LEMUR BIOLOGY

Hybrid male resulting from the cross of the hybrid
Lemur fulvus albocollaris × *L. f. rufus*, with *L. macaco*.
(See Chapter 3, p. 35.)

LEMUR BIOLOGY

Edited by

IAN TATTERSALL
Department of Anthropology
American Museum of Natural History
New York, New York
and
ROBERT W. SUSSMAN
Department of Anthropology
Washington University
St. Louis, Missouri

PLENUM PRESS • NEW YORK AND LONDON

Library of Congress Cataloging in Publication Data

Tattersall, Ian.
  Lemur biology.

  Includes bibliographies and index.
  1. Lemurs. I. Sussman, Robert W., 1941-   joint author. II. Title. [DNLM: 1.
Primates. QL737.P95 T221L]
QL737.P95T37   599'.81   74-28112
ISBN 0-306-30817-7

QL
737
.P95
T37

© 1975 Plenum Press, New York
A Division of Plenum Publishing Corporation
227 West 17th Street, New York, N.Y. 10011

United Kingdom edition published by Plenum Press, London
A Division of Plenum Publishing Company, Ltd.
Davis House (4th Floor), 8 Scrubs Lane, Harlesden, London, NW10 6SE, England

Printed in the United States of America

Contributors

JOHN BUETTNER-JANUSCH   Department of Anthropology, New York University, New York, New York.

NORMAN BUDNITZ   Department of Biology, Duke University, Durham, North Carolina.

KATHRYN DAINIS   Department of Biology, Duke University, Durham, North Carolina.

ISOBELLE DAVIDSON   Department of Anthropology, University of Pennsylvania, Philadelphia, Pennsylvania.

PHILIP D. GINGERICH   Museum of Paleontology, University of Michigan, Ann Arbor, Michigan.

JONATHAN E. HARRINGTON   Department of Biology, University of Missouri, St. Louis, Missouri.

RAYMOND HEIMBUCH   Department of Anthropology, Yale University, New Haven, Connecticut.

FRANCOISE K. JOUFFROY   Laboratoire d'Anatomie Comparée, Muséum National d'Histoire Naturelle, Paris, France.

GEORGES PARIENTE   Laboratoire d'Ecologie Générale, Muséum National d'Histoire Naturelle, Brunoy (Essonne), France.

J.-J. PETTER   Laboratoire d'Ecologie Générale, Muséum National d'Histoire Naturelle, Brunoy (Essonne), France.

A. PEYRIERAS   Laboratoire de Zoologie, O.R.S.T.O.M., Tananarive, Madagascar.

J. I. POLLOCK   Department of Anthropology, University College, London, England.

ALISON F. RICHARD   Department of Anthropology, Yale University, New Haven, Connecticut.

DAVID ROBERTS   School of Veterinary Medicine, University of Pennsylvania, Philadelphia, Pennsylvania.

YVES RUMPLER   Laboratoire d'Embryologie et de Cytogénétique, Ecole Nationale de Médécine, Tananarive, Madagascar.

ROBERT JAY RUSSELL   Department of Anatomy, Duke University, Durham, North Carolina.

ROGER SABAN   Laboratoire d'Anatomie Comparée, Muséum National d'Histoire Naturelle, Paris, France.

ALAIN SCHILLING   Laboratoire d'Ecologie Générale, Muséum National d'Histoire Naturelle, Brunoy (Essonne), France.

JEFFREY H. SCHWARTZ   Department of Anthropology, University of Pittsburgh, Pittsburgh, Pennsylvania.

ROBERT W. SUSSMAN   Department of Anthropology, Washington University, St. Louis, Missouri.

IAN TATTERSALL   Department of Anthropology, American Museum of Natural History, New York, New York.

Preface

The volume of studies on prosimian primates has, until recently, tended to lag well behind that of studies on the higher primates. This is so despite the fact that the considerable intrinsic interest of the living prosimians and the significance of their study for our understanding of the earlier stages of primate evolution have long been acknowledged by zoologists, paleontologists, and anthropologists alike. Among the prosimians, the Malagasy lemurs are of profound interest not only because they include the only extant diurnal forms, but also because it is only on Madagascar that the absence of competition with higher primates has allowed a surviving prosimian fauna to radiate, essentially unrestricted, into a broad spectrum of ecological zones. In contrast, the few extant prosimians of Africa and Asia occupy a relatively narrow range of "refuge" niches; although of considerable interest in themselves, they do not show the richness and variety of adaptation which make the Malagasy prosimian fauna such a fascinating object of study.

Over the past few years, however, there has been a considerable resurgence of interest in the prosimians in general, and in the lemurs in particular. The range of studies resulting from this rekindling of interest is wide, comprehending the systematics, evolution, anatomy, behavior, and ecology of these forms.

This volume constitutes a progress report on our knowledge of the lemurs. It is arbitrarily divided into five sections, but although each of these deals with a different aspect of the biology of the Malagasy prosimians, interpretations in each of these fields are dependent on knowledge of each of the others; we do not believe that a full understanding of any aspect of lemur biology can be obtained without crossing the boundaries of these conventional subdivisions. The book is not, however, an encyclopedic treatment of the primates of Madagascar; it cannot be, for lemur biology is a young subject which has yet to reach the stage of synthesis. Although for the sake of compactness some material which is readily available in English has been omitted, gaps in our coverage reflect to a large extent gaps in our knowledge. Similarly, the fact that our contributors are not invariably in agreement with one another reflects the fact that definitive answers to many of the most interesting and profound questions of lemur biology are not yet within our grasp.

The first section of the book includes a brief review of the history of discovery and study of the Malagasy prosimians, together with a short survey of the environmental background of Madagascar. The second section presents some new approaches to the analysis and interpretation of the relationships and evolution of the lemur fauna based on cytogenetic and dental information. The interested reader may wish to supplement this by consulting the recent articles by Martin (1972), Szalay and Katz (1973), and Tattersall and Schwartz (1974). The articles in the section on Morphology and Physiology summarize fairly comprehensively our current knowledge of these areas, although the recent contributions by Walker (1974) on the locomotor habits of the living and fossil lemurs, and of Stern and Oxnard (1973) on primate locomotion, should be mentioned. Section IV, Behavior and Ecology, presents data on a number of previously unstudied species, and on some species not previously reported upon in English. English language reports on a number of other species are available, however; these include *Microcebus murinus* (Martin, 1973), *Daubentonia madagascariensis* (Petter, 1965), *Lepilemur mustelinus* (Hladik and Charles-Dominique, 1974) and *Lemur mongoz* (Tattersall and Sussman, 1975). The final section contains a chapter which summarizes the problems of conservation of the Malagasy prosimian fauna and which reports the current status of these endangered animals.

Several chapters (6, 9, 11, and 15) were translated from the original French by Ian Tattersall in collaboration with Mlle. Christine Lémery, to whom, and to Mrs. Linda Sussman, we would like to express our deepest gratitude for invaluable assistance during the preparation of the book.

REFERENCES

Hladik, C. M., and Charles-Dominique, P., 1974, The behaviour and ecology of the sportive lemur (*Lepilemur mustelinus*) in relation to its dietary peculiarities, in: *Prosimian Biology* (R. D. Martin, G. A. Doyle, and A. C. Walker, eds.), Duckworth, London.

Martin, R. D., 1972, Adaptive radiation and behaviour of the Malagasy lemurs, *Phil. Trans. Roy. Soc. Lond. Soc. B* **185**:295–352.

Martin, R. D., 1973, A review of the behaviour and ecology of the lesser mouse lemur (*Microcebus murinus J.* F. Miller 1777), in: *Comparative Ecology and Behaviour of Primates* (R. P. Michael, and J. H. Crook, eds.), pp. 1–68, Academic Press, London.

Petter, J.-J., 1965, The aye-aye of Madagascar, in: *Social Communication among Primates* (S. A. Altmann, ed.), pp. 195–205, University of Chicago Press, Chicago.

Stern, J. T., Jr., and Oxnard, C. E., 1973, Primate locomotion: Some links with evolution and morphology, *Primatologia* **4**(11):1–93.

Szalay, F. S., and Katz, C. C., 1973, Phylogeny of lemurs, galagos and lorises, *Folia Primatol.* **19**:88–103.

Tattersall, I., and Schwartz, J. H., 1974, Craniodental morphology and the systematics of the Malagasy lemurs, *Anthropol. Papers Amer. Mus. Nat. Hist.* **52**(3):141–192.

Tattersall, I., and Sussman, R. W., 1975, Observations on the ecology and behavior of the Mongoose Lemur: *Lemur mongoz mongoz* Linnaeus (Primates, Lemuriformes) at Ampijoroa, Madagascar, *Anthropol. Papers Amer. Mus. Nat. Hist.* (in press).

Walker, A., 1974, Locomotor adaptations in past and present prosimian primates, in: *Primate Locomotion* (F. A. Jenkins, Jr., ed.), pp. 349–381, Academic Press, New York.

Contents

PART III
MORPHOLOGY AND PHYSIOLOGY

PART IV
BEHAVIOR AND ECOLOGY

PART I
INTRODUCTORY

History of Study of the Malagasy Lemurs, with Notes on Major Museum Collections[1]

1

History of Study of the Malagasy Lemurs, with Notes on Major Museum Collections[1]

JOHN BUETTNER-JANUSCH, IAN TATTERSALL,
AND ROBERT W. SUSSMAN

The history of Madagascar's contacts with Arab, European, and possibly Chinese explorers is a long one, and it is almost inconceivable that such early visitors should not have reported their impressions of the island's unique primates. However, the earliest extant record of these animals appeared as late as 1658, when Etienne de Flacourt published his *Histoire de la Grande Isle Madagascar*. This remarkable work, the fruit of several years' sojourn at Fort Dauphin, contains several recognizable descriptions of extant lemurs and a description of another animal which cannot be identified with any living species but which may correspond to one of the several recently extinct forms. One of de Flacourt's clearest portraits is of the sifaka, an animal sub-

[1]This article is based upon a chapter in Professor Buettner-Janusch's book *The Lemurs of Madagascar* (in preparation).

JOHN BUETTNER-JANUSCH Department of Anthropology, New York University, New York, New York 10003. IAN TATTERSALL Department of Anthropology, American Museum of Natural History, New York, New York 10024. ROBERT W. SUSSMAN Department of Anthropology, Washington University, St. Louis, Missouri 63130

sequently not described and possibly not even observed by another European until Grandidier named the species *Propithecus verreauxi* in 1867.

Although de Flacourt can thus claim the initial discovery by a European of these gentle, beautiful relatives of man, official zoological recognition was not bestowed upon them until the following century when Linnaeus established *Lemur*, in the form of *L. catta*, as one of his four primate genera in the tenth edition of his *Systema Naturae*. This species has subsequently and perhaps unfortunately become not only the type but the stereotype of the entire group.

The earliest notice of *Lemur*, however, antedates Linnaeus' great work. In 1754, the English naturalist Edwards began to issue a series of natural history plates, each accompanied by a text; these were published as a unit in 1758. Among the sections available in 1754 was Edwards' Figure 216, "The Mongoz," on which Linnaeus based *L. mongoz* in his twelfth edition of 1766. Edwards had found the animal living in 1752 at the house of the "obliging Mrs. Kennon, midwife to HRH the Princess of Wales," who informed him that it ate almost anything, including fish, and that it "had a great mind to catch birds." Edwards also illustrated "The Black Macauco" (Plate 217) from a living animal. Clearly a male *Lemur Macaco*, this was owned by a Mr. Critington, Clerk to the Society of Surgeons, who found that it ate cakes, bread and butter, and summer fruits. Edwards had one complaint about this lemur: "these animals differing from monkies in the parts which distinguish the sexes, they being more inward in the Macauco tribe, I cannot pronounce, of my own knowledge, of which gender it was." This is a lament which many later researchers have despairingly echoed.

In 1766, Buffon described and illustrated a small Malagasy mammal as *le rat de Madagascar*. More familiar now as *Microcebus murinus*, this "rat" appears to have been drawn from a living animal in the possession of the Comtesse de Marsan. How she obtained it has still to be explained.

The most significant eighteenth-century report dealing with observations on lemurs in their native habitat was published by Sonnerat, "Commissaire de la Marine, Naturaliste Pensionnaire du Roi, Correspondant de son Cabinet et de l'Academie Royale des Sciences de Paris, Membre de Celle du Lyon," in 1782. At the order of King Louis XVI, Sonnerat journeyed to China and the East Indies, and between 1774 and 1781 made numerous visits to the islands of the Indian Ocean. The substance of the two volumes of his *Voyage aux Indes Orientales et à la Chine* may seem fanciful and exaggerated to the modern reader, but much of the natural history is excellent. Thus his description of *Indri* is reasonably accurate, although his account of its habits, based upon secondhand information supplied by locals, is unreliable. Sonnerat was also the first to describe both the indriid *Avahi laniger*, and the aye-aye, *Daubentonia madagascariensis*. Sonnerat's engraving of *Avahi*, which he called *maquis à bourres*, does not represent the animal very well, but his illustration of the aye-aye accurately shows the thin, elongated third digit and the gnawing incisors. Nonetheless, Sonnerat's account of the behavior of the aye-aye was rather inaccurate; he described its movements through the trees as being squirrel-like, and, indeed, the animal was initially supposed to be related to squirrels. Thus Buffon and Gmelin (1828), who studied one of a pair of aye-ayes which Sonnerat had attempted to bring back to Paris but which had died onboard ship, believed that they were dealing with some kind of squirrel. In this they were followed by Cuvier, although the latter did find some primate characteristics in the skull.

Between 1792 and 1810, J. B. Audebert published a large atlas of the primates which included excellent engraved plates of a number of Lemuriformes. The engravings, with accompanying descriptions, were published, beginning in 1792, as separate *livraisons*; the entire set was finally issued complete with 1810 as the date of publication. Audebert's illustrations are generally accurate, since they were based on several collections of skins and skeletons, including his personal collection, those of several other private collectors, and those of the National Museum of Natural History in Paris (MNHN). For his descriptions of behavior and habitat Audebert used various earlier sources, including the reports of Flacourt and Sonnerat.

The ruffed lemur, *Lemur variegatus*, was known earlier than 1797, the year in which it was illustrated by Audebert, having been named (as *L. macacus variegatus*) by Kerr in 1792. Audebert's plate was drawn from a living animal which had for some time been in the possession of the Empress Josephine. The animal died in 1809 and was given to the MNHN, where it was later studied by Buffon, and where, faded but elegant, it is still to be seen. The Empress maintained a small menagerie at Malmaison, probably because fashion dictated that she show an interest in natural history. Geoffroy and Saint-Hilaire and Cuvier (1824) noted that a pair of *L. variegatus* kept there had copulated successfully and several births were recorded, one of twins; unfortunately, no data on gestation period or on events at birth were recorded.

The eighteenth century also witnessed the first rigorous dissections of lemurs, carried out by the French zoologist Daubenton, who was responsible for the anatomical studies of lemurs in Buffon's *Histoire Naturelle* (1789), and the English anatomist John Hunter, who dissected what was probably *Lemur mongoz*. Hunter's notes remained unknown until 1861, when Sir Richard Owen edited them for publication.

During the nineteenth century there was a rapid increase in the number of published descriptions of lemurs. The most notable are those of Etienne Geoffroy Saint-Hilaire (1812) and Isidore Geoffroy Saint-Hilaire (1851); of de Blainville, who first published a description of *Phaner* in his *Ostéographie* (1834–1841); of G. Fischer, whose *Anatomie der Maki* (1804) was a compendium of original observations made on museum specimens and material drawn from works published up to that time; of Cuvier and Laurillard (1850); and of J. E. Gray (1842 *et seq.*).

Beginning in 1819, Geoffroy and Saint-Hilaire and Cuvier published separate *livraisons* of the handsomely illustrated *Histoire des Mammiferes* (1824). Although the sections of this work devoted to lemurs are few, the "Maki à Front Blanc" was illustrated in *Livraison III*, which contains excellent renderings of a male and female *Lemur fulvus albifrons*. In 1816, the MNHN menagerie received two male "makis à front blanc," which were placed with females which did not look at all like them. On 23 December 1817 and during the following week, Geoffroy Saint-Hilaire and Cuvier observed the animals copulating. About 40 days thereafter, the mammary glands of a female began to swell, and on 13 April 1818 an infant female was born. A gestation period of about 120 days was thus recorded, which contemporary observations confirm. Geoffroy Saint-Hilaire and Cuvier were positive that the males and females were members of the same species, and in the light of their observations it is difficult to understand how the males and females of the sexually dichromatic *L. f. albifrons* were later mistakenly placed in separate species by Sclater (1880).

The second half of the nineteenth century produced growing discussion of the systematics and taxonomy of the lemurs, as well as descriptions of new specimens which, far too often, were presented as newly discovered species. Thus by the last quarter of the century the systematics and nomenclature of the Malagasy primates were in confusion despite numerous attempts to regularize them. Indeed, even today, unanimity is far from realization.

Also at this time, new observations of the lemurs in their natural habitat were being contributed by missionaries, European traders, French scientists, and professional collectors. This last category, in particular, did much to increase European knowledge of the lemurs. The collector Bernier, for instance, brought to Europe the first specimen of what is now known as *Lemur rubriventer*, while Goudot sent the first example of *Lepilemur* to France. The British collector Charles Telfair obtained a specimen of *Propithecus diadema*, the first member of the genus to arrive in Europe in the flesh, and the first confirmation of de Flacourt's account of the "white lemur." His compatriot Crossley brought back a large amount of material from Madagascar; this was sold to museums in England, France, and Germany. Crossley's collection contains two skins of the rarest of all lemurs, *Cheirogaleus* (or *Allocebus*) *trichotis*.

Accounts of lemurs in their natural habitat were also provided by members of the London Missionary Society during the second half of the nineteenth century. The missionaries' descriptions of the lemurs are often accurate, but the notions held by some of them of how to define species and evaluate reports of the habits of the animals are frequently rather idiosyncratic.

One of the most remarkable anatomical monographs published during this period is the famous work of Sir Richard Owen (1863) on the aye-aye. This work includes notes taken while the animal was kept in captivity by the Governor of Mauritius.

Among the French explorers, scientists, collectors, and government officials in Madagascar during the late nineteenth century were Coquerel, Verreaux, and Guinet, all of whom collected quantities of specimens of the Malagasy fauna, including lemurs. Coquerel's name is celebrated in several binomia and trinomia, as is Verreaux's. Two Dutch explorers and naturalists, Pollen and van Dam (1868), made large collections on Madagascar and the Comoros and produced some of the best accounts, up to that time, of the habitat, behavior, and geographical distributions of lemurs. These collections were given to the Rijksmuseum van Naturlijke Historie (RMNH) in Leiden, and were published by Schlegel (1866).

The most notable contribution of the late nineteenth century to the study of lemurs was, however, that of Alfred Grandidier. Grandidier, traveling widely across the island, made Madagascar his life's work. Twenty-seven volumes of his monograph *Histoire Physique, Naturelle et Politique de Madagascar* were published between 1875 and 1930, the last after his death in 1921, but the series is nonetheless incomplete. In particular, the section dealing with the lemurs and written in collaboration with Alphonse Milne-Edwards was never finished. A detailed account of the anatomy of Indriidae and *Hapalemur* was, however, published, together with a large number of plates illustrating the osteology, anatomy, and appearance of the other lemurines. Among Grandidier's numerous color and monochrome illustrations are some of the finest anatomical, osteological, and portrait plates ever published.

The names given by Grandidier to some of his specimens were, however, not those generally accepted, even given the confusion reigning at the time in lemur systematics. Either he was not aware of the nomenclature of the lemurs which had been developed by a number of his British and French colleagues, or he had decided to revise the taxonomy of the group but had no time before his death to provide written justifications. This was particularly unfortunate because some subsequent collectors, naturalists, and others not familar with the minutiae of lemuroid systematics used Grandidier's plates to establish identifications. The confusion that exists in some parts of the literature between *Lemur mongoz* and *Lemur fulvus*, for example, may be traced at least in part to the use of Grandidier's superb illustrations as a key to the identification of lemur skins or live specimens.

At this point, perhaps it is appropriate to turn to a brief consideration of the major collections of lemuroid material amassed during the nineteenth century and subsequently, and concentrated in the major natural history museums of Europe and North America. Unhappily, valuable materials have been destroyed, scattered, or allowed to deteriorate since enthusiastic eighteenth-century encyclopedists began to collect and itemize living organisms. Moths, mildew, and death-watch beetles have hardly been more destructive of such collections than war, social unrest, intellectual fashion, and governmental apathy and parsimony. It is fortunate, however, that despite the variety of forces inimical to their holdings, the great museums have managed to preserve quantities of material. Such material is indispensable to all who study Lemuriformes, for the changes of orientation in modern systematics have not eliminated the need to consult collections of lemur skins, skeletons, and wet specimens.

The collection of lemur skins and skeletons in the MNHN—the museum of Buffon, the two Cuviers, the two Geoffroys, Lacépède, and others—is one of the oldest, and contains many of the classic specimens described in the literature. Unfortunately, the documentation for many of these specimens is not particularly precise, and their usefulness is thus impaired for many kinds of studies. The lack of documentation for the collection, particularly its earliest specimens, was the result of the inadequate information provided by the collectors who brought the live animals to the menagerie. Nevertheless the collection has played an important role in the history of research on the Lemuriformes. Audebert (1810) contributed to the collection, and he used the contributions of others for the marvelous illustrations in his work on primates. Etienne Geoffroy Saint-Hilaire and Frederick Cuvier were able to base many sections of their *Histoire Naturelle des Mammifères* (1824) upon items in the collections and upon living animals in the menageries of the museum. A number of species of lemurs were bred at the menagerie early in the nineteenth century, and the earliest descriptions of the birth of a lemur and mother–offspring behavior are found in the reports of Geoffroy Saint-Hilaire and Cuvier (1824).

The collection of Lemuriformes in the British Museum of Natural History in London (BMNH) is probably the largest in the world, but it is rather poorly documented. It is possible to recover a reasonable amount of information from catalogues and labels in both the MNHN and the BMNH, but the modern researcher inevitably regrets that the importance of documenting the exact provenance of the specimens was not more commonly appreciated in the nineteenth century. The collection of the Tring Museum, now at the BMNH, includes many very fine specimens, but almost without exception they are of little use to modern students because they are not properly documented. Many specimens in the BMNH came from the London Zoological Gardens and no provenance other than Regent's Park is given.

The collection of skins and skulls of Lemuriformes in the BMNH includes examples, usually obtained by trade, from almost all the important collections made by other museums [MNHN, American Museum of Natural History (AMNH), National Museum in Stockholm] as well as several groups of specimens obtained specifically for the BMNH by collectors. Many skins are from the Archbold collection or are those that W. S. Webb collected during the 1939–1945 war. While Webb was stranded on the island of Madagascar, he managed, among other activities, to collect a number of important specimens which he later turned over to the BMNH. The documentation for his specimens is unique, because he recorded the latitude and longitude of the spots at which he captured or killed the animals.

A fine collection of lemurs has been assembled at the American Museum of Natural History in New York. The major and most useful part of the collection consists of the specimens collected by the Mission Zoologique Franco-Anglo-Americaine, often known to Americans as the Archbold expedition. The majority of the lemurs collected by this multinational group are in the AMNH. Some examples were kept by the BMNH and the MNHN. The Archbold specimens are among the best documented parts of the lemur collection in the latter two museums. Another group of lemur specimens at the AMNH came from the famous collection of Blüntschli, an anatomist at the University of Bern, Switzerland, whose collecting was primarily directed to making as complete a survey of the embryology of as many species of Lemuriformes as possible. His collection of embryological material, both slides and specimens, went to the Carnegie Institution of Washington, while the skeletons and skins were sold to the AMNH. Unfortunately, the excellent documentation which Blüntschli is known to have kept has not survived unimpaired.

Generally, the majority of the specimens found in the AMNH, BMNH, and MNHN are reasonably well documented. However, the documentation of most specimens of the early nineteenth century is poor. These specimens are the remains of famous private collections that were sold, usually by heirs, to dealers who then resold them to museums. Very seldom was an entire collection sold as a unit. Rather, bits and pieces were scattered as various museums required examples for exhibition or for study. The notes and documents that were parts of these collections have usually vanished. The collections of J. B. Audebert in the MNHN and of Crossley in the BMNH are known to have met this fate.

The excellent collection of Lemuriformes in the Rijksmuseum van Natuurlijke Historie in Leiden was started by Pollen and van Dam (1868). It was completed by J. Audebert, a resident of the island who was put under contract by Schlegel, director of the RMNH, to collect mammals and birds from Madagascar, the Comoros, and Mauritius (Gijzen, 1938). Stored away from the light, the collection of skins and skeletons of lemurs is in excellent condition today, and is one of the most useful in any major museum, especially since Audebert supplied rather more information than was common in those days about the localities in which he obtained his specimens.

There is a group of skins and skeletons of lemurs in the United States National Museum, Smithsonian Institution. Most of these were obtained by exchange with the AMNH or other institutions, or were obtained from the National Zoo after the animals had died. One group was collected by W. L. Abbott, and although there is no reason to doubt his documentation of them, it is likely that he obtained most of them in the seaports at which he touched, for his trip to Madagascar was part of an expedition that led him through East Africa and finally to India.

Other museums of natural history have examples of Lemuriformes, but the only other significant collections of which we are aware, besides that of the Museum of Comparative Zoology in Cambridge, Massachusetts, are in the zoological museum of the Humboldt-Universität in Berlin and in the National Museum of Natural History in Stockholm. None of the authors has yet been able to visit either of these museums. The Stockholm collection, fortunately, has been published by Kaudern (1910), and it is possible to make use of his descriptions.

Having outlined the impressive extent of the collections of lemur material available in the major museums, one might add that it is devoutly to be hoped that the era of the wholesale collection of lemur material is over. It was not unusual in the heyday of zoological collecting for several hundred animals to be shot during the course of a single expedition, and it is quite possible that such expeditions were responsible for the extirpation of entire local populations of lemurs.

Fortunately, however, collectors did not always confine themselves to the wholesale slaughter of animals. A. L. Rand of the Archbold Expedition, for instance, kept valuable records of the natural history of the animals in the regions he visited (Rand, 1935). Modern field research on the lemurs only began, however, in the late 1950s, when Petter carried out a survey in which he noted the ecology and behavior of a number of species (Petter, 1962). The first intensive field study on a particular species was that of Jolly (1966) on *Lemur catta* and *Propithecus verreauxi verreauxi*. However, even today only five of the approximately 20 lemur species have been intensively studied, and such studies have been carried out in very few sites, most of which are protected reserves (e.g., Berenty, Ankarafantsika, and Perinet). Thus the range of variation in behavior and ecology, even of these species, is generally not known. Precise geographical distributions are not known for any species.

As with primate field research generally, research on lemurs has only within the last 5 years changed its orientation from descriptive natural history to problem-oriented studies, in which specific questions concerning aspects of the behavior, ecology, and evolution of these animals are addressed. It is only as investigators move out into farther-flung areas of Madagascar, and particularly as their research becomes more problem-oriented, that we will begin to fully comprehend the flexibility and complexity of the behavior of these unique and fascinating animals.

REFERENCES

Audebert, J. B., 1810, *Histoire Naturelle des Singes et des Makis*, Desray, Paris.

Buffon, G. L. le C., 1766, *Histoire Naturelle, Générale et Particulière, avec le Description du Cabinet du Roi*, Vol. 3, Imprimerie Royale, Paris.

Buffon, G. L. le C., 1789, *Histoire Naturelle, Générale et Particulière*, Vol. 23, Imprimerie Royale, Paris.

Cuvier, G., and Laurillard, C. L., 1850, *Anatomie Comparée: Recueil des Planches de Myologie*, Dusacq, Paris.

de Blainville, H. M., 1841, *Ostéographie des Mammifères*, Paris.

de Flacourt, E., 1658, *Histoire de la Grande Isle Madagascar*, Pierre l'Amy, Paris.

Geoffroy Saint'Hilaire, E., 1812, Tableaux des quadrumanes ou des animaux composant la première ordre des mammifères, *Ann. Mus. Natl. Hist. Nat. Paris* **19**:156–170.

Geoffroy Saint-Hilaire, E., and Cuvier, F., 1824, *Histoire Naturelle des Mammifères, avec des Figures Originales, Coloriées, Dessinées d'Après des Animaux Vivants*, Paris.

Geoffroy Saint-Hilaire, I., 1851, *Catalogue Méthodique, Primates de la Collection des Mammifères*, Muséum National d' Histoire Naturelle, Paris.

Gijzen, A., 1938, *'S Rijks Museum van Natuurlijke Histoire, 1820–1915*, W. L. and J. Brusse's Uitgevers, Rotterdam.

Grandidier, A., 1867, Mammifères et oiseaux nouveaux découverts à Madagascar, *Rev. Mag. Zool.* (2nd. ser.) **19**:84–88.

Grandidier, A., 1875, *Histoire Physique, Naturelle et Politique de Madagascar*, Hachette et Cie, Paris.

Gray, J. E., 1842, Description of some new genre and fifty unrecorded species of mammalia, *Ann. Mag. Nat. Hist.* **8**:255–269.

Jolly, A., 1966, *Lemur Behavior*, University of Chicago Press, Chicago.

Kaudern, W., 1910, *Studien über die männlichen Geschlechtsorgane von Insektivoren und Lemuriden*, Hochschule, Stockholm.

Kerr, R., 1792, *Animal Kingdom, or the Zoological System of the Celebrated Sir Charles Linnaeus*, J. Murray and R. Faulden, London.

Owen, R. (ed.), 1861, *Essays and Observations on Natural History, Anatomy, Physiology, Psychology and Geology*, J. van Voorst, London.

Owen, R., 1863, *Monograph on the Aye-Aye (Chiromys madagascariensis Cuvier)*, Taylor and Francis, London.

Petter, J.-J., 1962, Recherches sur l'écologie et l'éthologie des lémuriens malgaches, *Mém. Mus. Natl. Hist. Nat. Paris* **27**:1–146.

Pollen, F. P. L., and van Dam, D. C., 1868, *Recherches sur la Faune de Madagascar et de ses Dépendences, d'Après les Découvertes de François P. L. Pollen et D. C. van Dam*, E. J. Brill, Leiden.

Rand, A. L., 1935, On the habits of some Madagascar mammals. *J. Mammal.* **16**:89–104.

Schlegel H., 1866, Contributions à la faune de Madagascar et des îles avoisinantes, d'après les découvertes et observations de Mm François Pollen et D. C. van Dam. *Ned. Tijdschr. Dierk.* **3**:73–89.

Sclater, P. L., 1880, Report on the additions to the Society's menagerie in May, 1880, and a description of a new lemur (*Lemur nigerrimus*), *Proc. Zool. Soc. Lond.*, no vol., 450–452.

Sonnerat, M., 1782, *Voyage aux Indes Orientales et à la Chine*, Froulé, Paris.

Notes on Topography, Climate, and Vegetation of Madagascar

2

IAN TATTERSALL AND
ROBERT W. SUSSMAN

The brief notes presented here are not meant to be exhaustive, but rather to provide some information on the topography, climate, and vegetation of Madagascar pertinent to various aspects of the ecology and distribution of the lemurs. More complete discussions may be found in the sources listed at the end of this chapter.

Madagascar, 995 miles (1600 km) long, with a maximum width of 360 miles (580 km) and covering an area of 230,000 square miles (590,000 km^2), is the world's fourth largest island, following Greenland, New Guinea, and Borneo. It lies in the southern Indian Ocean between 11°57' and 25°32' south latitude and thus falls almost entirely within the tropical zone. Its long axis is oriented NNE–SSW, and the width of the Mozambique Channel, which separates Madagascar from the eastern coast of Africa, varies from 220 miles (350 km) to 750 miles (1200 km).

IAN TATTERSALL   Department of Anthropology, American Museum of Natural History, New York, New York 10024. ROBERT W. SUSSMAN   Department of Anthropology, Washington University, St. Louis, Missouri 63130

TOPOGRAPHY

As Battistini (1972) has pointed out, one of the most striking aspects of Madagascar's topography is its asymmetry. There is an elevated central plateau running most of the length of the island, surrounded by coastal lowlands. The average elevation of the spine of the plateau is about 5000 ft (Fig. 1); the three highest peaks range from 8720 ft to 9468 ft in altitude. Most of the heavily dissected crystalline central highlands, however, lie to the east of the long axis of the island (Fig. 1). The eastward-facing slope is abrupt, limiting a narrow coastal strip, while the western slope of the plateau is much more gentle, bordering a broad belt of coastal lowlands enclosing two major sedimentary basins. Topographically, then, Madagascar falls into three major zones: the narrow eastern plain, including the steep escarpment which demarcates it to the west; the rugged high plateau; and the great sedimentary plains of the west and northwest.

CLIMATE

Madagascar's size is such as to give it the climatic characteristics of a small continent (Donque, 1972), and the variety of its relief gives rise to wide regional and local variations in climate. Broadly, two seasons can be distinguished: the austral winter (May–October) and the austral summer (November–April). Transitional periods separating these are short (Griffiths and Ranaivoson, 1972), each lasting under a month. It is tempting to characterize the summer as hot and wet and the winter as cooler and dry, but generalizations such as this tend to obscure the rather more complex realities of the situation.

Temperature

Although at any given latitude, the west coast is usually a few degrees warmer than the east coast, the mean annual temperature at sea level declines from about 77°F in the north to about 72°F in the south. Inland, temperatures are largely altitude dependent, declining by about 2.7°F per 1000 ft. Figure 2 gives the isotherms for mean annual temperature maxima and minima.

Throughout Madagascar, temperature minima occur in July or August, but the timing of the maxima is more variable. In the southwest and on the plateau, a single maximum temperature is reached, in the former between January and March and on the latter in November. Double maxima appear in the extreme northeast in April and December, and in the northwest in

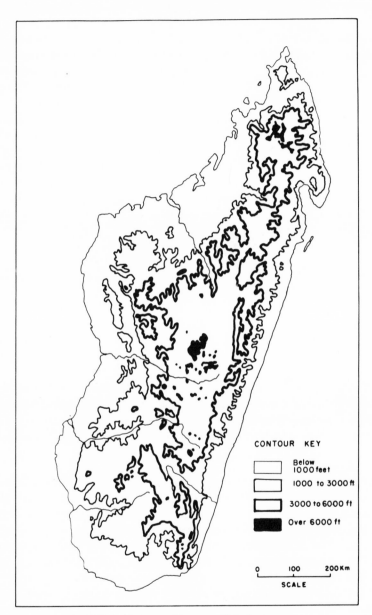

FIG. 1. Topographic map of Madagascar.

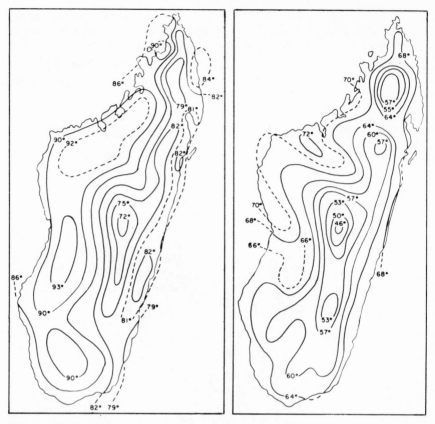

FIG. 2. Madagascar: mean annual maximum (left) and minimum isotherms. Calibrated in degrees Fahrenheit.

March–April and October–December. Along most of the east coast, a single maximum is achieved in January or February. Diurnal variation in temperature is smaller in the eastern, western, and northern coastal areas than along the southwestern coast and in the interior (Griffiths and Ranaivoson, 1972). In general, seasonal change in this variation is not marked.

Rainfall

Regionally, precipitation is highly variable. Annual total rainfall may vary between a maximum of almost 200 inches (5000 mm, Tamatave) and a minimum of 4.7 inches (120 mm, Morombe). Figure 3 shows the pattern of mean annual rainfall throughout the island.

In certain parts of the island it is impossible to distinguish meaningfully between wet and dry seasons. Thus between October and April, during which time 90% of the total annual precipitation falls on most of the plateau and in the western parts of the island, 50% or less falls in the east. Instead of a dry season along the eastern margin of the island and its immediate hinterland,

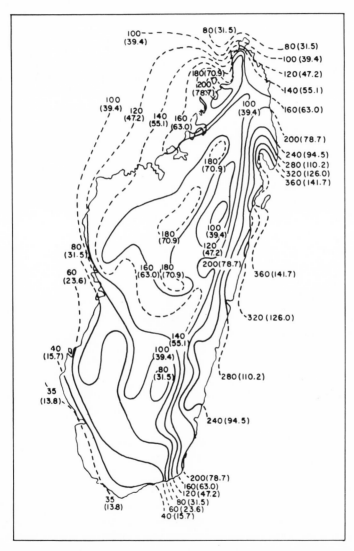

FIG. 3. Madagascar: mean annual rainfall. Primary figures in centimeters; those in parentheses in inches.

one can only note some reduction in rainfall during the months of September and October. It is not unusual on any day of the year to record rainfall in excess of 4 inches (100 mm) at stations on the east coast, while such an event will be rare in the west except during the period December–March (Griffiths and Ranaivoson, 1972).

Climatic Zones

Madagascar may, then, be divided into a number of climatic zones. The plateau, generally defined as that portion of the island's interior lying above 2300 ft (700 m), is generally temperate, although climatic conditions vary considerably with relief; there is a fairly clear distinction between wet and dry seasons. The eastern coastal strip, on the other hand, is hot and humid, and although rainfall maxima and minima can be distinguished, there is no clear dry season. The extreme southern area is, in contrast, hot but semiarid, with a minimal rainfall inconsistently spread over the year. Precipitation here tends to peak during certain limited periods, depending on location. Periods of drought may persist for several months. The warm western lowlands are more distinctly seasonal; there is a distinct dry period between May and October, and both temperature and rainfall show some decline with increasing latitude. The western coastal area north of Analalava, however, is characterized by high heat and humidity with abundant rainfall, of which 85% falls during the summer months.

VEGETATION

The vegetation of Madagascar has been described by Perrier de la Bathie (1921, 1936), Humbert (1955), and Humbert and Cours Darne (1965). The modern vegetal pattern of Madagascar is, of course, the product of centuries of human exploitation and interference. However, it is still unknown whether human exploitation is the direct cause of the small proportion of primary forests which exist on the island today (see Chapter 18).

The primary forests of Madagascar can be divided into two principal zones: the wet, evergreen vegetation zone (oriental zone) found in the east, the extreme north, the central plateau, and the high mountains; and the dry zone (occidental zone) of the western and southern regions of the island (Fig. 4).

Humbert subdivides the occidental zone into two domains: the southern and the western. The semiarid southern domain is characterized by rich thickets or forests of strongly endemic, bushy, xerophytic vegetation, with Didiereaceae and Euphorbiaceae predominant (Perrier de la Bathie, 1921; Hum-

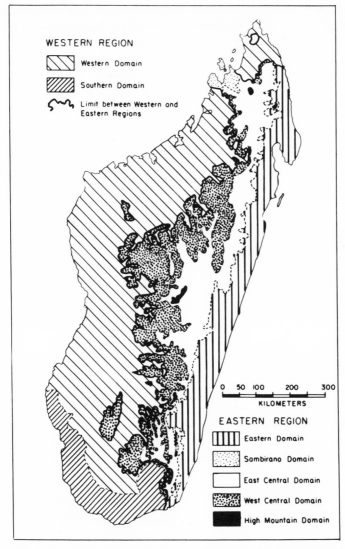

WESTERN REGION

Western Domain

Southern Domain

Limit between Western and
Eastern Regions

0   50   100   200   300
KILOMETERS

EASTERN REGION

Eastern Domain

Sambirano Domain

East Central Domain

West Central Domain

High Mountain Domain

FIG. 4. Phytogeographic zones of Madagascar, after the scheme of Humbert (1955).

bert, 1927). Deciduous forest similar to that of the western domain is also locally present in this part of Madagascar; it is found primarily along river margins, but under favorable edaphic conditions, it may penetrate even the driest areas (Koechlin, 1972).

Three types of forest have been distinguished by Perrier de la Bathie

(1936) in the western domain, their distribution depending largely on soil type (siliceous, calcareous, or rocky). The deciduous forests which flourish on the moist siliceous soils are generally characteristic of river valleys and are dominated by *Tamarindus indica* (kily). In the climax condition, they exhibit a continuous canopy 25 to 50 ft in height, with emergent trees reaching perhaps 65 ft; the forest floor is generally quite open. The drier calcareous soils support, predictably enough, a vegetation of a more xerophytic type in which no canopy or dominant tree species exists; the profile of these forests is much lower than that of those growing on the damper soils, and the undergrowth is dense. Where the soil is driest, xerophytic thickets, generally more impoverished in species than those of the south, are the rule. Forests of these different types may merge into each other over short distances where there are local changes in soil and moisture conditions; in general, however, there is a tendency toward increasing dryness from north to south, with a corresponding transition in predominant vegetal type.

The oriental zone is far more humid and exhibits a wider variety of forest types. The eastern domain is typified by dense rain forest (Koechlin, 1972) which flourishes to an altitude of about 2600 ft (800 m). This high, stratified, evergreen forest (reaching 100 ft) is composed of a vast number of species. There are a number of distinct forest strata, a relatively open floor, and abundant epiphytes. Much of the area originally covered by forest of this kind now supports dense secondary formations (*savoka*). The forest of the Sambirano domain is of similar structure but of slightly differing floral composition. Between 2600 ft and about 4300 ft (1300 m), the eastern portion of the central domain, more seasonal in climate, still supports in places a medium-altitude dense rain forest (Koechlin, 1972). As tall as the eastern rain forest, this type of formation is, however, less complexly stratified, presenting a single continuous stratum at a height of about 85–95 ft; below this is a substantial herbaceous or shrubby undergrowth. At altitudes in excess of 4300 ft and reaching up to about 6600 ft (2000 m), forest of this type gives way to a "lichen forest" composed of trees around 35–40 ft in height, with small, hard leaves and twisting trunks; these support abundant epiphytes, which also cover the mossy ground.

The western sector of the central domain, lying below 2600 ft, is warmer and drier than the eastern portion, but low minimum temperatures remain a limiting factor, and a low sclerophyllous forest with shrubby undergrowth is typical; in some valleys, however, a vegetation of rather damper aspect including a variety of eastern species is found. The high mountain domain is restricted to the few massifs which rise above 6600 ft (2000 m); its vegetation is adapted to the more severe climatic conditions prevailing in these areas and is usually bushy or herbaceous in nature.

Some of the forest types described above have almost entirely disappeared and are represented only by isolated remnant patches; this is particularly true of the forests of the central domain. Koechlin (1972) provides a brief discussion of the floral succession which follows the destruction of the various original climax forest types. Chauvet (1972) provides some depressing estimates of the various areas covered at present by aboriginal forest and of the current rates at which it is being destroyed. The problems of conservation are discussed in detail in Chapter 18.

REFERENCES

Battistini, R., 1972, Madagascar relief and main types of landscape, in: *Biogeography and Ecology in Madagascar* (R. Battistini and G. Richard-Vindard, eds.), pp. 1–25, W. Junk, The Hague.

Chauvet, B., 1972, The forest of Madagascar, in: *Biogeography and Ecology in Madagascar* (R. Battistini and G. Richard-Vindard, eds.), pp. 191–200, W. Junk, The Hague.

Donque, G., 1972, The climatology of Madagascar, in: *Biogeography and Ecology in Madagascar* (R. Battistini and G. Richard-Vindard, eds.), pp. 87–144, W. Junk, The Hague.

Griffiths, J. F., and Ranaivoson, R., 1972, Madagascar, in: *Climates of Africa* (J. F. Griffiths, ed.), pp. 461–498, (*World Survey of Climatology*, Vol. 10), Elsevier, Amsterdam.

Humbert, H., 1927, Destruction d'une flore insulaire par le feu. *Mém. Acad. Malgache* **5**:1–80.

Humbert, H., 1955, Les territoires phytogéographiques de Madagascar, leur cartographie, *Ann. Biol.* (3 sér.) **31**:195–204.

Humbert, H., and Cours Darne, G., 1965, *Notice de la Carte Madagascar*, Institut de la Carte Internationale du Tapis Végétal, Toulouse.

Koechlin, J., 1972, Flora and vegetation of Madagascar, in: *Biogeography and Ecology of Madagascar* (R. Battistini and G. Richard-Vindard, eds.), pp. 145–190, W. Junk, The Hague.

Perrier de la Bathie, H., 1921, La végétation malgache, *Ann. Mus. Colon. Marseille* (3 sér.) **9**:1–268.

Perrier de la Bathie, H., 1936, *Biogéographie des Plantes de Madagascar.* Soc. Edit. Géogr. Marit. et Colon., Paris.

PART II
SYSTEMATICS AND
EVOLUTION

The Significance of   3
Chromosomal Studies in the
Systematics of the Malagasy
Lemurs

YVES RUMPLER

Chromosomal studies of the Malagasy lemurs have been particularly fruitful in two areas of investigation where these animals pose various difficult problems: systematics and evolution. In this chapter, we will report on the significance of cytogenetics in the systematics of the lemurs.

In lemur taxonomy, the classical morphoanatomical and biogeographical criteria have occasionally proved inadequate, as the conflicting classifications proposed by different authors have shown. Much light can be shed, however, by integrating biochemical, physiological, and genetic considerations with data of a more classical type. Unfortunately, mere knowledge of the karyotypes of the lemurs has not always brought clarification and has sometimes even complicated classification. Such complication has been due to errors in establishing the karyotypes or in the zoological determination of the animals involved. This latter is difficult outside Madagascar, since scientists have to work on zoo animals whose exact location of capture is unknown.

YVES RUMPLER Laboratoire d'Embryologie et de Cytogénétique, Ecole Nationale de Médécine, B. P. 375, Tananarive, Madagascar

This has resulted in the publication of different karyotypes for the same species. Thus one might suppose the existence of chromosome polymorphism for *L. mongoz* ($2N$ = 60, Buettner-Janusch *et al.*, 1966; $2N$ = 58, Egozcue, 1967; $2N$ = 60, Rumpler and Albignac, 1969*d*), for *L. fulvus fulvus* ($2N$ = 48. Bender and Chu, 1963; $2N$ = 58, Egozcue, 1967; $2N$ = 60, Rumpler and Albignac, 1969*a*), and even for an animal which presents no geographical variation such as *L. catta* (10 SM, 44 A, Xac Yac: Bender and Chu, 1963; 14 SM, 40 A, Xac Yac: Egozcue, 1967; 8 SM, 46 A, Xm Yac: Rumpler, 1970). On the other hand, certain authors have tried to base their classifications exclusively on karyotypes and have put animals having identical or neighboring karyotypes into the same group while separating those with different karyotypes. This conception has led to the grouping of animals which are morphologically very different, such as *L. variegatus* and *L. macaco* (Egozcue, 1967; Eckhardt, 1972).

In the work reported here, we have tried to integrate cytogenetic results with information of a more classical type in an effort to avoid the traps mentioned above. We have therefore studied as large a number of individuals of each species as possible, and have paid particular attention to the strict zoological determination of animals of known location of capture. Moreover, we have taken karyotypes into account in determining the status of animals only where hybridization experiments were not possible. In fact, each time that conventional criteria did not allow us to decide whether an animal was to be classified in a separate species or merely in a separate subspecies, we tried hybridization, since it is the criterion of the fertility of hybrids that we regard as decisive. Only when such experiments of hybridization were not possible did we use the karyotype itself; we then classified in different species animals whose chromosomal formulas showed differences such as to suggest that fertile hybrids would not be obtained. We believe that chromosomal differences between two groups do not signify *a priori* that two species are present. Even numerically important differences between two karyotypes cannot be significant in terms of a chromosomal barrier if they only slightly affect the process of meiosis. Thus centric fusions are hardly selective, because they do not prevent the preferential formation of balanced germ cells, as has been shown in the pig (MacFee and Banner, 1969), in *Leggada* (Matthey, 1964), and in the hybrids *Mus poschiavinus* × *Mus musculus* (Gropp *et al.*, 1970), in contrast to the more dramatic effects of pericentric and paracentric inversions or translocation of chromosomal fragments, as demonstrated in studies of human pathology (Turpin and Lejeune, 1964).

MATERIALS AND METHODS

We have karyotyped all the Malagasy lemurs from leukocyte cultures. Most slides were stained with Giemsa or Unna blue. Thermic denaturation

(Dutrillaux and Lejeune, 1971) and observation under ultraviolet light (Caspersson *et al.*, 1970) were only used in certain cases. A summary of the results of our karyotyping is given in Table I. Hybridization experiments have been tried in a number of cases: in the genus *Lemur*, between *L. macaco* and the different subspecies of *L. fulvus*, and between *L. mongoz* and *L. coronatus*; in the genus *Hapalemur*, between the different subspecies of *H. griseus*; and in the genus *Propithecus*, between the different subspecies of *P. verreauxi*. Finally, we have been able to study the karyotypes of natural hybrids, caught in the wild, between the different subspecies of *Lepilemur septentrionalis*. The karyotypes of the various hybrids are summarized in Tables II–V.

DISCUSSION

The last full-scale classification of the lemurs was that of Hill (1953), which distinguished three families: Lemuridae, Indriidae, and Daubentoniidae. We have modified this in some respects. We believe (1) that the genus *Phaner* should be isolated in its own subfamily, Phanerinae (Rumpler and Rakotosamimanana, 1971), and (2) that *Lepilemur* should be raised to its own subfamily, Lepilemurinae (Rumpler and Rakotosamimanana, 1971; Rumpler, 1974).

Similarly, our hybridization experiments have allowed us to clarify the taxonomy of the genera *Lemur*, *Hapalemur*, and *Lepilemur*. Thus in the genus *Lemur*, according to Hill (1953), six species exist: *L. catta*, *L. variegatus*, *L. rubriventer*, *L. mongoz* (with two subspecies: *L. m. mongoz* and *L. m. coronatus*), *L. fulvus* (with six subspecies: *L. f. fulvus*, *L. f. rufus*, *L. f. albifrons*, *L. f. sanfordi*, *L. f. collaris*, and *L. f. flavifrons*). Morphological and biogeographical (Petter, 1962) studies have suggested that *L. macaco* and the various *L. fulvus* subspecies belong to a single species, *L. macaco*, which thus includes seven different subspecies. Our chromosomal studies did not permit us to decide between the two taxonomies because some animals which are very distinct morphologically such as *L. mongoz* and *L. fulvus* have the same karyotype, whereas some morphologically similar animals such as *L. f. fulvus* and *L. f. collaris* have different karyotypes. Moreover, the karyotypes of all the animals of this group derive from that of *L. fulvus* ($2N = 60$) principally through centric fusion. These transformations from the basic karyotype are not highly selective; i.e., the existence of these different karyotypes does not allow a prejudgment on whether or not fertile hybrids might be obtained. We therefore carried out hybridization experiments between these animals and have placed those producing fertile hybrids in the same species. Our results have led to the following conclusions:

1. The different subspecies of *L. fulvus* ($2N = 60$) regularly give fertile hybrids. Hill's (1953) grouping of them into the same species is correct.

TABLE I. Known Karyotypes of the Malagasy Lemurs

Species	Number of specimens examined M	F	2N	Chromosomes[a] M	S	A	X	Y	Reference
Microcebus m. murinus	—	1	66	—	2	64	—	—	Chu and Bender (1962)
M. m. murinus	2	—	66	—	—	64	M	A	Rumpler and Albignac (1973a)
M. m. murinus	—	1	66	—	—	—	—	—	de Boer (1973)
M. m. rufus	1	—	66	—	—	64	M	A	Rumpler and Albignac (1973a)
M. coquereli	4	1	66	—	—	64	M	A	Rumpler and Albignac (1973a)
Cheirogaleus major	—	1	66	—	2	64	A	A	Chu and Bender (1962)
C. major	2	1	66	—	—	64	M	A	Rumpler and Albignac (1973a)
C. medius	3	1	66	—	—	64	M	A	Rumpler and Albignac (1973a)
Phaner furcifer	3	3	46	4	12	28	M	A	Rumpler and Albignac (1973a)
Lemur f. fulvus	1	1	48	10	6	30	A	A	Chu and Swomley (1961)
L. f. fulvus	Sex unknown		58	—	—	—	—	—	Chiarelli (see Chu and Bender, 1962)
L. f. fulvus	2	1	58	—	4	52	A	A	Egozcue (1967)
L. f. fulvus	4	3	60	—	4	54	A	A	Rumpler and Albignac (1969b)
L. f. rufus	—	2	60	—	4	54	A	A	Chu and Swomley (1961)
L. f. rufus	2	4	60	—	4	54	A	A	Rumpler and Albignac (1969a)
L. f. albifrons	1	2	60	—	4	54	A	A	Rumpler and Albignac (1969c)
L. f. albifrons	—	1	60	—	4	54	A	A	Chu and Swomley (1961)
L. f. sanfordi	1	1	60	—	4	54	A	A	Rumpler and Albignac (1969c)
L. f. collaris	2	2	52	6	6	38	A	A	Rumpler and Albignac (1969b)
L. f. albocollaris	2	2	48	6	10	30	A	A	Rumpler and Albignac (1969b)
L. macaco	1	1	44	12	8	22	A	A	Chu and Bender (1961)
L. macaco	3	3	44	12	8	22	A	A	Egozcue (1967)
L. macaco	2	3	44	12	8	22	A	A	Rumpler and Albignac (1969a)
L. rubriventer	—	1	50	2	12	36	—	—	Rumpler and Albignac (1971)
L. catta	2	—	56	6	4	44	A	A	Chu and Swomley (1961)
L. catta	—	2	56	8	6	40	A	A	Egozcue (1967)
L. catta	11	13	56	2	6	46	M	A	Rumpler (1970)
L. catta	1	1	56	2	6	46	M	A	Hayata et al. (1971)
L. mongoz	2	3	60	—	4	54	A	A	Chu and Swomley (1961)
L. mongoz	1	7	58	—	4	52	A	A	Egozcue (1967)
L. mongoz	2	1	60	—	4	54	A	A	Rumpler and Albignac (1969c)
L. mongoz	1	1	60	—	4	54	A	A	Hayata et al. (1971)
L. mongoz	—	1	60	—	4	54	A	A	de Boer (1973)
L. coronatus	2	1	46	6	12	26	A	A	Rumpler and Albignac (1969c)
Hapalemur g. griseus	1	1	54	4	6	42	A	A	Chu and Swomley (1961)
H. g. griseus	6	5	54	4	6	42	A	A	Rumpler and Albignac (1973b)
H. g. alaotrensis	1	1	54	4	6	42	A	A	Rumpler and Albignac (1973b)
H. g. occidentalis	—	1	58	2	4	52	—	—	Chu and Swomley (1961)
H. g. occidentalis	2	1	58	2	4	50	A	A	Rumpler and Albignac (1973b)
H. simus	2	1	60	—	4	54	M	A	Rumpler and Albignac (1973b)
Varecia variegata	1	—	46	14	4	26	M	A	Chu and Swomley (1961)
V. variegata	2	4	46	8	10	26	S	A	Rumpler and Albignac (1971)
Lepilemur ruficaudatus	7	9	20	2	16	—	M	A	Rumpler et al. (1972)

TABLE I. (continued)

Species	Number of specimens examined M	F	Chromosomes [a] 2N	M	S	A	X	Y	Reference
L. leucopus	2	2	26	4	14	6	M	A	In preparation
L. mustelinus	3	—	34	—	6	26	S	A	In preparation
L. rufescens	3	2	22	4	14	2	A	A	In preparation
L. dorsalis	8	7	26	8	10	6	M	A	In preparation
L. s. septentrionalis	1	5	34	—	6	26	M		Rumpler and Albignac (in press)
L. s. ankaranensis	2	2	36	—	4	30	M		Rumpler and Albignac (in press)
L. s. andrafiamenensis	2	—	38	—	2	34	M		Rumpler and Albignac (in press)
L. s. sahafarensis	2	—	36	—	4	30	M		Rumpler and Albignac (in press)
Avahi laniger	1	—	[64]	—	2	60	M	A	Rumpler and Albignac (1973c)
Propithecus v. verreauxi	1	—	48	—	—	—	—	—	Chu and Bender (1962)
P. v. verreauxi	2	2	48	14	16	16	S	A	Rumpler and Albignac (1973c)
P. v. deckeni	1	2	48	—	—	—	—	—	Chu and Bender (1962)
P. v. deckeni	1	1	48	14	16	16	S	A	Rumpler and Albignac (1973c)
P. v. coronatus	1	1	48	14	16	16	S	A	Rumpler and Albignac (1973c)
P. v. coquereli	1	—	48	14	16	16	S	A	Rumpler and Albignac (1973c)
P. d. diadema	3	—	42	18	14	8	M	A	Rumpler and Albignac (1973c)
P. d. perrieri	1	—	42	18	14	8	M	A	Rumpler and Albignac (1973c)
Indri indri	1	—	40	12	20	6	M	A	Rumpler and Albignac (1973c)
Daubentonia madagascarensis	1	1	30	6	18	4	M	A	In preparation

[a]M, Metacentric chromosome; S, submetacentric chromosome; A, acrocentric chromosome
Data in brackets are tentative.

2. The classification of *L. f. albocollaris*[1] ($2N = 48$) in the species *L. fulvus* can be discussed from the point of view of the hybrids which we obtained (crossbreedings 12 and 13 from Table II). The F_1 hybrids between *L. f. rufus* ($2N = 60$) and *L. f. albocollaris* ($2N = 48$) have a diploid number of $2N = 54$ (Fig. 1). At meiosis, the chromosomes of the hybrid F_1 form 18 bivalents and six trivalents in the germ cells. The six trivalents can spread out in a balanced way, producing germ cells with seven different haploid numbers be-

[1] The cytogenetic study of the animals of this group, however, allowed us to isolate a new type, *L. f. albocollaris*, characterized by its karyotype: $2N = 48$. Morphologically, the male is different from that of *L. f. collaris* in the possession of white instead of rufous cheeks, whereas the females of both types have reddish cheeks. The first animals we studied were females, and were temporarily identified as *L. f. collaris* (red beard) type b or type II, in contrast to *L. f. collaris* sensu stricto, which we called *L. f. collaris* (red beard) type c or type III (Rumpler and Albignac, 1969b).

TABLE II. Number and Karyotypes of the Hybrids Between Different *Lemur* Subspecies

	Number			Chromosomes[a]				
Hybrids	M	F	2N	M	S	A	X	Y
L. f. fulvus × *L. f. rufus*	2	2	60	—	4	56	A	A
L. f. fulvus × *L. f. albifrons*	1	—	60	—	4	56	A	A
L. f. fulvus × *L. f. sanfordi*	1	—	60	—	4	56	A	A
L. m. macaco × *L. f. albifrons*	1	2	52	6	6	38	A	A
L. m. macaco × *L. f. fulvus*	2	3	52	6	6	38	A	A
L. m. macaco × *L. f. rufus*	3	3	52	6	6	38	A	A
L. m. macaco × *L. f. albocollaris*	3	2	46	10	8	26	A	A
L. f. albocollaris × *L. f. rufus*	2	2	54	4	6	42	A	A
L. f. collaris × *L. macaco*	—	1	48	9	7	30	A	A
L. f. collaris × *L. f. fulvus*	—	1	56	3	5	46	A	A
Hybrid 1 × *hybrid 2*	1	1	60	—	4	56	A	A
Hybrid 8 × *L. m. macaco*	1	—	50	6	8	34	A	A
Hybrid 8 × *L. f. fulvus*	1	—	54	4	6	42	A	A

[a]M, Metacentric chromosome; S, submetacentric chromosome; A, acrocentric chromosome.

tween 24 and 30 (Table III). Crossbreeding 12, which has produced an animal of $2N = 50$, is the result of fertilization of an ova $N = 28$ by a spermatozoon $N = 22$ from *L. macaco* ($2N = 44$). Crossbreeding 13, which produced an animal of $2N = 54$, is the result of the fertilization of an ova $N = 24$ by a spermatozoon $N = 30$ from *L. fulvus* ($2N = 60$) (Fig. 1). The success of these two fertile crossbreedings out of three attempts would have been highly improbable if the segregation of the trivalents had taken place completely by chance, because then there would only have been one balanced germ cell out of 3^6, i.e., one out of 729. The preferential segregation of the trivalents in favor of balanced germ cells with such a high frequency shows that the chromosomal barrier is penetrable, and one must agree with Hill that *L. f. albocollaris* should be subsumed into the species *L. fulvus*.

3. The status of *L. f. collaris* ($2N = 52$) is more difficult to determine at the present stage of our research. This animal is easily crossbred with *L. f. fulvus* ($2N = 60$) to give hybrids of $2N = 56$, and with *L. f. albocollaris* ($2N = 48$) to give hybrids of $2N = 50$, but we have not been able to test the fertility of F₁ individuals, which have not yet reached sexual maturity. The chromosomal barrier between *L. f. collaris* ($2N = 52$) and *L. fulvus* ($2N = 60$) is weaker than in the previous case because the forms differ only in four pairs of chromosomes. The barrier is still smaller between *L. f. collaris* and *L. f. albocollaris* since these differ only in two pairs of chromosomes. It seems logical to keep *L. f. collaris* in the species *L. fulvus*, at least until the fertility of the hybrids can be examined.

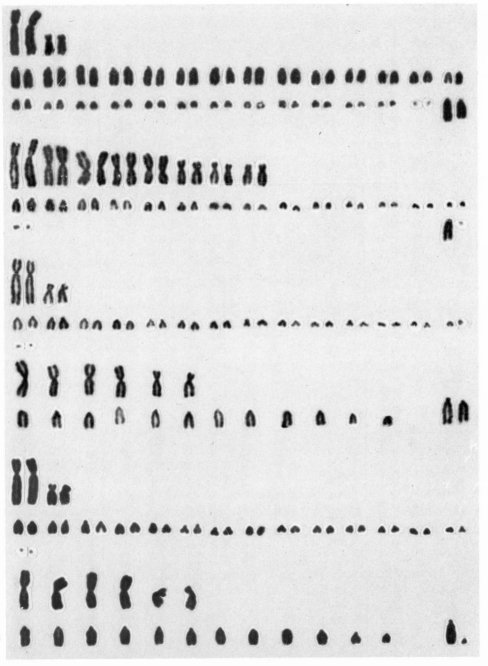

FIG. 1. Karyotypes of (a) *Lemur fulvus rufus* females, (b) *L. f. albocollaris* male, (c) female hybrid between the above two, (d) a male hybrid resulting from the fertilization of the hybrid F_1 by a *L. f. fulvus* male. Note that the karyotype of the second hybrid is similar to that of the F_1 hybrid (ignoring the sex chromosomes).

TABLE III. Production of the Seven Haploid Numbers of the Balanced Germ Cells from the Hybrid Between *L. f. rufus* and *L. f. albocollaris* and the Fecundation of Two of These by *L. f. fulvus* and *L. m. macaco* Spermatozoons[a]

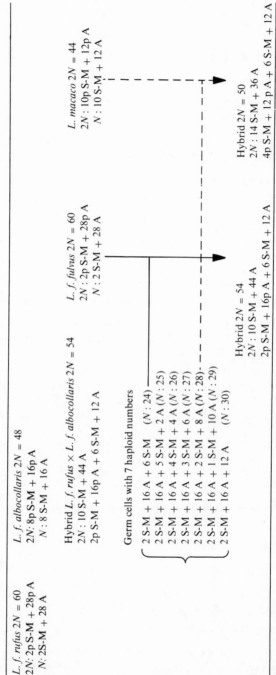

L. f. rufus 2N = 60
2N: 2p S-M + 28p A
N: 2S-M + 28 A

L. f. albocollaris 2N = 48
2N: 8p S-M + 16p A
N: 8 S-M + 16 A

Hybrid *L. f. rufus* × *L. f. albocollaris* 2N = 54
2N: 10 S-M + 44 A
2p S-M + 16p A + 6 S-M + 12 A

Germ cells with 7 haploid numbers

2 S-M + 16 A + 6 S-M   (N: 24)
2 S-M + 16 A + 5 S-M + 2 A (N: 25)
2 S-M + 16 A + 4 S-M + 4 A (N: 26)
2 S-M + 16 A + 3 S-M + 6 A (N: 27)
2 S-M + 16 A + 2 S-M + 8 A (N: 28)
2 S-M + 16 A + 1 S-M + 10 A (N: 29)
2 S-M + 16 A + 12 A   (N: 30)

L. f. fulvus 2N = 60
2N: 2p S-M + 28p A
N: 2 S-M + 28 A

L. macaco 2N = 44
2N: 10p S-M + 12p A
N: 10 S-M + 12 A

Hybrid 2N = 54
2N: 10 S-M + 44 A
2p S-M + 16p A + 6 S-M + 12 A

Hybrid 2N = 50
2N: 14 S-M + 36 A
4p S-M + 12 p A + 6 S-M + 12 A

[a]Note that the hybrid resulting from the fecundation of the first type of balanced germ cells has the same diploid number, 2N = 54, as the hybrid *L. f. rufus* × *L. f. albocollaris*. M, Metacentric chromosome; S, submetacentric chromosome; A, acrocentric chromosome; 2p, two pairs; p. pair.

4. The form *L. macaco* seems to be clearly different from the others. Although the procuring of F_1 hybrids between *L. macaco* and the other groups of *L. fulvus* is easy, we have not yet been able to obtain either the F_2 or the backcross. The hybrids between *L. macaco* and *L. fulvus albocollaris* have not reproduced, although they have been through several reproductive seasons (in each of the original crossbreedings, the frequency of reproduction was 0/5). This absence of fertility shows that there is a break between *L. macaco* and *L. fulvus* and suggests the maintenance of *L. macaco* as a separate species. It will be possible to reach a more definite conclusion only when we have more hybrids and F_1 crosses.

5. Still in the genus *Lemur*, we consider *L. mongoz* and *L. coronatus* to be separate species. These two animals have very different karyotypes, although it seems that that of *L. coronatus* derives from $2N = 60$ only through centric fusion. Even though this numerical difference does not argue in favor of two different species, the absence of F_1 hybrids between the two, in addition to morphological considerations (Petter, 1962; Mahé, 1972; Rakotosamimanana and Rumpler, 1970), argues strongly for separation.

In the genus *Hapalemur*, cytogenetic studies and hybridization experiments have similarly allowed us to clarify the systematics. There exist, in the species *H. griseus*, three subspecies (Rumpler and Albignac, 1973): *H. g. griseus* ($2N = 54$), living in the east; *H. g. alaotrensis* ($2N = 54$), recently discovered in the area of Lake Alaotra by Petter and Peyrieras (1970, Chapter 15); and *H. g. occidentalis* ($2N = 58$), living in the north and west. The two subspecies sharing $2N = 54$, however, are morphologically very different; moreover, *H. g. griseus* is small, half the weight of *H. g. alaotrensis*. *H. g. occidentalis*, confused by certain biologists with the hypothetical *Hapalemur olivaceus*, has a body size very little different from that of *H. g. griseus*, although its karyotype differs through the centric fusion of two pairs of chromosomes. The two subspecies are able to crossbreed, but we have not yet been able to test the fertility of the hybrids ($2N = 56$), which are still

TABLE IV. Number and Karyotypes of the Hybrids Between the Different Subspecies of *Hapalemur griseus*

Hybrids	Number of specimens			Chromosomes[a]				
	M	F	2N	M	S	A	X	Y
H. g. griseus × *H. g. alaotrensis*	2	1	54	4	6	42	A	A
H. g. griseus × *H. g. occidentalis*	1	—	56	3	5	46	A	A

[a]M, Male; F, female; M, metacentric chromosome; S, submetacentric chromosome; A, acrocentric chromosome.

immature (Table IV). The existence of a chromosome difference as weak as this is not *a priori* an obstacle to obtaining fertile hybrids, especially since it is due to centric fusion. As for *H. simus*, its karyotype differentiates it clearly from *H. griseus*, in particular through the pericentric inversion of an X chromosome. Its place in a separate species is therefore justified.

The taxonomy of *Lepilemur* has similarly been very controversial. For Hill (1953), the genus includes two species: *L. mustelinus* and *L. ruficaudatus*, the latter with two subspecies: *L. r. ruficaudatus* and *L. r. leucopus*. For Petit (1933) and for Webb (1946), on the other hand, the species *L. ruficaudatus* contains a third subspecies, *L. r. dorsalis*. Petter and Petter-Rousseaux (1960)

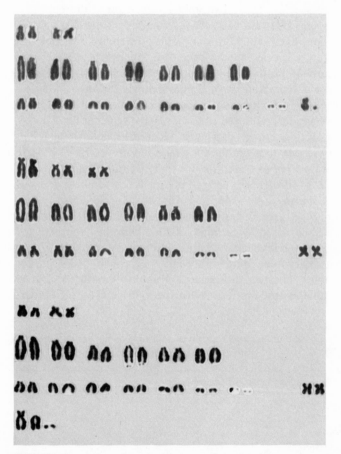

FIG. 2. Karyotypes of (a) *Lepilemur s. sahafarensis*, (b) *L. s. septentrionalis*, (c) a fertile hybrid between the two first. Two un-paired acrocentric chromosomes correspond to the unpaired sub-metacentric chromosome of the hybrid.

FIG. 3. Hybrid male resulting from the cross of the hybrid *Lemur fulvus albocollaris* × *L. f. rufus* with *L. macaco*. The coloration of the pelage is reddish-yellow, with white ears and lighter patches above the eyes. Pelage coloration in this animal is very different from both the various subspecies of *Lemur fulvus* and the hybrids between *L. fulvus* and *L. f. albocollaris* and between *L. fulvus* and *L. macaco*.

thought there was only one species, *L. mustelinus*, with five subspecies: *L. m. mustelinus*, *L. m. ruficaudatus*, *L. m. leucopus*, *L. m. dorsalis*, and *L. m. microdon*. The cytogenetic study of *Lepilemur* from different areas in the island has revealed the existence of very dissimilar karyotypes (Table I), resulting from chromosomal transformations more complex than in *Lemur*. The rearing of these animals is difficult and we have not been able to induce

TABLE V. Number and Karyotypes of the Hybrids Between the Different Subspecies
of *Lepilemur septentrionalis*

Hybrids	Number of specimens			Chromosomes[a]			
	M	F	2N	S	A	X	Y
L. s. ankaranensis × *L. s. andrafiamenensis*	1	1	37	3	32	M	A
L. s. sahafarensis × *L. s. septentrionalis*	2	5	35[b]	5	28	M	A

[a]M, Male; F, female; S, submetacentric chromosome; A, acrocentric chromosome; M, metacentric chromosome.

[b]Among the five caught females, three were pregnant, providing the fertility of these hybrids.

breeding even in couples of the same species. Because of the impossibility of
performing hybridization experiments between the various forms of *Lepilemur*, we have been led to propose a systematic revision based solely on the
karyotype and we have considered as separate species those animals whose
karyotypes permit us to infer the absence of fertile hybrids. We have been
able in this way to distinguish seven species of *Lepilemur: L. leucopus*, which
lives in the region of Fort Dauphin; *L. ruficaudatus*, in the west from Sakaraha to Antsalova (Rumpler *et al.*, 1972); *L. rufescens*, in the west from
the north of Antsalova to Ankarafantsika; *L. dorsalis*, in the Sambirano region and Nosy-Be; *L. mustelinus*, in the east from Perinet to Foulepointe; *L. microdon*; and *L. septentrionalis* (Fig. 2) in the north beyond
Ambilobe, which includes four subspecies deriving from each other by centric
fusion, *L. s. andrafiamenensis, L. s. ankaranensis, L. s. sahafarensis*, and
L. s. septentrionalis (Rumpler and Albignac, in press). We have provisionally
kept the species *L. microdon*, although its karyotype is still unknown, in order to reflect the views of Hill and Petter.

The karyotypes of the various *Lepilemur* species differ considerably
from those of other Lemurinae. If one views the karyotypes of all the Malagasy lemurs as coming from a single primitive formula similar to that of the
Cheirogaleinae, very complex chromosomal transformations must have
taken place to give the karyotypes of *Lepilemur* species. These great chromosomal differences, allied to the existence of several other morphological,
ecoethological, hematological, and biochemical characteristics, have led us to
classify the genus in a separate subfamily: Lepilemurinae (Rumpler and
Rakotosamimanana, 1971; Rumpler, 1974).

In the Indriidae, the cytogenetic results are in agreement with classical
taxonomy. The existence in the genus *Propithecus* of two dissimilar karyotypes confirms classification of the genus into two different species: *P. verreauxi* and *P. diadema*. The karyotype of *P. diadema* did not derive directly
from that of *P. verreauxi*; rather, both come from a common, more primitive,
karyotype through centric fusion and pericentric inversion not always affect-

ing the same chromosomes (Rumpler and Albignac, 1973c). Subspecies exist in these two species with karyotypes which are identical (Table I). The production of hybrids between the subspecies of *P. verreauxi* is not surprising, but their fertility had only recently been confirmed.

In the Cheirogaleidae, *Phaner furcifer* is the only animal to have a divergent chromosomal complement. Its fundamental number is different (NF = 62) from that of Cheirogaleinae (NF = 66) and from that of the Lemurinae (NF = 64). This means that chromosomal transformations more complex than simple centric fusion have occurred in the differentiation of its karyotype. These cytogenetic considerations, plus the discovery of new morphological characteristics such as the presence of a scent-marking gland in the neck (Rumpler and Andriamiandra, 1971), the specific dermatoglyphics (Rumpler and Rakotosamimanana, 1971), and the skull characteristics (Mahé, 1972), have led us to put *Phaner* in a separate subfamily (Rumpler and Rakotosamimanana, 1971).

In view of the results of our cytogenetic studies and hybridization experiments, we would modify Hill's (1953) classification of the Malagasy lemurs as shown below:

Family Cheirogaleidae Rumpler and Albignac, 1973
   Subfamily Cheirogaleinae Gregory, 1915
      Microcebus E. Geoffroy, 1828
        M. murinus (Miller, 1777)
        M. coquereli (Grandidier, 1867)
      Cheirogaleus E. Geoffroy, 1812
        C. major E. Geoffroy, 1812
        C. medius E. Geoffroy, 1812
      Allocebus Petter-Rousseaux and Petter, 1967
        A. trichotis (Günther, 1875)
   Subfamily Phanerinae Rumpler and Rakotosamimanana, 1971
      Phaner Gray, 1870
        P. furcifer (Blainville, 1841)
Family Lemuridae Gray, 1821
   Subfamily Lemurinae Mivart, 1864
      Lemur Linnaeus, 1758
        L. fulvus E. Geoffroy, 1812
          L. f. fulvus E. Geoffroy, 1812
          L. f. rufus Audebert, 1800
          L. f. albifrons E. Geoffroy, 1796
          L. f. sanfordi Archbold, 1932
          L. f. collaris E. Geoffroy, 1812
          L. f. albocollaris ssp. nov.
        L. macaco Linnaeus, 1766
          L. m. macaco Linnaeus, 1766

L. m. flavifrons (Gray, 1867)
L. mongoz Linnaeus, 1766
L. coronatus Gray, 1842
L. catta Linnaeus, 1758
L. rubriventer I. Geoffroy, 1850
Hapalemur I. Geoffroy, 1851
  H. griseus (Link, 1795)
    H. g. griseus (Link, 1795)
    H. g. alaotrensis ssp. nov.
    H. g. occidentalis ssp. nov.
  H. simus Gray, 1870
Varecia Gray, 1870
  V. variegata (Kerr, 1792)
    V. v. variegata (Kerr, 1792)
    V. v. ruber (E. Geoffroy, 1812)
Subfamily Lepilemurinae Rumpler and Rakotosamimanana, 1971
  Lepilemur I. Geoffroy, 1851
    L. dorsalis Gray, 1870
    L. ruficaudatus A. Grandidier, 1867
    L. rufescens Lorenz-Liburnau, 1898
    L. mustelinus I. Geoffroy, 1851
    L. leucopus Forsyth-Major, 1894
    L. microdon Forsyth-Major, 1894
    L. septentrionalis Rumpler and Albignac, 1974
      L. s. andrafiamenensis Rumpler and Albignac, 1974
      L. s. septentrionalis Rumpler and Albignac, 1974
      L. s. sahafarensis Rumpler and Albignac, 1974
      L. s. ankaranensis Rumpler and Albignac, 1974
Family Indriidae Burnett, 1828
  Avahi Jourdan, 1834
  Avahi laniger (Gmelin, 1788)
  Propithecus Bennett, 1832
    P. diadema Bennett, 1832
    P. verreauxi A. Grandidier, 1867
  Indri E. Geoffroy and Cuvier, 1795
Family Daubentoniidae Gray, 1870
  Daubentonia E. Geoffroy, 1795

REFERENCES

Bender, M. A., and Chu, E. H. Y., 1963, The chromosomes of the primates, in: *Evolutionary and Genetic Biology of Primates*, Vol. 1 (J. Buettner-Janusch, ed.), pp. 261–310, Academic Press, New York.

Buettner-Janusch, J., Swomley, B. A., and Chu, E. H. Y., 1966, Les nombres chromosomiques de certains lémuriens de Madagascar, *Publ. ORSTOM Sec. Biol.* **2**:2–7.

Caspersson, T., Zech, L., and Johansson, C., 1970, Analysis of human metaphase chromosome set by aid of DNA binding fluorescent agents, *Exp. Cell Res.* **62**:490–492.

Chu, E. H. Y., and Bender, M. A., 1961, Chromosome cytology and evolution in primates, *Science* **133**:1399–1405.

Chu, E. H. Y., and Bender, M. A., 1962, Cytogenetics and evolution of primates, *Ann. N.Y. Acad. Sci.* **102**:253–266.

Chu, E. H. Y., and Swomley, B. A., 1961, Chromosomes of lemurine lemurs, *Science* **133**:1925–1926.

de Boer, L. M., 1973, Studies on the cytogenetics of prosimians, *J. Hum. Evol.* **2**:271–278.

Dutrillaux, B., and Lejeune, J., 1971, Sur une nouvelle technique d'analyse du caryotype humaine, *C. R. Acad. Sci. (Paris)* **272**:2638–2640.

Eckhardt, R. B., 1972, A chromosome arm number index and its applications to the phylogeny and classification of Lemurs, *Am. J. Phys. Anthropol.* **31**:85–88.

Egozcue, J., 1967, Chromosome variability in the Lemuridae, *Am. J. Phys. Anthropol.* **26**:341–348.

Gropp, A., Tettenborn, U., and von Lehmann, E., 1970, Chromosomen Variation vom Robertson'schen Typus bei der Tabakmaus, *M. proschiavinus* und ihren Hybriden mit der Laboratoriummaus, *Cytogenetics* **9**:9–23.

Hayata, I., Sonta, S., Itoh, M., and Kondo, N., 1971, Notes on the karyotypes of some prosimians, *Lemur mongoz, Lemur catta, Nycticebus coucang* and *Galago crassicaudatus, Jap. Genet.* **46**:61–64.

Hill, W. C. O., 1953, *Primates: Comparative Anatomy and Taxonomy*, Vol. 1, The University Press, Edinburgh.

MacFee, A. F., and Banner, M. W., 1969, Inheritance of chromosome number in pigs, *J. Reprod. Fert.* **18**:9–14.

Mahé, J., 1972, Thèse: Crâniométrie des Lémuriens–Analyses Multivariables–Phylogénie (personal communication).

Matthey, R., 1964, Evolution chromosomique et spéciation chez les *Mus* du sous-genre *Leggada* Gray, 1837, *Experientia* **20**:657–666.

Petit, G., 1933, Le genre *Lepidolemur* et sa répartition géographique, *C.R. Soc. Biogéogr. Paris* **X**:33–37.

Petter, J. J., 1962, Recherches sur l'écologie et l'éthologie des lémuriens malgaches, *Mém. Mus. Natl. Hist. Nat. (Paris)* **A27**:1–146.

Petter, J.J. and Petter-Rousseaux, A., 1960, Remarque sur la systématique du genre *Lepilemur, Mammalia* **24**:76–86.

Petter, J. J., and Peyrieras, A., 1970, Observations éco-éthologiques sur les lémuriens malgaches du genre *Hapalemur, Terre Vie* **24**(3):356–382.

Rakotosamimanana, B., and Rumpler, Y., 1970, Etude des dermatoglyphes palmaires et plantaires de quelques lémuriens malgaches, *Bull. Assoc. Anat.* **148**:493–510.

Rumpler, Y., 1970, Etude cytogénétique de *Lemur catta, Cytogenetics* **9**:239–244.

Rumpler, Y., 1974, Contribution of cytogenetics to a new classification of the malagasy Lemurs, in: *Prosimian Biology.* (R. D. Martin, G. A. Doyle, and A. C. Walker, eds.), Duckworth, London.

Rumpler, Y. and Albignac, R., 1969a, Etude cytogénétique de deux lémuriens: *L. macaco macaco* Linnaeus, 1766 et *L. fulvus rufus* Audeburt, 1800, et d'un hybride *macaco macaco* × *fulvus rufus, C.R. Soc. Biol.* **163**:1247–1250.

Rumpler, Y., and Albignac, R., 1969b, Existence d'un variabilité chromosomique intraspécifique chez certains lémuriens, *C. R. Soc. Biol.* **163**:1989–1992.

Rumpler, Y., and Albignac, R., 1969c, Etude cytogénétique de quelques hybrides intraspécifiques et interspécifiques de lémuriens, *Ann. Univ. Besançon Med.* **6**:15–18.

Rumpler, Y., and Albignac, R., 1969d, Evolution chromosomique des lémuriens malgaches, *Ann. Univ. Madagascar Med. et Biol.* **12**:123–131.

Rumpler, Y., and Albignac, R., 1971, Etude cytogenetique du *Varecia variegata* et du *Lemur rubriventer, C. R. Soc. Biol.* **165**: 741–747.

Rumpler, Y., and Albignac, R., 1973a, Cytogenetic study of the endemic Malagasy lemur subfamily Cheirogaleinae Gregory, 1915, *Am. J. Phys. Anthropol.* **38**:261–264.

Rumpler, Y., and Albignac, R., 1973b, Cytogenetic study of the endemic Malagasy lemur: Hapalemur, I. Geoffroy, 1851, *J. Hum. Evol.* **2**:267–270.

Rumpler, Y., and Albignac, R., 1973c, Cytogenetic study of the Intrüdae Burnett 1828, a malagasy endemic Lemur's family (International Congress of Anthropological and Ethnological Sciences—Conference and Workshop on Comparative Kariology of Primates) August, 1973, Detroit and Chicago (in press).

Rumpler, Y. and Albignac, R., (in press), Intraspecific variability in a Lemur from the North of Madagascar: *Lepilemur Septentrionalis*, species nova, *Am. J. Phys. Anthropol.*

Rumpler, Y., and Andriamiandra, A., 1971, Etude histologique des glandes de marquage du cou des lémuriens malgaches, *C. R. Soc. Biol.* **164**:436–441.

Rumpler, Y., and Rakotosamimanana, B., 1971, Coussinets plantaires et dermatoglyphes de répresentants de lémuriformes malgaches, *Bull. Assoc. Anat.* **54**:493–510.

Rumpler, Y., Albignac, R., and Rumpler-Randriamonta, N., 1972, Etude cytogénétique de *Lepilemur ruficaudatus, C. R. Soc. Biol.* **166**:1208–1211.

Turpin, R., and Lejeune, J., 1964, *Les Chromosomes Humains*, Gauthier-Villars, Paris.

Webb, C. S., 1946, Some Madagascan animals, *Zoo Life London* **i**:57–58 (cited by Hill, W. C. O., 1953).

Development and Eruption of the Premolar Region of Prosimians and Its Bearing on Their Evolution

4

JEFFREY H. SCHWARTZ

Among the complexes most frequently called on in the establishment of relationships within and between various mammal groups, dentition stands out as perhaps the most important traditional avenue of comparative investigation. That dental morphology has been so greatly relied on in paleontological studies is not surprising, as teeth survive the processes of fossilization far better and much more frequently than do other parts of the skeleton. The application to paleontology (and neontology as well) most commonly undertaken is the comparative study of occlusal morphology, whereby identifications of taxa are based on such criteria as the number of cusps, their disposition within the tooth, the presence of cingula, and the presence or absence of such structures as conules, stylids, and lophs. However, when one is dealing with such final stages of development, one is faced solely with static morphologies which may on the surface appear to be similar but which in fact may only represent secondary and convergent modifications of the final stages of processes

JEFFREY H. SCHWARTZ Department of Anthropology, University of Pittsburgh, Pittsburgh, Pennsylvania 15260

which are different on a more fundamental level. This is equally the case with other morphologies such as the ear region. In other words, crown and root morphology represent only one, the last, of many levels of dental organization.

The basic levels at which one may study the dentition include (1) processes and mechanisms (such as budding) involving the dental lamina, (2) sequences of dental development, (3) sequences of eruption, and (4) crown and root morphology. Each such level of dental organization is determined by a different genetic substrate (Butler, 1939, 1956, 1963; Schwartz, 1974a). Since all mammals, regardless of differences at other levels, share the same pattern of budding from the dental lamina, i.e., the interaction of neural crest cells with the mesenchyme and ectomesenchyme resulting in discrete waves of incisor, premolar, and molar budding (Osborn, 1970, 1973), it would appear that this is the most stable genetic substrate. On the other hand, morphology is often subject to quite rapid change; the fossil record abounds with examples of convergence which reflect similar modes of locomotion or dietary habits in phylogenetically unrelated forms. Thus since crown morphology, and therefore tooth morphogenesis, can be altered rapidly by crowding or inhibition within the jaw (Butler, 1939, 1956; Osborn, 1971) as well as by selection pressures on the genetic fields which govern such morphology (Butler, 1939, 1963; Patterson, 1956; Van Valen, 1970), it follows that the genetic substrate controlling this level is relatively unstable and most likely to change. It is reasonable, then, to expect a continuum between processes of the dental lamina and tooth morphology, with levels closer to tooth morphology being less genetically stable and more subject to change than levels closer to lamina processes, affording significant taxonomic and phylogenetic information (Schwartz, 1974a). Within these intermediate levels, one may also distinguish the genetic substrate governing the order of dental development as more stable than that which determines the sequence of eruption (Schwartz, 1974a).

Given the above, the purpose of this chapter is to present data on the development and eruption of the premolar region of the prosimians in the hope that this new set of characters will be helpful in determining phylogenetic relationships, particularly among the Malagasy lemurs. Although it might be preferable at times to consider total sequences of dental development and eruption (e.g., see Schwartz, 1974a; Tattersall and Schwartz, 1974), the data presented here on the premolar region are sufficiently sensitive for discussion down to the generic level. This is in striking contrast to certain other sets of teeth which present characters common to all prosimians. For instance, the molars invariably develop and erupt in the order M1–M2–M3, and although the order of development and eruption of the upper incisors may be variable, as a set their appearance is coordinated with that of the toothcomb; the toothcombs of different taxa develop and erupt consistently, with the central teeth (incisors) appearing as a unit well ahead of the intially retarded and visibly distinct lateral teeth (canines) (Schwartz, 1974c).

A total of 212 prosimians was studied, with sample sizes for subfamilies as follows: Cheirogaleinae (n = 16), Lemurinae (n = 72), Indriinae (n = 27), Daubentoniinae (n = 2), Lorisidae (n = 78), and Tarsiinae (n = 14); in addition, two specimens of *Archaeolemur majori* (PM 1935-419 and PM not registered) and one of *Hadropithecus stenognathus* (PM 1935-407) were analyzed. All specimens were radiographed.

Abbreviations used in the text and figures are as follows:

AMNH   American Museum of Natural History
PM   Museum National d'Histoire Naturelle, Paris
RM   Rijksmuseum van Natuurlijke Historie, Leiden

THE MALAGASY LEMURS

As used here, "Malagasy lemur" does not carry any taxonomic or phylogenetic implications; it is merely a reference to those prosimians which happen to occur on the island of Madagascar. Since a reclassification of the Malagasy lemurs seems warranted (Tattersall and Schwartz, 1974), reference will be made only to subfamilies and the genera and species contained within each. An exception is Indriidae, which unquestionably contains Indriinae, Archaeolemurinae, and Palaeopropithecinae (Tattersall, 1973a, b). It should also be mentioned that for those prosimians possessing three pairs of premolars (deciduous and permanent) in each jaw, the homologies accepted here are dp2, dp3, and dp4, and P2, P3, and P4. The indriines are recognized as possessing two pairs of both deciduous and permanent upper premolars (most likely dp^{3-4} and P^{3-4}), two pairs of permanent lower premolars (probably P_{2-3}), but four pairs of deciduous lower premolars (not necessarily homologous with the ancestral primate dp1–4) (Schwartz, 1974b, c); for convenience, the notation Pa (anterior), Pp (posterior), and $dp^{a,p}$ will be employed. The dentition of *Daubentonia* will be discussed separately.

Cheirogaleinae

Three specimens of *Phaner furcifer* were available for study. Data on the appearance of the deciduous premolars are lacking, but it was noted that the permanent premolars develop and erupt in the order P2–P4–P3 in both upper and lower jaw .

The data on *Microcebus* are better. R. D. Martin (personal communication), who has a breeding colony of mouse lemurs, reports that the order of

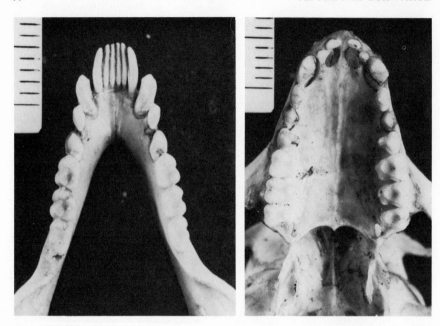

FIG. 1. *Microcebus coquereli* (RM K). Maxilla shows P² and P⁴ erupted and P³ only partially erupted; mandible shows P₂ fully, P₄ almost fully erupted, dp₃ still in place with the underlying tip of P₃ visible

eruption of the upper and lower deciduous premolars is dp2–dp4–dp3. From museum specimens of both *M. murinus* and *M. coquereli*, (n = 5), it was found that the order in which the permanent premolars developed and erupted was P2–P4–P3 (Fig. 1).

No specimens of *Cheirogaleus* were found which shed any light on the eruption of the deciduous dentition. The permanent premolars, however, were found to develop and erupt in the same order as those of *Phaner* and *Microcebus* (Fig. 2). The seven specimens of *Cheirogaleus* included *C. major* and *C. medius*.

Aside from my study, the only reported eruption data on the cheirogaleines comes from the work of Bennejeant (1936) in which the sequence P2–P3–P4 was presented for the entire subfamily solely on the basis of a radiograph of *Microcebus*. This is clearly incorrect.

Lemurinae

The deciduous premolars of *Lepilemur mustelinus* (n = 11) were found to erupt in the order dp2–dp4–dp3, in both the upper and lower jaws (Fig.

3A). In all specimens with the deciduous teeth present, the upper incisors (un-replaced in the adult) were also present. It is interesting to note that in some cases the upper deciduous incisors persisted throughout the eruption of the permanent tooth comb and would thus have had an unusually long functional life. The permanent premolars erupt in the order P4-P3-P2 (Fig. 3C, D), a condition markedly different from that in all cheirogaleines. If one did not see this order, but only had radiographs of the developing premolars, one would infer that the order of eruption would be P2-P4-P3 (Fig. 3B); this is because the crown of P2 develops and begins to calcify quite rapidly and must migrate and approach the alveolar margin well before P4 and P3 have reached their final stages of calcification. Lamberton (1938) gives the order $P^4-P^3-P^2$ for eruption in the upper jaw of *Lepilemur* but incorrectly reports $P_4-P_2-P_3$ for the lower.

Hapalemur griseus (n = 9) was found to have the eruption sequence dp2-dp3-dp4, for both upper and lower jaws. The permanent premolars develop and erupt in the order P4-P3-P2 (Fig. 4), which in the latter process is similar to *Lepilemur* but as a continuum is distinct.

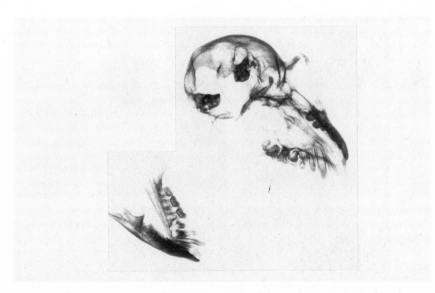

FIG. 2. *Cheirogaleus major* (PM 1882-1561). Radiograph of skull (P^2 is fully erupted, dp^4 is in place with P^4 fully developed, and dp^3 is in place but P^3 is only partially developed) and mandible (P_2 is fully erupted, P_4 is beginning to erupt, and dp_3 is in place with P_3 partially developed.

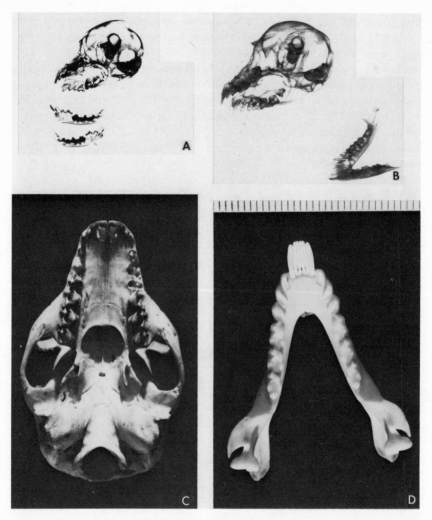

FIG 3. (A) *Lepilemur mustelinus* (PM 1962–2721). Radiograph of skull and mandible with dp2 fully, dp4 almost fully, and dp3 partially erupted. (B) (PM 1962–2717). Radiograph of skull and particularly the mandible showing development of permanent premolars in the sequence P2–P4–P3. (C) (PM 1962–2733). Occlusal view of skull. Note that P⁴ and P³ have erupted and P² is just beginning to break the alveolar surface. (D) (PM not registered). Occlusal view of mandible. Both P₄ and P₃ have erupted while P₂ is just partially erupted.

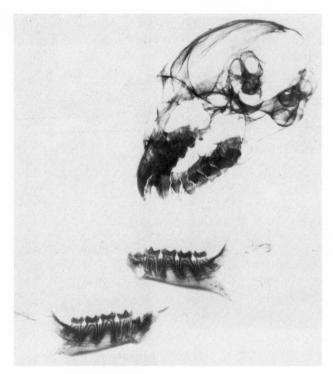

FIG. 4. *Hapalemur griseus* (PM 1880–2511). Radiograph of skull (with P^4 fully, P^3 almost fully, and P^2 beginning to erupt) and mandible (with P_4 fully and P_3 beginning to erupt while P_2 is in the final stages of development); differentiation between the two anterior premolars is based on the fact that the root system and crown of P_3 are complete while those of P_2 are not.

Contrary to what might be expected, the species of the genus *Lemur* do not present a homogeneous picture. The following points emerge:

1. *L. catta* ($n = 12$): The deciduous premolars erupt in the order dp2–dp3–dp4 (Fig. 5A), as in *Hapalemur*. The permanent premolars develop and erupt in the sequence P4–P3–P2 (Fig. 5B), again as in *Hapalemur*, but unlike other lemurines or any cheirogaleines.

2. *L. fulvus* ($n = 16$): The only data available pertain to the permanent premolars, which develop and erupt in the order P2–P4–P3, similar to the cheirogaleines and, in development, to *Lepilemur* (Fig. 5C).

3. *L. mongoz* ($n = 9$): The only data on the deciduous dentition came from one specimen, in which the lower premolars erupt in the order dp2–dp3–dp4; the upper premolars have not begun to erupt. Although

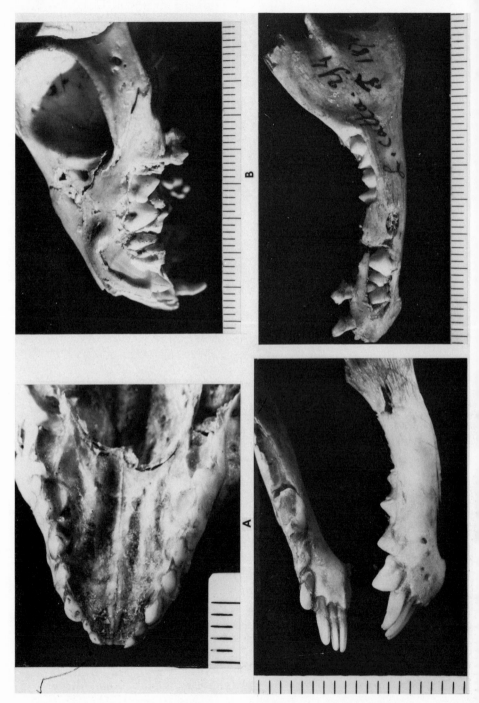

FIG. 5. (A) *Lemur catta* (RM h). Maxilla and mandible showing eruption sequence dp2–dp3–dp4. (B) *L. catta* (PM 1871–254). Above: Dissection of left maxilla showing P⁴ most developed and closest to breaking the alveolar margin, P³ less developed, and P² even less so. Below: Dissection of left mandible shows the same condition; in cases such as this, relative development of the root system is most significant. (C) *L. fulvus* (RM h). Above: Partial dissection of left maxilla. P² is almost fully erupted as evidenced by the cingulum being almost fully below the alveolar margin, P⁴ is only partially below the alveolar margin, and dp³ is still in place with P³ still deep in the bone. Below: Occlusal view of mandible showing that P₂ is fully erupted, P₄ is almost completely in place, and dp₃ is still present with the preeruption perforation of the bone indicating the presence of P₃.

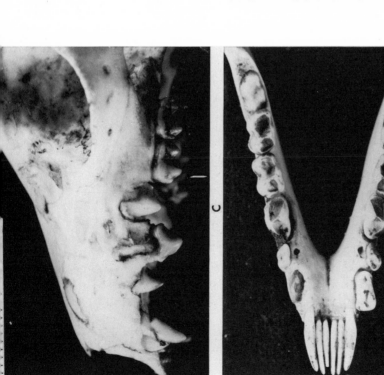

relative times of eruption may vary, it was found among most other prosimians, for both the deciduous and permanent premolars, that the sequences of eruption are the same in both the upper and lower jaws (see Table I); this is probably also the case with *L. mongoz*. The permanent set is well documented: the order of development and eruption is P2–P4–P3 (Fig. 6D). Bennejeant (1936) reported an eruption sequence of P4–P3–P2 for all Lemurinae, based on a specimen of *L. mongoz*. Either his analysis was incorrect or the individual was not *L. mongoz*; if not, it was probably *L. catta*.

4. *L. rubriventer* (*n* = 2): The deciduous premolars of both specimens erupted in the order dp2–dp3–dp4 (Fig. 6B). Although no specimens with the permanent premolars erupting were found, it was noted that P2 was the first to develop, suggesting the order P2–P4–P3 as seen in all other *Lemur* species with the exception of *L. catta*.

5. *L. macaco* (*n* = 9): The deciduous premolars were found to erupt in the order dp2–dp3–dp4 (Fig. 6A). The permanent premolars develop in the order P2–P4–P3, but completed eruption was not noted. However, as P2 is the first to erupt initially, it seems reasonable to expect that P4 would appear next, followed by P3.

6. *L. (Varecia) variegatus* (*n* = 4): No specimens were found with the deciduous teeth erupting. The permanent premolars develop and erupt in the order P2–P4–P3 (Fig. 6C).

Indriinae

Propithecus (*n* = 21) is among the best-represented prosimians in any museum collection. The upper posterior deciduous premolar develops and erupts prior to the anterior (Fig. 7A). No examples of development or eruption of any of the four lower deciduous premolars were noted, because even in the most juvenile specimens they were always in place. Besides the posterior-to-anterior direction of appearance of the upper deciduous premolars, *P. verreauxi* and *P. diadema* share a similarly directed order of development and eruption of the permanent premolars, i.e., Pp–Pa (Fig. 7C).

No specimens of *Avahi* (*n* = 6) were found which showed the deciduous premolars in a state either of development or of eruption. The permanent premolars, however, were noted to develop and erupt in the order Pp–Pa (Fig. 7B).

Although there were only two specimens of juvenile *Indri* available for study, one clearly showed the upper deciduous premolars erupting in a posterior-to-anterior direction while the other showed the same order for the permanent set, i.e., Pp–Pa. Lamberton (1938) has also reported this anteriorly directed sequence of premolar eruption for *Indri*.

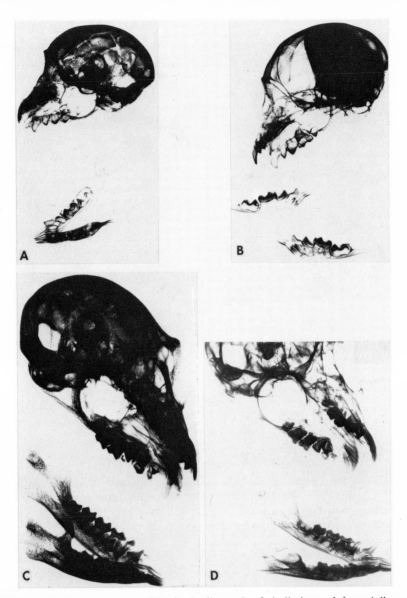

FIG. 6. (A) *Lemur macaco* (RM j). Radiograph of skull shows dp^2 partially and more erupted than dp^3 with dp^4 still in stages of eruption; radiograph of mandible shows dp$_2$ erupted and dp$_3$ more developed than dp$_4$. (B) *L. rubriventer* (PM 1871–245). Radiograph of maxilla shows dp^2 and dp^3 fully and dp^4 partially erupted; radiograph of mandible shows eruption in the sequence dp$_2$–dp$_3$–dp$_4$. (C) *L. variegatus* (PM A–12.938). Radiograph of mandible shows P$_2$ partially erupted, P$_4$ fully developed and about to erupt, and the crown of P$_3$ only slightly calcified; the maxilla is not as clear, but it appears that the development and eruption would be as in the mandible. (D) *L. mongoz* (RM e). Radiograph of mandible shows the same as for *L. rubriventer;* in the maxilla P^2 has been lost from the specimen and P^4 is more advanced in eruption than P^3 as evidenced by the retention of dp^3.

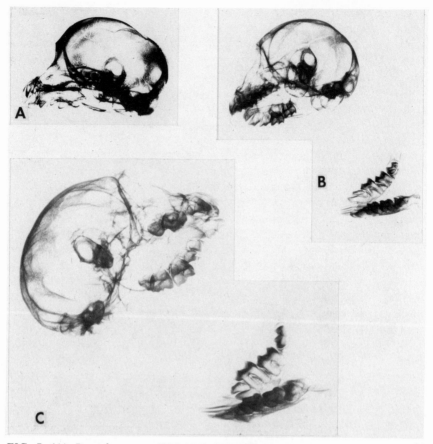

FIG. 7. (A) *Propithecus* sp. (PM 1962-228). Radiograph of skull showing dp$^{(p)}$, erupting before dp$^{(a)}$. (B) *Avahi laniger* (PM 1871-228). Radiograph of skull and mandible showing P(p) erupting before P(a). (C) *P. verreauxi* (PM 1909-267). Radiograph of skull and mandible showing same eruption sequence as *Avahi*.

From the study of a specimen of *Propithecus diadema*, Bennejeant (1936) stated that the premolar eruption sequence of Indriidae (Indriinae) was posterior-to-anterior.

Daubentoniinae

The dentition of *Daubentonia madagascariensis*, the single extant representative of this subfamily, is rather difficult to describe. The deciduous upper dentition consists of a large, curved anterior tooth with its point of growth posterior to the premaxillary suture but growing through the premaxilla; this tooth is retained throughout life and therefore, if the definition of a milk tooth were to be strictly adhered to, would be represented in the deciduous rather than the permanent formula, as would the molars. Traditionally, this tooth is represented in the formula for the permanent dentition; to convey the notion of continuous growth, this tooth should be represented and bracketed in both the permanent and deciduous formulas. Furthermore, this tooth has traditionally been identified as an incisor, essentially as a result of comparison with rodents (e.g., Le Gros Clark, 1962; James, 1960). However, this animal is a primate, not a rodent, and should be interpreted in context. It has been argued elsewhere (Schwartz, 1974c) that anterior tooth loss in primate evolution would affect the incisors rather than the canines. Thus this anterior tooth in *Daubentonia*, which has its origin of growth in the maxilla (Schwartz, 1974a), is better considered a canine. The remainder of the deciduous upper dentition consists of three premolars, with the first immediately posterior to the canine and the other two well back in the maxilla; the first two are small and peglike and the last is somewhat "molariform" (Fig. 8). Reference to another deciduous tooth, small and peglike and located in the premaxilla, has been made by Remane (1960) and Tomes (1914); the vacated alveolus of such a tooth (deciduous incisor) appears to be present in the specimen illustrated here.

The deciduous lower dentition consists of a large, continuously growing canine (Schwartz, 1974a) and two premolars well back in the mandible; the anterior deciduous premolar is small and peglike, while the posterior is much larger and somewhat "molariform" (Fig. 8). These two lower deciduous premolars are strongly reminiscent of the indriine posterior lower deciduous premolars (Schwartz, 1974a; Tattersall and Schwartz, 1974) and thus may help corroborate Gregory's (1915) suggestion that daubentoniines were derived from an early indriid stock. This relationship may be further appreciated by the fact that although there is only one pair of permanent (upper) premolars in *Daubentonia*, these replace the most posterior of the deciduous set, suggesting a vestigial retention of an anteriorly directed premolar eruption sequence.

DISCUSSION

The data presented here can provide information on a variety of different levels. First, as has been argued elsewhere (Schwartz, 1974a, b), evolutionary pressures affecting the deciduous dentition need not be the same as those affecting the permanent dentition. As is evident from the data of this study, the eruption sequence of the permanent premolars need not be the same as

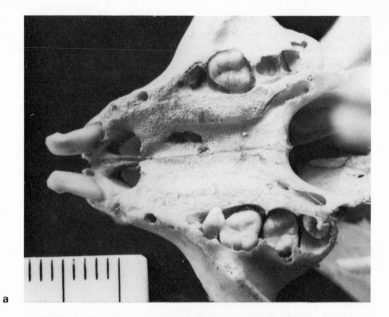

a

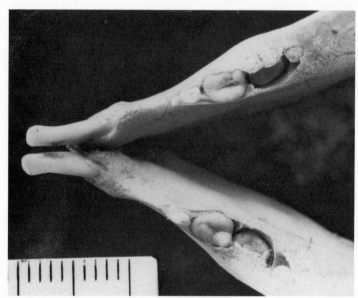

b

FIG. 8. *Daubentonia madagascariensis* (RM b). Skull and mandible showing some deciduous premolars (upper posterior and lower two posterior) and alveoli for others (upper extreme anterior and middle); this illustration was selected as it shows the possible closed alveolus of a deciduous tooth posterior in the premaxilla.

that of the deciduous, while the order of development of the permanent premolars may differ from that in which they erupt (e.g., as in *Lepilemur*). Second, because different states of evolution may be reflected, the two dentitions
(and even the two levels of organization, development and eruption) may be
studied as independent entities.

It seems reasonable to propose that primitively the deciduous and permanent dentitions would have reflected the same evolutionary pressures, and
that, moreover, developmental and eruption sequences would have been the
same. It is possible that subsequent change from the primitive character state
would have first been realized in modification of the order of appearance of
the deciduous dentition and only subsequently in that of the permanent. The
mechanisms of speciation and adaptive radiation, however, are not in total
accord with a model of uniform and universal gradual change (e.g., see Eldredge and Gould, 1972, but also see the argument of Gingerich, 1974). Dif-
ferentials in selection pressures might preclude an observable transitional
stage: e.g., the evolutionary loss of a premolar might not be preceded by an
observable diminution in size of that tooth (Schwartz, 1974*b*). Thus any
model of change from the primitive sequence must include the possibility of
differential, or punctuated, change. This might take various forms: for instance, the deciduous dentition might retain the primitive sequence, while the
permanent became derived; or the deciduous and permanent dentitions might
both represent derived, but differently so, character states; or the developmental level of a sequence might remain primitive while that of eruption
became derived.

Since the sequences of development and eruption are not known for the
"ancestral" primate stock, one cannot know directly the primitive character
state and thus cannot discern at this level which of the prosimians have retained the primitive condition or possess derived characters. However, it is
usually possible to infer primitiveness vs. derivedness of character states on
the basis of communality of possession, both within and between taxa (Hennig, 1966). Furthermore, given the incompleteness of the lemur fossil record,
it is necessary to analyze the few pertinent fossils and the extant forms together, as urged by Schaeffer *et al.* (1972).

As Table I shows, more prosimians, including the lorisids and *Tarsius*
(Fig. 9), share the sequence of development P2–P4–P3 than the sequence P4–
P3–P2, or at least an anteriorly directed order of appearance; similarly, the
most frequent sequence of eruption is P2–P4–P3, not P4–P3–P2. It is therefore possible to suggest that not only is the order P2–P4–P3 primitive for
each level of dental organization (i.e., development and eruption) but primitive for the whole continuum of dental sequencing. That this order may be
primitive not only for the Malagasy lemurs but also for prosimians and possibly primates in general is further implied by its frequent occurrence in fossil
and extant insectivores (Slaughter *et al.*, 1974) and its possession by the

TABLE I. Development and Eruption Sequences of the Deciduous and Permanent Premolars of Prosimians[a]

Taxon	dp2–dp4 –dp3	dp2–dp3 –dp4	dp(p)– dp(a)	P2–P4– P3	P4–P3– P2	Pp–Pa
Tarsius				x/xd,e		
Loris				x/xd,e		
Nycticebus				x/xd,e		
Perodicticus	x/xe			x/xd,e		
Arctocebus[b]						
Galago demidovii	x/xe			x/xd,e		
G. senegalensis	x/xe			x/xd,e		
G. crassicaudatus	x/xe			x/xd,e		
G. alleni				x/xd,e		
Euoticus				x/xd,e		
Microcebus[c]	x/xe			x/xd,e		
Cheirogaleus				x/xd,e		
Phaner				x/xd,e		
Lemur macaco		x/xe		x/xd,e		
L. fulvus				x/xd,e		
L. mongoz		?/xe		x/xd,e		
L. rubriventer[b]		x/xe				
L. variegatus				x/xd,e		
L. catta		x/xe			x/xd,e	
Hapalemur		x/xe			x/xd,e	
Lepilemur	x/xe			x/xd	x/xe	
Megaladapis[d]				x/xd	x/xe	
Adapis[e]			x/xd?,e			
Notharctus[f]			x/xd,e		x/xd,e	
Archaeolemur[d]					x/xd,e	
Hadropithecus[d]					x/xd,e	
Avahi						x/xd,e
Indri			x/			x/xd,e
Propithecus			x/			x/xd,e

[a] x/x, Observed in upper/lower jaws; ?, inferred data (see text); d, development; e, eruption.
[b] Only data on permanent: P2 develops first.
[c] Data on deciduous dentition courtesy of R. D. Martin.
[d] From Lamberton (1938) and Schwartz (1974*a*).
[e] From Stehlin (1912).
[f] From Gregory (1920).

Oligocene primates *Apidium* and *Parapithecus* (Conroy *et al.*, 1975). Thus those prosimians which possess levels of dental organization or a continuum other than P2–P4–P3 may be considered as derived in this character (Table I). These forms include Indriinae, Archaeolemurinae (Lamberton, 1938; Schwartz, 1974*a*), *Lemur catta, Hapalemur, Lepilemur, Megaladapis*, and the adapids (Gregory, 1920, for *Notharctus*; Stehlin, 1912, for *Adapis parisiensis*).

Lepilemur and *Megaladapis* represent what may be termed a "transitional" stage between the sequences P2–P4–P3 and P4–P3–P2; i.e., in development, the primitive prosimian sequence persists while in eruption the derived character state is exhibited (Schwartz, 1974*a*). Although we may consider each of the levels of organization as an independent character, we may also view the entire continuum as a unit, and as such, the transition P2–P4–P3 (in development) through P4–P3–P2 (in eruption) should be considered to represent a derived condition. Since both *Lepilemur* and *Megaladapis*, and only these genera, share this uniquely derived continuum of premolar appearance, it is most probable that these two form a close sister group, having a common ancestor not shared by any other primate (see Hennig, 1966). In converting a possible P2–P4–P3 eruption sequence to one of P4–P3–P2, one of three mechanisms might be employed: (1) late in development, there is a marked increase in growth of P4 and then of P3; (2) there is an inhibition and retardation in growth of P2, allowing P4 and P3 to develop and erupt at their normal rates; or (3) a combination of both (1) and (2). Considering the conservative nature of evolutionary change, mechanism (2) is probably responsible for the change.

Differing from the primitive level of organization and the continuum seen in most prosimians (i.e., the cheirogaleines, most *Lemur* species, *Varecia*, and the lorisids and *Tarsius*) and from the derived developmental–eruption continuum shared by *Lepilemur* and *Megaladapis* are the derived character complexes of *Lemur catta, Hapalemur*, Indriinae, Archaeolemurinae (unfortunately data are lacking on the palaeopropithecines), *Daubentonia*, and the adapids (at least *Adapis parisiensis* and species of *Notharctus*). Regardless of the number of premolars, all are characterized by development and eruption proceeding in a posterior-to-anterior direction. The mechanism by which change from the primitive P2–P4–P3 sequence occurred is best viewed as an extention of mechanism (2) above, i.e., an inhibition and retardation in growth of P2, allowing P4 and P3 to develop and erupt at their normal rates. If the sequences of premolar development and eruption were the only characters available for determining phylogenetic relationships, those possessing the derived posterior-to-anterior continuum of appearance would have to be considered as forming a sister group apart from the other prosimians. In the light of other information, this hypothesis is needlessly

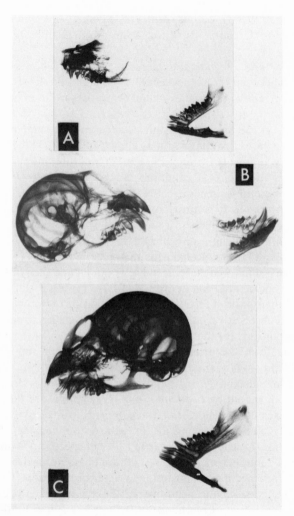

FIG. 9. (A) *Loris tardigradus* (AMNH 240827). Radiograph of skull and mandible with P2 erupted, dp3 and dp4 present, and (particularly visible in the mandible) the substantial development of P4 relative to P3. (B) *Nycticebus coucang* (PM A-2.830). Radiograph of skull and mandible showing condition similar to *Loris*. (C) *Perodicticus potto* (AMNH 51024). Radiograph of skull shows P^2 partially erupted and presence of dp^4 and dp^3; mandible shows P_2 erupted, presence of dp_4 and dp_3 with small light area below dp₄ indicating growth of P_4. (D) *P. potto* (AMNH 51023). Radiograph of skull and mandible with dp2 fully, dp4 partially, and dp3 beginning

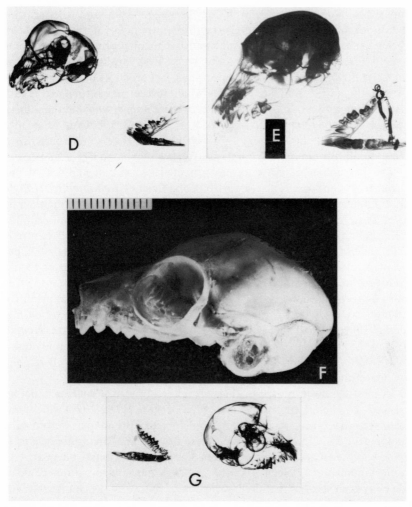

to erupt. (E) *Galago crassicaudatus* (PM A-3.014). Radiograph of skull and mandible showing P2 starting to erupt, presence of dp3 and dp4 with P4 partially developed; this is the same as was found in other species of *Galago*. (F) *G. crassicaudatus* (AMNH 116259). Skull showing dp^2 fully, dp^4 almost fully, and dp^3 partially; the same eruption sequence was found for the mandible; this eruption sequence was found for other species of *Galago*. (G) *Tarsius spectrum* (RM 1892d). Radiograph of skull and mandible showing eruption P2–P4–P3; the eruption in the mandible is slightly delayed relative to that in the maxilla.

complex and in many cases these similarities must be considered as parallel developments.

On the basis of other criteria, Tattersall (1973*a*, *b*) has convincingly argued for a close relationship between the indriines and archaeolemurines; comparisons of craniodental morphology between indriids and other prosimians indicate that in most characters the former are highly derived (Tattersall and Schwartz, 1974). Since Indriinae and Archaeolemurinae, although possessing different numbers of permanent premolars, both present a posterior-to-anterior direction of premolar appearance, it is probable that this is a characteristically indriid character complex which further serves to unite the family. If the suggestion of close affinities between *Daubentonia* and the indriids is correct, then the retention of a replacing P⁴ (a possible remnant of an anteriorly directed order of premolar appearance) rather than a more mesially located premolar would represent a derived character shared by the two.

The possession of an anteriorly directed sequential complex of permanent premolar appearance (a derived condition) by *Lemur catta*, *Hapalemur*, and the adapids presents an interesting problem in determining their affinities. It is generally accepted that "lemur" is equated with the "primitive primate condition" (e.g., Le Gros Clark, 1962; Simons, 1972) in all, or almost all, characters, and that since the adapids share many characters (albeit primitive) with the "lemurs" the former must in some way represent the ancestral prosimian condition (e.g., Gregory, 1915, 1920, 1922; Simons, 1972). It must be recognized, however, that it is the sharing of *derived*, not primitive, characters which is the criterion for determining phylogenetic relationships (Hennig, 1966; Schaeffer *et al.*, 1972). Thus the sharing by *L. catta*, *Hapalemur*, and the adapids of the primitive bulla and carotid circulation in no way proves their close affinities but merely indicates that they have all retained these ancestral characters.

Interestingly, we may cite the very characters of dental morphology presented elsewhere (Gingerich, Chapter 5; Gregory, 1915, 1920) which were used to show the primitive and generalized nature of the adapids (particularly the adapines), to suggest that, in fact, these Eocene prosimians are related to but a few of the extant Malagasy lemurs. Gregory has strongly advocated the extreme similarities between *Adapis* and *Lepilemur* in occlusal morphology, while Gingerich stresses that the positioning of the cusps, the configuration of the shearing crest, and the nature of the cingulum are more appropriate to a comparison between *Adapis* and *Hapalemur*. In fact, it is not surprising that *Adapis* is similar in molar morphology to both *Lepilemur* and *Hapalemur*, for the latter two are themselves quite similar in such characters (Tattersall and Schwartz, 1974). However, the features of similarity in molar morphology are not only distinct from *Lemur* (and *Varecia*) but are derived (Tattersall and Schwartz, 1974). It is suggested, then, that as the sharing of a derived molar morphology by Adapidae (or at least *Adapis*), *Hapalemur*, and *Lepile-*

mur indicates close affinities between them, so may the derived nature of premolar appearance (with *Lepilemur*, and thus *Megaladapis*, retaining a more primitive character complex) be viewed as a specialization of this group.

This, then, leaves *Lemur catta* to be considered. Although distinct from other *Lemur* species (and *Varecia*) in its derived order of premolar development and eruption, *L. catta* is definitely united with them on the basis of dental morphology (e.g., simple premolars, the protocone the largest of the three distinct cusps of the trigon, anterior displacement of the lingual aspect of the molars) and general configurations of the mandible and skull (Tattersall and Schwartz, 1974). They also share the derived sequence of eruption of the deciduous premolars. Analysis of complete sequences of development and eruption (Schwartz, 1974a) indicates that there is a morphocline proceeding from *Varecia* to other *Lemur* species to *L. catta*, with *L. catta* being the most derived; this is certainly true of premolar development and eruption. Perhaps further investigation of the characters of *L. catta* will warrant its separation, at least at the subgeneric level.

Sequences of development and eruption of the deciduous premolars are dp2–dp4–dp3 and dp2–dp3–dp4 (Table I). Unfortunately, the data are less abundant than those for the permanent dentition; inferences as to the primitive character complex are thus more tenuous. What information there is suggests the significance of the observation that not only the lorisids, but *Microcebus* and *Lepilemur* as well, share the same sequence (dp2–dp4–dp3), while it is only among the lemurines that one finds dp2–dp3–dp4. Although it would be equally simple to derive one from the other (inhibition of dp4 in the sequence dp2–dp4–dp3 would give dp2–dp3–dp4, and inhibition of dp3 in dp2–dp3–dp4 would effect the change), it appears significant that the order 2–4–3 is found in both the permanent and deciduous sets while other sequences are only evident in one dentition or the other. The presence of the same sequence of premolar appearance in both dentitions would appear, in accordance with the suggestion offered earlier, to argue that dp2–dp4–dp3 is the primitive sequence since it mirrors that of the permanent dentition.

CONCLUSION

As expected when dealing with a small number of characters, the number of definitive statements on phylogenetic relationships which can be made is limited. Although the majority of prosimians share the same developmental and eruptional sequence of the permanent premolars (and, when the data are available, of the deciduous premolars), this sequence, P2–P4–P3, is primitive for the group and thus cannot be used for determining relationships. Statements concerning the relationships between various groups (e.g.,

Tarsius, Lorisidae, Cheirogaleinae, Lemurinae) are therefore not possible on the basis of this evidence. However, the likelihood that P2–P4–P3, in development and eruption, was present in the last common ancestor of the prosimians is nonetheless of interest. With the derived (for Prosimii) sequences, it is possible to suggest that (1) an anteriorly directed sequence of premolar appearance is characteristic of the indriids, to which *Daubentonia* may very well be related, as a group; (2) *Lemur catta* is derived compared to *Varecia* and other *Lemur* species, from which some taxonomic separation may be warranted; (3) *Lepilemur* and *Megaladapis* form a close sister group; (4) the *Lepilemur–Megaladapis* sister group may be part of a larger group with *Hapalemur* and the adapids; and (5) the possession of the derived sequence by representatives of both subfamilies of adapids suggests that none is ancestral to any modern lemuriform.

ACKNOWLEDGMENTS

I would like to thank Drs. Roger Saban (Laboratoire d'Anatomie Comparée, Paris), François Petter (Laboratoire de Mammifères et Oiseaux, Paris), Ms. Prue Napier [Department of Mammals, British Museum (Natural History)], Drs. A. M. Husson (Department of Mammals, Rijksmuseum van Natuurlijke Historie, Leiden), Sydney Anderson (Department of Mammals, American Museum of Natural History), Richard Thorington, Jr. (National Museum of Natural History), and Mr. Charles W. Mack (Museum of Comparative Zoology, Harvard University) for kindly allowing me to study all specimens in their charge. To Dr. R. D. Martin, I am indebted for exchange of data as well as discussion. I thank Ms. Theya Molleson [Subdepartment of Anthropology, British Museum (Natural History)], Mr. J. Simons and his assistant Mr. G. J. de Jing (Zoological Laboratory, University of Leiden), and Dr. Jean-Pierre Gasc (Muséum National d'Histoire Naturelle) for assistance in radiographic work. Special thanks go to Drs. Eric Delson and D. A. Hooijer for helpful critical discussion, and to Dr. Ian Tattersall I am most grateful for his unbounded patience and devotion of time to discussion preceding and during the preparation of this chapter.

REFERENCES

Bennejeant, C., 1936, *Anomalies et Variations Dentaires chez les Primates*, 285 pp., Clermont-Ferrand, Paris.
Butler, P. M., 1939, Studies of the mammalian dentition: Differentiation of the post-canine dentition, *Proc. Zool. Soc. Lond. Ser. B* **109**:1–36.
Butler, P. M., 1956, The ontogeny of molar pattern, *Biol. Rev.* **31**(1):30–70.
Butler, P. M., 1963, Tooth morphology and primate evolution, in: *Dental Anthropology.* (D. Brothwell, ed.), pp. 1–14, Pergamon Press, London.

Conroy, G. C., Schwartz, J. H., and Simons, E. L., Dental eruption patterns in Parapithecidae (Primates, Anthropoidea). *Folia primat.*, in press.

Eldredge, N., and Gould, S. J., 1972, Punctuated equilibria: An alternative to phyletic gradualism, in: *Models in Paleobiology.* (T. J. M. Schopf, ed.), pp. 82–115, Freeman, Cooper, San Francisco.

Gingerich, P. D., 1974, Stratigraphic record of early Eocene *Hyopsodus* and the geometry of mammalian phylogeny, *Nature* **248**:107–109.

Gregory, W. K., 1915, 1. On the relationship of the Eocene lemur *Notharctus* to the Adapidae and other primates. 2. On the classification and phylogeny of the Lemuroidea, *Bull. Geol. Soc. Am.* **26**:419–446.

Gregory, W. K., 1920, On the structure and relations of *Notharctus*, an American Eocene primate, *Mem. Am. Mus. Nat. Hist. (n.s.)* **3**(2):49–243.

Gregory, W. K., 1922, *The Origin and Evolution of the Human Dentition*, 548 pp., Williams and Wilkins, Baltimore.

Hennig, W., 1966, *Phylogenetic Systematics*, 263 pp., University of Illinois Press, Urbana.

James, W. W., 1960, *The Jaws and Teeth of Primates*, 199 pp., Pitman Medical Publishing Co., London.

Lamberton, C., 1938, Dentition lactéale de quelques Lémuriens subfossiles malgaches, *Mammalia* **2**(2):57–80.

Le Gros Clark, W. E., 1962, *The Antecedents of Man*, 2nd ed., 388 pp., Edinburgh University Press, Edinburgh.

Osborn, J. W., 1970, New approach to Zahnreihen, *Nature* **225**(5230):343–346.

Osborn, J. W., 1971, The ontogeny of tooth succession in *Lacerta vivipara* Jacquin (1787), *Proc. Roy. Soc. Lond. Ser. B* **179**:261–289.

Osborn, J. W., 1973, The evolution of dentitions, *Am. Sci.* **5**(2):548–599.

Patterson, B., 1956, Early Cretaceous mammals and the evolution of mammalian molar teeth, *Fieldiana Geol.* **13**:1–105.

Remane, A., 1960, Zähne and Gebiss, *Primatologia* **3**(2):637–846.

Schaeffer, B., Hecht, M. K., and Eldredge, N., 1972, Phylogeny and paleontology, in: *Evolutionary Biology*, Vol. 6. (T. Dobzhansky, M. K. Hecht, and W. C. Steere, eds.), pp. 31–46, Appleton-Century-Crofts, New York.

Schwartz, J. H., 1974a, Dental development and eruption in the Prosimians and its bearing on their evolution, Ph.D. thesis, Columbia University, University Microfilms, Ann Arbor, Mich.

Schwartz, J. H., 1974b, Premolar loss in primates: A re-investigation, in: *Prosimian Biology.* (R. D. Martin, G. A. Doyle, and A. C. Walker, eds.), Duckworth, London.

Schwartz, J. H., 1974c, Observations on the dentition of the Indriidae, *Am. J. Phys. Anthropol.* **41**(1):107–114.

Simons, E. L., 1972, *Primate Evolution*, 322 pp., Macmillan, New York.

Slaughter, B. H., Pine, R. H., and Pine, N. E., 1974, Eruption of cheek teeth in Insectivora and Carnivora, *J. Mammal.* **55**(1):118–125.

Stehlin, H. G., 1912, Die Säugetiere des schweizerischen Eocäns: Kritischer Katalog der Materialien, 7ter Theil, erste Hälfte *Adapis, Abh. Schweiz. Pal. Ges.* **38**:1165–1298.

Tattersall, I., 1973a, Subfossil lemuroids and the "adaptive radiation" of the Malagasy lemurs, *Trans. N.Y. Acad. Sci. (ser. II)* **35**(4):314–324.

Tattersall, I., 1973b, Cranial anatomy of the Archaeolemurinae (Lemuroidea, Primates), *Anthropol. Pap. Am. Mus. Nat. Hist.* **52**(1):1–110.

Tattersall, I., and Schwartz, J. H., 1974, Craniodental morphology and the systematics of the Malagasy lemurs (Prosimii, Primates), *Anthropol. Pap. Am. Mus. Nat. Hist.* **52**(3):141–192.

Tomes, C. S., 1914, *A Manual of Dental Anatomy*, 7th ed., 616 pp., Blakiston, Philadelphia.

Van Valen, L., 1970, An analysis of developmental fields, *Develop. Biol.* **23**:456–477.

Dentition of *Adapis* 5
parisiensis and the Evolution
of Lemuriform Primates

PHILIP D. GINGERICH

The fossil record of Madagascar unfortunately sheds no light on the origin of its unique lemur fauna, nor does it assist in determining the relationships of the Malagasy lemurs to other primates. The known fossil forms are at most 3000–4000 years old (Tattersall, 1973a), which makes them virtual contemporaries rather than ancestors of the living species. The only adequately known fossil primates possibly related to the ancestry of the lemurs are found in sediments of Eocene age in North America, Europe, and Asia. The European genus *Adapis* includes species which deserve special attention in this regard.

A major problem in any consideration of lemur origins is biogeographical. Madagascar has apparently been an island since the Late Cretaceous, though previously it was joined on the west to Africa and on the east to India. (Cracraft, 1973, provides a good review of recent geophysical evidence bearing on this continental separation.) When and how did the ancestral lemurs get to Madagascar? Virtually all recent authors identify Africa as the source of the ancestral lemur stock, and their efforts to date the lemur invasion of

PHILIP D. GINGERICH   Museum of Paleontology, The University of Michigan, Ann Arbor, Michigan 48104

Madagascar involve balancing a number of factors. The Mozambique Channel is believed to have widened gradually during the early Tertiary; thus the earlier the lemur invasion, the easier it would have been. On the other hand, the lemurs are advanced forms compared to the known Paleocene primates, suggesting that they probably did not evolve before the Eocene. For these reasons, the time of invasion of Madagascar by the ancestral lemur stock is usually considered to have been Early Eocene (McKenna, 1967; Cooke, 1968; Fooden, 1972; Tattersall, 1973b). Charles-Dominique and Martin (1970) suggest the possibility that numerous lemur types were present in the African region in the Paleocene, which they believe was possibly before Madagascar separated from Africa. Walker (1972) concludes that ancestral lemurs could have rafted across the Mozambique Channel from Africa until about the end of the Eocene, by which time the channel was probably about 240 km wide.

There are a number of problems with this theory. As Simons has pointed out (1972, p. 169), if Africa is the continent of origin of the lemur fauna, it is curious that lemurs have not been found in the primate-rich Oligocene and Miocene sediments of Africa. The paleogeography of Madagascar in the early Tertiary is not completely agreed on, but it appears certain that in the Eocene a considerable distance of ocean (100–200 km minimum) separated Africa and Madagascar. Prevailing winds and ocean currents make chance crossing of the Mozambique Channel extremely improbable today, and presumably had approximately the same effect in the Eocene. Finally, a satisfactory account of the origin of the lemur dentition has never been given. If the procumbent incisors and canines of lemurs are primitive, it is surprising that they do not appear in any of the Eocene lemuroids of Europe or North America. If they are not primitive, it is necessary to explain how the procumbent incisors and canines of lemurs isolated on Madagascar came to be shared with the lorises of Africa and Asia, which first appear in the Miocene. The purpose of this chapter is to present new observations on the dentition of Eocene lemuroids, and to discuss their implications for the origin of the lemur fauna of Madagascar.

DENTITION OF *ADAPIS*

All of the adequately known Paleocene primates are members of extinct side branches of early primate evolution, and the fossil record of lemuriform primates begins in the Eocene. All Eocene lemuroids are classified in the family Adapidae. Adapids share two diagnostic cranial characters with the living lemurs: they have a free ectotympanic annulus within the auditory bulla, and the internal carotid artery divides into two branches, the promontory and the stapedial. Among fossil primates, this combination of characters is known only in the Adapidae.

Phylogeny of Adapidae

An outline phylogeny of adapid evolution is presented in Fig. 1. The earliest adapids, species of *Pelycodus*, first appear at the beginning of the Eocene in England (Blackheath beds), France (*Lignites de Soissonais*), and North America (Clark Fork beds). In North America, *Pelycodus* can be traced upward through the strata of the Lower Eocene and the advanced forms are placed in the genera *Notharctus* and *Smilodectes*. A skull of the Middle Eocene species *Notharctus "osborni"* is illustrated in Fig. 2, and the skull of *Pelycodus* undoubtedly was very similar. A Princeton University partial skull of *Pelycodus* (PU No. 14515) confirms that this primitive form had a fully developed postorbital bar as in *Notharctus*. The principal morphological difference between *Pelycodus* and *Notharctus* is the presence in the latter of well-developed mesostyles and hypocones on the upper molars.

In France, the *Sables à Unios et Térédines* are stratigraphically higher in the Lower Eocene than the *Pelycodus*-bearing *Lignites de Soissonais*, and yield a more advanced adapid, *Protoadapis*, a genus which persisted through the remainder of the Eocene (Russell *et al.*, 1967). Early species of *Protoadapis* resemble *Pelycodus*, differing chiefly in the replacement of the postprotocingulum by a hypocone on the upper molars, and reduction of the paraconid on the lower molars. Molars of *Protoadapis* are more sharply crested and have more open trigonids than do those of *Pelycodus*. Both *Pelycodus*

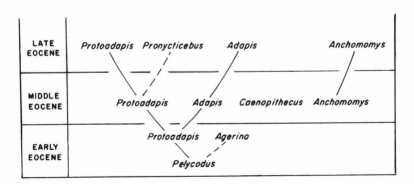

FIG. 1. Phylogeny of European genera included in the primate family Adapidae. *Adapis parisiensis* is a Late Eocene species of *Adapis*.

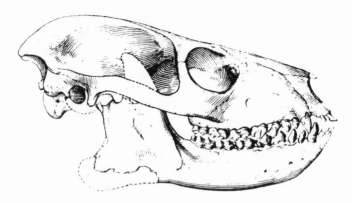

FIG. 2. Skull of *Notharctus "osborni,"* a closely related descend-ant of the earliest adapid *Pelycodus*. Drawing natural size, taken from Gregory (1920). This specimen may in fact be a female *Notharctus tenebrosus*.

and *Protoadapis* have small, vertically implanted incisors, large protruding canines, and premolars showing little molarization.

The Late Eocene *Pronycticebus* is known from a single skull and asso-ciated right mandible. The lower teeth of *Pronycticebus* are virtually identical to those of *Protoadapis*; however, the upper molars are relatively broader and have a more strongly developed hypocone. As its name implies, when first de-scribing the skull of *Pronycticebus*, Grandidier (1904) considered it to be pos-sibly related to the living lorisoid *Nycticebus*. Restudy by Le Gros Clark (1934), Simons (1962), and Szalay (1971) has led to agreement that *Pronycti-cebus* is an adapid. Simons dissected the bulla of *Pronycticebus* and demon-strated that this genus has a free tympanic ring, although the ring is located nearer to the external auditory meatus than is typical of lemuroid primates. This Simons (1962) interpreted as a possible indication of near fusion of the ring to the lateral wall of the bulla, suggesting that perhaps *Pronycticebus* is related in some way to lorisoid origins. Szalay (1971) interpreted the same evidence differently, as he believes the free tympanic ring of lemuroids to be derived from the lorisoid condition (Szalay, 1972). While Szalay's inter-pretation of tympanic evolution in Primates is theoretically possible, it is im-probable and not supported by either ontogenetic development or the fossil record, a point that will be discussed in detail elsewhere.

The adapids *Agerina*, *Caenopithecus*, and *Anchomomys* (Fig. 1) are poorly known anatomically and as yet contribute little to our knowledge of primate evolution. They do, however, indicate that the European Eocene ra-diation of Adapidae was a broad one, with a minimum of four genera present in the Middle and Late Eocene.

The remaining adapids, species of *Adapis*, have the greatest bearing on the origin of lemurs. These include some of the best-known fossil primates anatomically. Six species of *Adapis* are presently recognized (Schmidt-Kittler, 1971). These are, in the Middle Eocene: *A. sciureus*, *A. ruetimeyeri*, and *A. priscus*, and in the Late Eocene: *A. ulmensis*, *A. magnus*, and *A. parisiensis*. The type specimen of the genus *Adapis* was first figured and described by Cuvier in 1812 in the first edition of his *Recherches sur les Ossemens Fossiles*. The specimen is a badly crushed skull (Fig. 3), and it is not surprising that Cuvier was never able to determine its true affinities. In the second edition (1822) of his *Ossemens Fossiles*, Cuvier first named the specimen *Adapis*, a name sometimes used at that time for hyraxes, and classified it as a small pachyderm. No specific name was proposed for the taxon until publication of the section of de Blainville's *Ostéographie* dealing with *Adapis*. Cuvier's specimen was refigured by de Blainville (1849), who referred to it as *Adapis parisiensis*.

In 1859, Gervais described additional specimens as *Aphelotherium duvernoyi*, which he classified with *Adapis parisiensis* as an omnivorous pachyderm. The true affinities of these specimens were not recognized until Delfortrie (1873) described a nearly complete skull, which he named *Palaeolemur*

FIG. 3. Badly crushed palate and right mandible of the type specimen of *Adapis parisiensis*. This specimen was first figured and described by Cuvier in 1812. As with most early lithographs, Cuvier's figure is reversed left for right. Specimen is in Muséum National d'Histoire Naturelle in Paris.

betillei and correctly identified as a lemur-like primate. Gervais (1873) immediately recognized the identity of *Adapis parisiensis, Aphelotherium duvernoyi,* and *Palaeolemur betillei,* the first name having priority.

The following year Filhol (1874) described a new skull of a much larger primate as *Adapis magnus.* Gervais (1876) subsequently placed this species in a new genus *Leptadapis.* The canine teeth of *Adapis magnus* are significantly different in conformation from those of *A. parisiensis* (see below), and the name *Leptadapis* should probably be retained, at least as a subgenus. *Adapis parisiensis* and *A. (L.) magnus* are both now known from numerous skulls and jaws which were described in detail by Stehlin (1912) in a monograph on *Adapis.* The other four recognized species of *Adapis* were named by Stehlin (1912, 1916) and Schmidt-Kittler (1971), and remain inadequately known.

The phylogeny of Eocene adapids presented in Fig. 1 traces the evolution of *A. parisiensis* at the generic level from the earliest known ancestor *Pelycodus,* through the intermediate genus *Protoadapis,* leading in the Middle Eocene to species placed in *Adapis. Adapis parisiensis* is an advanced member of the genus, with important specialization of the lower canines, and first appears in the Late Eocene.

Dentition of *Adapis parisiensis*

The following description and discussion is based on the dentition of *A. parisiensis.* Comparison is made throughout with the dentition of the living lemur *Hapalemur griseus. Hapalemur* is approximately the same size as *A. parisiensis,* which eliminates problems of allometry when comparing the two. Considered together, the dentitions of all of the living lemurs exhibit a range of variation in morphology in which *Hapalemur* is, in most respects, intermediate. Because of its relatively generalized nature, the dentition of *Hapalemur* may closely approximate the condition of the ancestral lemur stock. Its resemblances to *Adapis* are thus of particular interest and importance.

Adapis parisiensis has a dental formula of $I_2^2C_1^1P_4^4M_3^3$. Its incisors above and below are broad and spatulate (Figs. 4 and 5), resembling those of anthropoid primates. The upper and lower canines are of medium size, protruding significantly beyond the premolars. The four premolars become progressively larger and more molariform, with the last being morphologically virtually molars. The upper and lower molars are all sharply crested. Small hypocones are present on M^1 and M^2, and M_3 is elongated, with a well developed hypoconulid, as in most primitive primates.

In *Hapalemur,* on the other hand, the upper incisors are very small teeth, placed close to the canine. The lower incisors are very narrow and styliform, forming part of a dental scraper or tooth comb. The upper canine in *Hapalemur* is a large sharply pointed tooth, but the lower canine is small and styli-

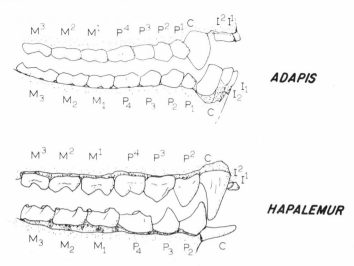

FIG. 4. Dentition of *Adapis parisiensis* and *Hapalemur griseus*, in lateral view. Note particularly the incisiform lower canine of *Adapis*, and the caniniform P_2 and procumbent lower canine of *Hapalemur*. Both to same scale, twice natural size. Drawing of *Adapis* based on specimens in Muséum National d'Histoire Naturelle in Paris and the Harvard Museum of Comparative Zoology.

form like the incisors and forms the lateral member of the dental scraper. The most anterior lower premolar (P_2) is enlarged and caniniform, and shears against the back of the upper canine. As in *Adapis*, P_4^4 are very molariform. In *Hapalemur*, the upper molars lack hypocones and M_3 does not have the hypoconulid lobe seen in *Adapis*.

Evolution of the Anterior Dentition

One of the most important characters distinguishing the lemurs and lorises from other primates is the tooth scraper formed by the lower incisors and canines. In the Indriidae, the scraper is formed by a single pair of lower incisors and the canines (or possibly two pairs of incisors and no canines), and was undoubtedly derived from the condition seen in the Lemuridae, where two pair of incisors plus the canines form the scraper (see *Hapalemur*, Fig. 5). The scraper of Lorisidae is virtually identical to that of the Lemuridae and undoubtedly shares common ancestry with it, as Martin (1972, p. 320) notes. In having the tympanic ring fixed in the lateral wall of the auditory bulla rather than free within it as in lemurs, the Lorisidae represent a "derived"

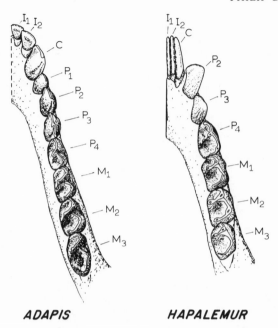

ADAPIS   **HAPALEMUR**

FIG. 5. Lower dentition of *Adapis parisiensis* and *Hapalemur griseus*, in occlusal view. Note incisor–canine functional unit, and the open, crested premolars and molars in both species. Both to same scale, twice natural size. Drawing of *Adapis* based on specimens in Muséum National d'Historie Naturelle, Paris.

group with respect to the ancestral lemur stock. The Lemuridae are primitive in both of these important characters and probably reflect the ancestral condition.

Two questions must be answered to explain the origin of the lemur dental scraper: what is its evolutionary history, and what is its functional basis? Regarding the evolutionary history of the dental scraper, it was previously thought that the pronounced procumbency of the lower incisors and canines in the tree shrews indicated a phase of evolutionary development preceding the more highly specialized dental comb of the living lemurs and lorises (Le Gros Clark, 1962). Van Valen (1965), Martin (1966, 1968), Campbell (1966), and others have recently restudied the relationships of the tree shrews, and concluded that the similarities between living tree shrews and Primates are based exclusively on retention of ancestral placental mammal characteristics and the convergent development of certain features in the two groups. If tree shrews are related to Primates at all, it would seem to be at the microsyopoid level and, as Martin (1968, p. 390) notes, the "lemur-like" proclivity of the lower canines in some tree shrews represents a separate development convergently resembling (and not very closely at that) this character in lemurs.

An answer to the question of when the canine became incorporated into a functional unit with the incisors in the ancestral lemur/loris is provided by *Adapis*. *Adapis parisiensis* has spatulate, nearly vertically implanted lower incisors as is typical of the other adapids, and of anthropoids as well, but its canines differ in one important respect from those of all the other Eocene adapids. The lower canines of *A. parisiensis* are functionally incisors; the six lower anterior teeth in this species form a single functional unit (see Figs. 4 and 5, this chapter; Figs. 246 and 247 in Stehlin, 1912). Once the lower canines and incisors came to function as a unit, it is not difficult to understand how all six teeth were modified similarly to form a dental scraper.

Recent field studies have determined that a primary function of the tooth scraper in the wild is to scrape resin or gum from the bark of trees. This behavior has been reported in *Euoticus* by Charles-Dominique (1971), and in *Phaner* by Petter *et al.* (1971). The indriids include a significant amount of bark in their diet (Petter, 1962), which is pried loose with the dental scraper (Richard, 1973). It is entirely plausible that repeated use of the lower anterior teeth to scrape bark would, by natural selection, result in the modification of even vertically oriented incisors to form the procumbent scraper seen in lemurs and lorises. Incorporation of the lower canines into the scraper of lemurs and lorises suggests that the lower canines of the ancestral lemur/loris were functionally incisors. *Adapis parisiensis* is the only fossil primate known which had this configuration of canines, and it is thus the best candidate yet known for the common ancestor of lemurs and lorises.

One often cited character which supposedly bars *Adapis* from lemur ancestry is its fused mandibular symphysis. The reason for symphyseal fusion in Mammalia is not completely understood, but it is apparent that fusion of the mandibular rami is correlated with the development of sectorial incisors such as those seen in *Adapis* and in anthropoid primates. With modification of the sectorial incisors of *Adapis* to form a dental scraper, symphyseal fusion would no longer be necessary, and consequently the mandibles need no longer fuse during ontogeny. This is, if true, a relatively simple example of neoteny.

Evolution of the Cheek Teeth

No living lemurs or lorises have more than three premolars, while *Adapis* has four. Loss of one or more premolars is well documented in a number of fossil lineages. Loss of P_1 would pose no problem in the transition from *Adapis* to the ancestral lemur/loris, except that in *Adapis* there is already a well-developed functional relationship between the lower first premolar and the upper canine. The lower first premolar (P_1) wears against the back of the upper canine in *Adapis*. The function of this wear in Anthropoidea is usually assumed to be to sharpen the upper canine (Every, 1970; Zingeser, 1969). It is also possible that a food-shearing mechanism is involved, and sharpening is not the principal function. Whatever the function of the lower premolar–upper canine relationship, the specimen illustrated in Fig. 6 shows how P_1 might have been lost without disturbing the premolar–canine wear pattern. In this specimen, P_1 is reduced and the upper canine wore against the back of the lower canine, P_1, and the front of P_2 simultaneously. In *A. parisiensis*, progressive enlargement of P_2 and reduction of P_1[1] would presumably have led to the stage illustrated by Fig. 6, and finally to that illustrated by *Hapalemur* (Fig. 4).

The cheek teeth of *Hapalemur griseus* exhibit a number of additional important derived characters which resemble, in detail, characters of the cheek teeth of *A. parisiensis*. P_3 in *Hapalemur* shows a peculiar pattern of ridging on the posterior surface of its principal cusp, which is very similar to that of *A. parisiensis*. *Hapalemur* has well-developed metastylids on P_4 and M_1, which is also a characteristic of *A. parisiensis*. These metastylids are very posteriorly placed, forming pseudoentoconids, in *Lepilemur*. Both *Adapis* and *Hapalemur* have sharply crested cheek teeth. On the lower molars of both genera, the paraconids are lost, being replaced by almost identical, sloping anterior cingulids. The result is a very open trigonid, and the talonids of the lower molars are also open on the lingual side. A functional analysis of tooth wear in *Adapis* and definitions of the terms used to describe *Adapis* molars were presented in an earlier paper (Gingerich, 1972).

One significant structural difference in the upper molars of *Adapis* and *Hapalemur* should be noted. The upper first and second molars of *A. parisiensis* usually have moderately developed hypocones, which are entirely lacking in *Hapalemur*. However, hypocones are well developed on the upper mo-

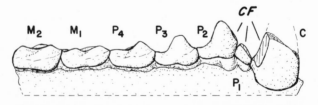

FIG. 6. Right mandible of *Adapis* (*Leptadapis*) *magnus*, in lateral view. *CF*, Wear facets formed by occlusion of upper canine. Apex of lower canine is broken. Twice natural size. Specimen in Muséum National d'Histoire Naturelle in Paris.

lars of indriids and lorisids, and may well have been present in the common ancestor of the lemurs and lorises.

The molars of *A. parisiensis* and *H. griseus* are so similar that if they were found together in the same fossil bed they would undoubtedly be placed together in the same primate family and possibly in the same genus. This could not be said for any of the other lemurs or lorises, except possibly *Lepilemur,* but it indicates the close relationship of the advanced adapid *A. parisiensis* to the living lemurs and lorises.

EVOLUTION OF LEMURIFORM PRIMATES

The phylogeny of Eocene adapids is outlined in Fig. 1. As was mentioned earlier, very little is known of the ancestor of *Pelycodus*. The presence of a free tympanic ring in adapids is a character linking them to Paleocene primates of the superfamily Microsyopoidea rather than Plesiadapoidea, and new specimens of the microsyopoid *Purgatorius unio* discovered by Clemens (1974) might be close to the ancestry of *Pelycodus*. Primitive members of the primate family Omomyidae have cheek teeth resembling those of *Purgatorius* and *Pelycodus* (Gingerich, 1973*a*), but enlargement of the anterior teeth and specializations of the ear region make it unlikely that omomyids are closely related to the origin of either Lemuriformes or Anthropoidea. Anthropoid primates were almost certainly derived from a primitive adapid (Gingerich, 1973*b*).

The dental scraper of lemurs and lorises is plausibly interpreted as a derivative of the anterior dentition of *A. parisiensis*, and the close resemblance of the cheek teeth of this species to those of the lemur *H. griseus* is additional evidence that *Adapis* is very closely related to the ancestry of the lemurs and lorises. Thus the phylogenetic relations of Lemuriformes (including Lorisidae) to other primates, and among themselves, are probably as represented in Fig. 7.

The proposed derivation of Lemuriformes from the advanced adapid *A. parisiensis* or a closely related contemporary species requires a reconsideration of the time and continent of origin of the ancestral lemur stock. The evolution of the lower canines in adapids can be traced in some detail through the Eocene. The first appearance of canines which are functionally a unit with the lower incisors is in the Late Eocene *A. parisiensis*, and it is unlikely that the tooth scraper of the lemur/loris ancestor evolved before the Early Oligocene, i. e., about 36–37 million years B.P.

At the beginning of the Oligocene, Walker (1972) estimates the Mozambique Channel to have been 250 km wide at its narrowest point, and estimates that the probability of animals making a chance crossing is about what it is

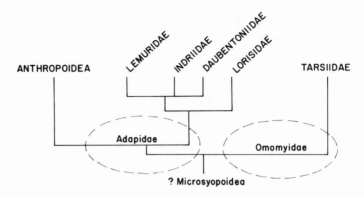

FIG. 7. Diagrammatic representation of the probable phyletic relationships of living primates. Dashed ovals show approximate relationships of the primitive Eocene families Adapidae and Omomyidae.

today—unlikely. This factor, plus the absence of lemurs in the Oligocene and later fossil faunas of Africa, casts some doubt on the African origin of the lemur fauna of Madagascar.

The possibility of an Asian origin of the lemur fauna of Madagascar is enhanced by the presence in Asia of lemuriform fossil primates, and by the patterns of ocean currents and wind circulation. *Lantianius xiehuensis*, named and described by Chow (1964) from the ?Late Eocene of China, is an adapid.[1] *Amphipithecus mogaungensis* from the Late Eocene of Burma may well, as Szalay (1970) believes, be a lemuroid (but see also Simons, 1971). Simons (1972) noted that *Pondaungia cotteri*, also from the Late Eocene of Burma, could, on the basis of tooth structure, be related to the adapid *Pelycodus*, though it also resembles some Anthropoidea. Finally, *Indraloris lulli* described by Lewis (1933) and Tattersall (1968) from the Late Miocene of India is clearly related to the adapid–lemurid–lorisid group, though its precise affinities cannot yet be determined. These fossil forms, from a continent with an as yet poorly sampled Tertiary mammalian microfauna, suggest the possibility of a large Eocene and Oligocene radiation of Asian lemuriform primates.

Walker (1972) showed that the present fauna of Madagascar is more similar to the Miocene African fauna than to the present-day African fauna. A similar comparison could be made with the Miocene fauna of Asia. In fact, the Early Miocene was a time of great faunal interchange between Africa and Asia (Coryndon and Savage, 1973), and as Lorisidae first appear in Africa in the Early Miocene (Walker, 1974), it is plausible that they first emigrated

---

[1]Recent careful study of Chow's figures and description suggests that *Lantianius* is an early artiodactyl and not an adapid primate.

from Asia at this time. Of the three subfamilies of Viverridae in Madagascar, two are endemic and the other is most closely related to Asian forms (G. Petter, 1962). F. Petter (1962) suggests that the cricetid rodents of Madagascar are derived from African Oligocene forms. Cricetids are known from the Oligocene of Asia (Romer, 1966, p. 395), but not that of Africa (Wood, 1968). Thus an Asian origin may be the best explanation for the carnivores and rodents of Madagascar, as well as for the lemurs.

Fortunately, the paleogeography of the Indian Ocean at the beginning of the Oligocene has recently been reconstructed by geophysicists (McKenzie and Sclater, 1971), and a part of this reconstruction is reproduced in Fig. 8. Apart from Africa, the Asian continent is the only plausible area of origin for the ancestral lemur stock. It should be noted that at the beginning of the Oligocene, India and Ceylon were still south of the equator (Fig. 8). The Indian Ocean at this time was already large, and would almost certainly have contained a strong subtropical gyre and a strong south equatorial current. This current, together with southeast trade winds, could have carried rafts of drifting vegetation, including small mammals, from India to Madagascar. By the end of the Oligocene, India would presumably have moved entirely into the northern hemisphere, thereby greatly reducing the chances of subsequent rafting of mammals from Asia to Madagascar. Though the distance from

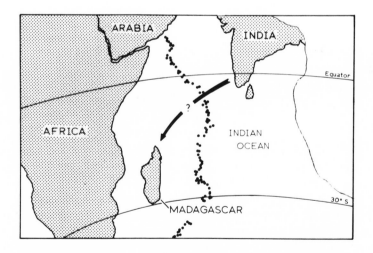

FIG. 8. Paleogeography of Indian Ocean at the beginning of the Oligocene, 36 million years B.P., showing suggested emigration of the ancestral Malagasy lemur stock from Asia rather than Africa. Solid circles are earthquake epicenters marking the boundary between the African and Indian crustal plates. Based on McKenzie and Sclater (1971).

Asia to Madagascar is considerably more than would be necessary for transport from Africa, the currents and winds appear to be more favorable, a point Matthew (1915) noted many years ago.

CONCLUSIONS

To summarize, the anterior dentition of *A. parisiensis* is unique among Eocene lemuriform primates in having the lower canines reduced in size and functioning as a unit with the lower incisors, a condition prestaging the incisor–canine dental scraper of lemurs and lorises. The detailed resemblances of the cheek teeth of *A. parisiensis* to those of *H. griseus* further support a close phyletic relationship of *Adapis* and the living Lemuriformes. It is postulated that increased use of the anterior lower teeth to procure resin or bark by *Adapis* or a close relative led to the well-developed anterior tooth scraper characteristic of both lemurs and lorises.

The phylogeny of Eocene lemuroids is relatively well known, and it is unlikely that the derived dental characters shared by *Adapis* and living lemurs evolved before the Late Eocene, or that the tooth scraper evolved before the Early Oligocene. For faunal and paleogeographic reasons, an Asian origin of lemurs and lorises in the Early Oligocene is suggested, with initial colonization of Madagascar by lemurs in the Oligocene presumably preceding the Miocene immigration of lorisids into Africa.

This evolutionary and migrational scheme is in a number of respects more plausible than those previously advanced to explain the origin of the lemur fauna of Madagascar; however, the conclusions must remain tentative until new fossil evidence from Africa and Asia is found to confirm them.

ACKNOWLEDGMENTS

I have profited from numerous discussions with Drs. Elwyn Simons, Friderun Simons, David Pilbeam, and Ian Tattersall. Dr. Alison Richard kindly permitted me to cite her observations on incisor function in *Propithecus*. I am grateful to Dr. D. E. Russell in Paris and Mr. Pierre Louis in Reims for the opportunity to study adapids from the French Eocene. Dr. Russell kindly provided casts of the specimens illustrated in Figs. 4–6. In addition, I thank Drs. G. L. Jepsen and V. J. Maglio of Princeton University, and Dr. F. A. Jenkins of Harvard University for the loan of *Adapis* specimens in their care.

REFERENCES

de Blainville, II, M. D., 1849, *Ostéographic des Mammifères: Des Anoplothériums,* 155 pp., Bertrand, Paris.

Campbell, C. B. G., 1966, Taxonomic status of tree shrews, *Science* **153**:436.

Charles-Dominique, P., 1971, Eco-éthologie des prosimiens du Gabon, *Biol. Gabonica* **7**(2):121–228.

Charles-Dominique, P., and Martin, R. D., 1970, Evolution of lorises and lemurs, *Nature* **227**:257–260.

Chow, M., 1964, A lemuroid primate from the Eocene of Lantian, Shensi, *Vertebr. Palasiat.* **8**(3):257–263.

Clemens, W. A., 1974, *Purgatorius,* an early paromomyid primate (Mammalia), *Science* **184**: 903–905.

Cooke, H. B. S., 1968, The fossil mammal fauna of Africa, *Quart. Rev. Biol.* **43**(3):234–264.

Coryndon, S. C., and Savage, R. J. G., 1973, The origin and affinities of African mammal faunas, *Spec. Pap. Palaeontol.* **12**:121–135.

Cracraft, J., 1973, Continental drift, paleoclimatology, and the evolution and biogeography of birds, *J. Zool. Lond.* **169**:455–545.

Cuvier, G., 1822, *Recherches sur les Ossemens Fossiles,* Dufour et d'Ocagne, Paris. (*Adapis* named in Vol. 3, p. 265.)

Delfortrie, E., 1873, Un singe de la famille des lémuriens, *Actes Soc. Linn. Bordeaux* **29**:87–95.

Every, R. G., 1970, Sharpness of teeth in man and other primates, *Postilla* **143**:1–30.

Filhol, H., 1874, Nouvelles observations sur les mammifères, lémuriens et pachylémuriens, *Ann. Sci. Geol. (Paris)* **5**(4):1–36.

Fooden, J., 1972, Breakup of Pangaea and isolation of relict mammals in Australia, South America, and Madagascar, *Science* **175**:894–898.

Gervais, P., 1859, *Zoologie et Paléontologie Françaises,* 544 pp., Bertrand, Paris.

Gervais, P., 1873, Remarques au sujet du genre *Palaeolemur, J. Zool. (Paris)* **2**:421–426.

Gervais, P., 1876, *Zoologie et Paléontologie Générales,* Bertrand, Paris.

Gingerich, P. D., 1972, Molar occlusion and jaw mechanics of the Eocene primate *Adapis, Am. J. Phys. Anthropol.* **36**:359–368.

Gingerich, P. D., 1973*a,* First record of the Palaeocene primate *Chiromyoides* from North America, *Nature* **244**:517–518.

Gingerich, P. D., 1973*b,* Anatomy of the temporal bone in the Oligocene anthropoid *Apidium* and the origin of Anthropoidea, *Folia Primatol.* **19**:329–337.

Grandidier, G., 1904, Un nouveau lémurien fossile de France, le *Pronycticebus gaudryi, Bull. Mus. Natl. Hist. Nat. (Paris)* **10**:9–13.

Gregory, W. K., 1920, On the structure and relations of *Notharctus, Mem. Am. Mus. Nat. Hist.* **3**:49–253.

Le Gros Clark, W. E., 1934, On the skull structure of *Pronycticebus gaudreyi, Proc. Zool. Soc. Lond.* **1934**:19–27.

Le Gros Clark, W. E., 1962, *The Antecedents of Man,* 388 pp., University Press, Edinburgh.

Lewis, G. E., 1933, Preliminary notice of a new genus of lemuroid from the Siwaliks, *Am. J. Sci.* **26**:134–138.

Martin, R. D., 1966, Tree shrews: Unique reproductive mechanism of systematic importance, *Science* **152**:1402–1404.

Martin, R. D., 1968, Towards a new definition of Primates, *Man* **3**:377–401.

Martin, R. D., 1972, Adaptive radiation and behaviour of the Malagasy lemurs, *Phil. Trans. Roy. Soc. Lond.* **264**:295–352.

Matthew, W. D., 1915, Climate and evolution, *Ann. N.Y. Acad. Sci.* **24**:171–318.

McKenna, M. C., 1967, Classification, range, and deployment of the prosimian primates, *Colloq. Int. Cent. Natl. Rech. Sci. (Paris)* **163**:603–610.

McKenzie, D., and Sclater, J. G., 1971, The evolution of the Indian Ocean since the Late Cretaceous, *Geophys. J. Roy. Astr. Soc.* **25**:437–528.

Petter, F., 1962, Monophylétisme ou polyphylétisme des rongeurs malgaches, *Colloq. Int. Cent. Natl. Rech. Sci. (Paris)* **104**:301–310.

Petter, G., 1962, Le peuplement en carnivores de Madagascar, *Colloq. Int. Cent. Natl. Rech. Sci. Paris* **104**:331–342.

Petter, J.-J., 1962, Recherches sur l'écologie et l'éthologie des lémuriens malgaches, *Mém. Mus. Natl. Hist. Nat., Sér. A* **27**:1–146.

Petter, J.-J., Schilling, A., and Pariente, G., 1971, Observations éco-éthologiques sur deux lémuriens malgaches nocturnes: *Phaner furcifer* et *Microcebus coquereli, Terre Vie* **25**:287–327.

Richard, A., 1973, Social organization and ecology of *Propithecus verreauxi* Grandidier, Ph.D. Thesis, London University.

Romer, A. S., 1966, *Vertebrate Paleontology*, 3rd ed., 468 pp., University of Chicago Press, Chicago.

Russell, D. E., Louis, P., and Savage, D. E., 1967, Primates of the French Early Eocene, *Univ. California Publ. Geol. Sci.* **73**:1–46.

Schmidt-Kittler, N., 1971, Eine unteroligozäne Primatenfauna von Ehrenstein bei Ulm, *Mitt. Bayer. Staatssamml. Paläeontol. Hist. Geol.* **11**:171–204.

Simons, E. L., 1962, A new Eocene primate genus, *Cantius*, and a revision of some allied European lemuroids, *Bull. Brit. Mus. (Nat. Hist.) Geol.* **7**:1–36.

Simons, E. L., 1971, Relationships of *Amphipithecus* and *Oligopithecus, Nature* **232**:489–491.

Simons, E. L., 1972, *Primate Evolution*, 322 pp., Macmillan, New York.

Stehlin, H. G., 1912, Die Säugetiere des schweizerischen Eocaens, *Adapis, Abh. Schweiz. Paläeontol. Ges.* **38**:1163–1298.

Stehlin, H. G., 1916, Die Säugetiere des schweizerischen Eocaens, *Caenopithecus*, etc., *Abh. Schweiz. Paläeontol. Ges.* **41**:1299–1552.

Szalay, F. S., 1970, Late Eocene *Amphipithecus* and the origins of catarrhine primates, *Nature* **227**:355–357.

Szalay, F. S., 1971, The European adapid primates *Agerina* and *Pronycticebus, Am. Mus. Novit.* **2466**:1–19.

Szalay, F. S., 1972, Cranial morphology of the Early Tertiary *Phenacolemur* and its bearing on primate phylogeny, *Am. J. Phys. Anthropol.* **36**:59–76.

Tattersall, I., 1968, A mandible of *Indraloris* (Primates, Lorisidae) from the Miocene of India, *Postilla* **123**:1–10.

Tattersall, I., 1973a, A note on the age of the subfossil site of Ampasambazimba, Miarinarivo Province, Malagasy Republic, *Am. Mus. Novit.* **2520**:1–6.

Tattersall, I., 1973b, Subfossil lemuroids and the "adaptive radiation" of the Malagasy lemurs. *Trans. N.Y. Acad. Sci.* **35**:314–324.

Van Valen, L., 1965, Treeshrews, primates, and fossils, *Evolution* **19**:137–151.

Walker, A., 1972, The dissemination and segregation of early primates in relation to continental configuration, in: *Calibration of Hominoid Evolution* (W. W. Bishop and J. A. Miller eds.), pp. 195–218. Scottish Academic Press, Edinburgh.

Walker, A., 1974, A review of the Miocene Lorisidae of East Africa, in: *Prosimian Biology* (R. D. Martin, G. A. Doyle, and A. Walker, eds.), Duckworth, London.

Wood, A. E., 1968, The African Oligocene Rodentia, *Bull. Peabody Mus. Nat. Hist.* **28**:23–105.

Zingeser, M. R., 1969, Cercopithecoid canine tooth honing mechanisms, *Am. J. Phys. Anthropol.* **31**:205–214.

PART III
MORPHOLOGY AND PHYSIOLOGY

Structure of the Ear Region in Living and Subfossil Lemurs

6

ROGER SABAN

This study is limited to the examination of the bony ear, including the interior of the temporal bone (Figs. 1 and 2). Consisting as it does of three parts (external, middle, and inner ears), this region is one of cavities of varied dimension and complexity (Lamberton, 1941; Saban, 1956, 1963). Evolutionary modification is particularly evident in the external and middle ears, whereas the inner ear is relatively homogeneous within the lemurs. In the region with which we are concerned, the following are encountered during the course of primate evolution: (1) the appearance of the external auditory meatus by the extension of the tympanic bone (external ear), and (2) the disappearance of the petrosal auditory bulla and the progressive pneumatization of the middle ear and the surrounding area (squama temporalis and the petromastoid region), which eventually reaches the inner ear as the mastoid process appears (Saban, 1964b). The auditory region is crossed by numerous canals associated with the passage of nerves, vessels, and even muscles: These canals permit the reconstruction of vascular relationships among the fossil forms, an important point because primate evolution is characterized by the developing dominance of the carotid over the vertebral system in the arterial irrigation of the brain, and of the internal jugular over the external jugular system in the venous drainage.

ROGER SABAN   Laboratoire d'Anatomie Comparée, 55 Rue de Buffon, 75005 Paris, France

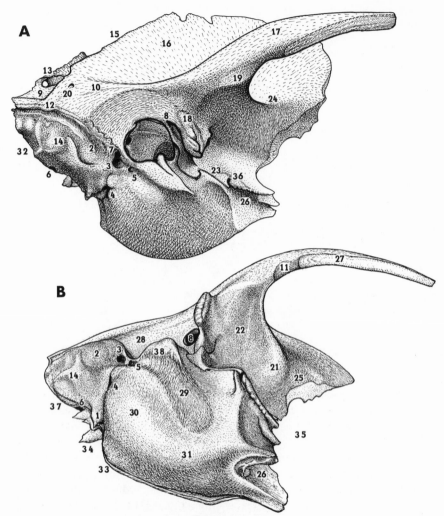

FIG. 1. *Lemur.* Right temporal bone, exocranial aspect. (A) Lateral view. (B) Ventral view. 1, Ostium introitum; 2, mastoid process; 3, stylomastoid foramen; 4, carotid foramen; 5, hyoid fossa; 6, mastoid foramen; 7, posterior canal for chorda tympani nerve; 8, postglenoid foramen; 9, incisura parietalis; 10, linea temporalis; 11, masseteric fossa; 12, posterior petrosquamous suture; 13, petrosal endocranial aspect with aqueduct of Verga; 14, petromastoid portion; 15, squamous suture; 16, squama temporalis; 17, zygomatic process; 18, postglenoid process; 19, zygomatic tubercle; 20, suprasquamous foramen; 21, temporal condyle; 22, mandibular fossa; 23, Glaserian fissure; 24, infratemporal crest; 25, infratemporal plane; 26, auditory tube; 27, temporozygomatic suture; 28, external auditory foramen; 29, groove for stylohyoid muscles; 30, groove for digastric muscle; 31, auditory bulla; 32, petro-occipital suture; 33, petrobasilar canal; 34, jugular process; 35, incisura sphenoidalis; 36, petrotympanic groove; 37, jugular surface; 38, external auditory meatus.

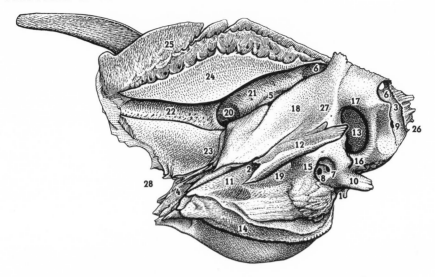

FIG. 2.—*Lemur*. Right temporal bone, endocranial aspect. 1, Opening for cochlear aqueduct; 2, hiatus of the facial canal; 3, sigmoid groove; 4, entocarotid canal; 5, opening for stapedial canal; 6, aqueduct of Verga; 7, foramen singulare of Morgagni; 8, opening of the saccule; 9, opening for sigmoidoantral vein canal; 10, jugular process; 11, Gasserian fossa; 12, tentorium osseum; 13, fossa subarcuata; 14, groove for ventral petrosal sinus; 15, internal auditory meatus; 16, opening for vestibular aqueduct; 17, superior semicircular canal; 18, eminentia arcuata; 19, trigeminal notch; 20, postglenoid foramen; 21, groove for petrosquamous sinus; 22, vascular groove; 23, internal petrosquamous suture; 24, squama temporalis; 25, squamous suture; 26, mastoid foramen; 27, groove for dorsal petrosal sinus; 28, incisura sphenoidalis.

THE EXTERNAL EAR

The Tympanic Bone and the External Auditory Meatus

Among the living lemurs, the tympanic appears as an almost complete ring, comparable to the ring of the human fetus. Open superiorly and furrowed internally by the tympanic sulcus into which the tympanic membrane inserts, this ring (Fig. 3) thus forms two horns (anterior and posterior). Free inside the auditory bulla, the ring attaches, via its two horns, to the horizontal portion of the temporal squama. The anterior horn (grooved on its internal face, Fig. 3B) together with the squama forms the petrotympanic groove. This, itself a portion of the anterior canal of the chorda tympani nerve, is pen-

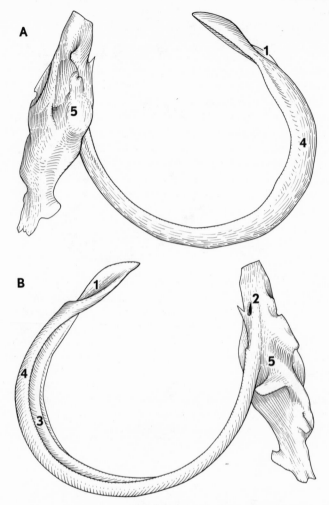

FIG. 3.—*Lemur*. Right tympanic ring. (A) External view. (B) Internal view. 1, Petrotympanic groove; 2, posterior canal for chorda tympani nerve; 3, tympanic sulcus; 4, anterior horn; 5, posterior horn.

etrated by the anterior process of the malleus. In this region, the chorda tympani borders on the anterior ligament of the malleus and the anterior tympanic artery. The posterior horn possesses a bony downward projection crossed by the posterior canal of the chorda tympani. Above, this projection lies firmly against the postmeatal portion of the squama (posterior tympanosquamous suture), and, below, it lies against the petrosal bone (posterior petrotympanic suture), to which it joins between the hyoid fossa and the stylomastoid foramen (end of the third portion of the facial canal). The projection tends to spread out among Indriidae as the edge of the external face of the ring widens. On the other hand, *Daubentonia* is characterized by the narrowing of the horns, giving the ring a U-shaped aspect, and by the disappearance of the projection of the posterior horn and the extension of the anterior horn into the long anterior process which penetrates into the auditory tube (Fig. 4). Except in *Microcebus* (Fig. 5), the tympanic ring does not reach the wall of the bulla. The external auditory meatus is thus formed (Saban, 1971) by the annular cartilage, which attaches directly to the external auditory foramen since there is no bony tube. The annular membrane covering the cartilaginous conduit thickens slightly ("Annulusmembran" of Stehlin, 1912), and extends to the tympanic ring, which with the tympanic membrane separates the external auditory meatus from the tympanic cavity. In front of this is a depression (meatal recess), which is particularly well developed in *Daubentonia* (Fig. 6).

Among the subfossil forms, the tympanic ring remains in *Lemur* (*Pachylemur*), very similar to that in living Lemurinae. In *Mesopropithecus* and the archaeolemurines, it is very delicate, with a mixture of *Indri-* and *Propi-*

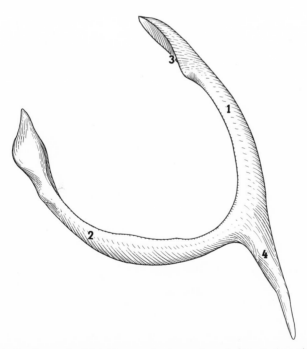

FIG. 4. *Daubentonia*. Right tympanic ring, external view. 1, Anterior horn; 2, posterior horn; 3, petrotympanic groove; 4, anterior process.

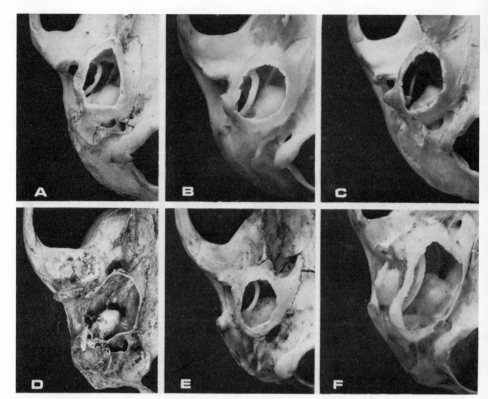

FIG. 5. Right auditory bullae of lemurs. Inferior view with tympanic ring in position (obstructing bone removed). A, *Lemur*; B, *Lepilemur*; C, *Hapalemur*; D, *Pachylemur jullyi*; E, *Cheirogaleus*; F, *Microcebus*; G, *Phaner*; H, *Indri*; I, *Propithecus*; J,

thecus-like characters. Although the ring is held away from the wall of the bulla, it is always, among these forms, positioned inside the opening of the meatus. Its lowest point, located above the level of the floor of the external auditory meatus, would certainly not permit the formation of a meatal recess. However, the anterior horn in *L. (P.) jullyi*, as in *Lepilemur*, shows the beginnings of development of an anterior process (Fig. 7), which becomes important in *Hadropithecus* and *Daubentonia*. The external auditory foramen is thus constituted, as among the living forms, of the squama above and the petrosal below. Usually circular in form, it nonetheless shows, in *Daubentonia*, an anterior flattening due to the rearward movement of the postglenoid process. This corresponds to the anteroposterior elongation of the glenoid fossa for reasons connected with mastication.

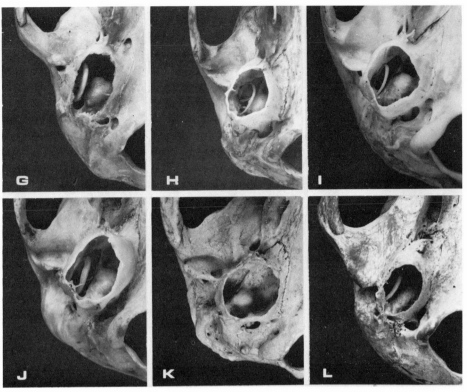

Avahi; K, *Mesopropithecus*; L, *Daubentonia*. Photography by Laboratoire d'Anatomie Comparée du Muséum National d'Histoire Naturelle, Paris.

In the other subfossil indriines (*Archaeoindris* and *Palaeopropithecus*), as well as in *Megaladapis*, the tympanic extends toward the exterior, forming, with the temporal squama, a long, downwardly inclined, bony external auditory meatus (Figs. 8 and 9A). It is important to note that among these large forms all such characters are accentuated. The external auditory foramen through which the meatus opens is dorsoventrally flattened in *Archaeoindris* because of the thickening of the posterior root of the zygomatic arch, which overhangs it. On the other hand, it tends to subcircularity in *Palaeopropithecus*. In both genera the tympanic stretches downward in an important folded layer, although this is more compressed in *Palaeopropithecus*, lying between the postglenoid and paroccipital processes. At the base of the mastoid process the posterior petrotympanic suture encloses the mastoid foramen

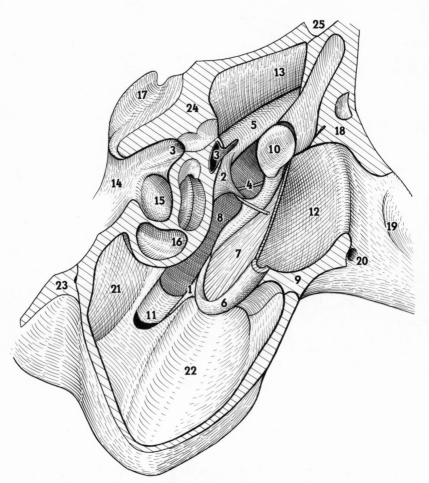

FIG. 6. *Daubentonia*. Cross-section of right temporal (anterior portion). 1, Anterior process of tympanal; 2, tensor tympani muscle; 3, facial canal (first portion); 4, chorda tympani nerve; 5, stapedial canal; 6, tympanic ring; 7, tympanic membrane; 8, eardrum region; 9, annulus membrane; 10, malleus; 11, foramen pneumaticum (auditory tube); 12, recessus meatus; 13, epitympanic recess; 14, internal auditory meatus; 15, opening of the central canal of the cochlea; 16, cochlea (first turn of spiral canal); 17, tentorium osseum; 18, squama temporalis; 19, zygomatic process; 20, postglenoid foramen; 21, hypotympanic sinus (diverticle D$_2$); 22, hypotympanic sinus (diverticle D$_1$); 23, basioccipital bone; 24, petrosal bone; 25, groove for petrosquamous sinus.

and the posterior canal for chorda tympani. This may correspond to the primitive bony projection of the posterior horn of the ring, which among living forms represents the beginning of exteriorization of the tympanic. The anterior edge of the tympanic joins the postglenoid process; a large venous canal communicating with the postglenoid canal appears in the suture. In *Archaeoindris*, the downward spreading of the tympanic creates a lateral wall, an exaggeration of the vaginal process of human anatomy, which totally encloses the styloid process and the tympanic canal; in *Palaeopropithecus*, it constitutes only a partial covering. In ventral view, this wall forms almost a right angle with the portion of the tympanic covering the tympanic cavity from the jugular foramen to the Glaserian fissure, and ends in a long muscular process, digitated in *Archaeoindris*. The enormous development of the external wall and of the muscular process seems to indicate a very powerful musculature (partially styloid and pharyngeal) in the latter genus.

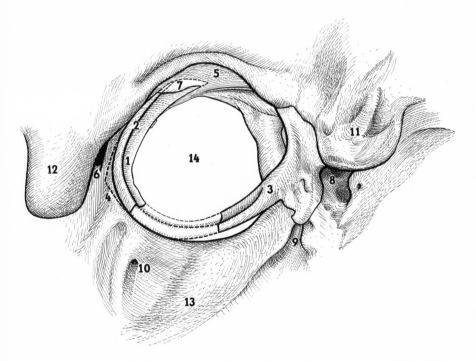

FIG. 7. *Lemur (Pachylemur) jullyi*. Left external auditory meatus. 1, Tympanic sulcus; 2, tympanic ring (anterior horn); 3, tympanic ring (posterior horn); 4, anterior process of tympanic; 5, horizontal portion of squama; 6, postglenoid foramen; 7, petrotympanic groove; 8, stylomastoid foramen; 9, hyoid fossa; 10, canal for deep auricular artery; 11, mastoid process; 12, postglenoid process; 13, auditory bulla; 14, external auditory meatus.

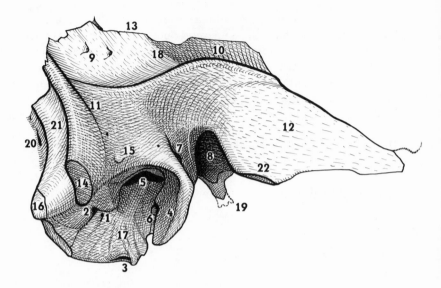

FIG. 8. *Archaeoindris fontoynonti.* Right temporal bone. Exocranial aspect, lateral view. 1, Posterior canal for chorda tympani nerve; 2, stylomastoid foramen; 3, hyoid fossa; 4, postglenoid process; 5, external auditory meatus; 6, postglenoid foramen; 7, accessory postglenoid foramen; 8, mandibular fossa; 9, suprasquamous foramen; 10, squama temporalis; 11, posterior petrosquamous suture; 12, zygomatic process; 13, squamous suture; 14, mastoid process; 15, suprameatal spine; 16, paroccipital process; 17, tympanic; 18, linea temporalis; 19, muscular process; 20, mastoid foramen; 21, petromastoid portion; 22, zygomatic tubercle.

The tympanic bone in *Megaladapis* resembles, in its exaggerated morphology and bony outgrowths, that of the two other fossils; however, the postglenoid process is part of the external wall, providing itself with a large postglenoid canal within the suture (Fig. 9A). Viewed ventrally, the tympanic, covering the petrosal, shows a slight protrusion at the level of the tympanic cavity, by the bulla. In these three subfossil forms, the tympanic constitutes a long bony meatus which narrows in its deep portion, and terminates, as in modern higher primates, in a small meatal recess in front of the ring. Inside the tympanic cavity (Fig. 9), the ring, small in size and slightly grooved by the tympanic sulcus, is prolonged by a long and wide anterior process which, as in *Daubentonia*, utilizes the auditory tube. The lateral wall is found in no other primate, but it may be comparable to that seen in the Oligocene insectivore *Apternodus* (Schlaikjer, 1933, 1934).

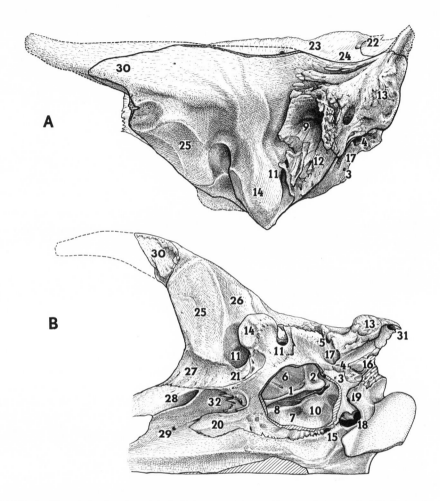

FIG. 9. *Megaladapis edwardsi.* Left temporal bone. Exocranial aspect. (A) Lateral view. (B) Ventral view (tympanic bone partially removed). 1, Anterior process of tympanic; 2, tympanic ring; 3, carotid foramen; 4, hyoid fossa; 5, posterior canal for chorda tympani nerve; 6, hypotympanic sinus (diverticulum D_1); 7, hypotympanic sinus (diverticulum D_2); 8, longitudinal septum; 9, external auditory meatus; 10, cochlea; 11, postglenoid foramen; 12, canal for deep auricular artery; 13, mastoid process; 14, postglenoid process; 15, petrobasilar canal; 16, digastric fossa; 17, stylomastoid foramen; 18, jugular foramen; 19, ostium introitum; 20, apex of pyramid; 21, petrotympanic groove; 22, suprasquamous foramen; 23, squama temporalis; 24, linea temporalis; 25, temporal condyle; 26, mandibular fossa; 27, infratemporal plane; 28, oval foramen; 29, alisphenoid canal; 30, zygomatic process; 31, mastoid foramen; 32, auditory tube.

THE MIDDLE EAR

The middle ear, enclosed within the petrosal, is formed of several inter-communicating cavities and connects with the exterior. Limited by the auditory bulla, it is situated more deeply within the petrosal in the three large sub-fossil lemurs (*Archaeoindris, Palaeopropithecus*, and *Megaladapis*), in which the bulla disappears and pneumatization increases.

Regression of the Auditory Bulla

Generally spherical or ovoid, the bulla (Figs. 1–7), of which the wall represents an expansion of the petrosal, is situated between the occipital and the alisphenoid. Except in *L.* (*Pachylemur*) and Archaeolemurinae, its wall contacts the external pterygoid plate. It extends anteriorly up to the postglenoid process and the oval foramen. On its wall it bears the scars of the stylohyoid and digastric muscles; these are most pronounced among living Indriinae and *Daubentonia*, and among the extinct *L.* (*Pachylemur*) and Archae-olemurinae. In association with the special morphology of the tympanic and the formation of an external auditory meatus, the bulla is reduced to a slight protrusion (Fig. 9) in *Megaladapis*, and completely disappears in *Archaeoindris* and *Palaeopropithecus* (Fig. 10). The bulla, which envelopes the tympanic cavity, presents at its borders various important vascular and nervous foramina; the consistent locations of these are directly related to the endocra-nium in all genera, living and extinct (Figs. 1, 5, 9, and 10). Anteroexternally, the postglenoid foramen transmits a large emissary vein (multiple foramina connected by a system of canals in Archaeolemurinae and the three large genera); anterointernally is the oval foramen, with, in *Archaeoindris* and *Palaeopropithecus*, a distinct foramen spinosum indicative of the possible presence of a middle meningeal artery, and, in Cheirogaleinae, the middle lacerate foramen for the "anterior" carotid (Saban, 1963). Posteroexternally, the stylomastoid foramen, the termination of the facial canal, transmits the facial nerve and the stylomastoid vessels, and opens between the external auditory foramen, the mastoid process, and the hyoid fossa, in which the styloid process originates. Secondary orifices may appear for the posterior canal of the chorda tympani (living and extinct indriids, *Megaladapis*) or for the vascular supply of the stapedius muscle (living lemurids). Posterointernally, the jugular foramen (internal jugular vein, cranial nerves IX, X, XI) lies in the petro-occipital suture, facing the condylar foramen. This communicates with the inferior petrosal sinus, which runs along a canal formed by the expansion of the basioccipital (petrobasilar canal: *Lemur, Hapalemur, Phaner, L. (Pachylemur), Indri, Propithecus, Archaeolemur,* and sometimes *Megaladapis*); this

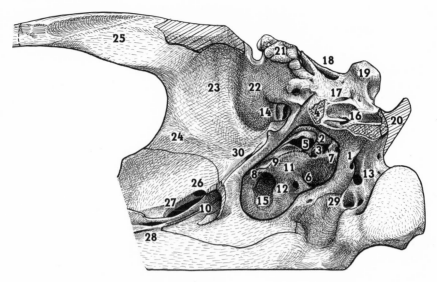

FIG. 10. *Palaeopropithecus maximus*. Left temporal bone. Ventral view (tympanic partly removed). 1, Carotid foramen; 2, facial canal (third portion); 3, hypotympanic sinus (diverticulum D_3); 4, styloid process; 5, epitympanic recess; 6, cochlear aqueduct; 7, carotid groove; 8, longitudinal septum; 9, entocarotid groove, 10, auditory tube; 11, cochlea; 12, internal auditory meatus; 13, jugular foramen; 14, postglenoid foramen; 15, hypotympanic sinus (diverticulum D_2); 16, stylomastoid foramen; 17, posterior canal for chorda tympani nerve; 18, external auditory meatus; 19, mastoid process; 20, paroccipital process; 21, postglenoid process; 22, mandibular fossa; 23, temporal condyle; 24, infratemporal plane; 25, zygomatic process; 26, foramen spinosum; 27, oval foramen; 28, alisphenoid canal; 29, condylar foramen; 30, petrotympanic groove.

canal forks in *Archaeoindris* and possesses secondary orifices in this genus and *Megaladapis*. In the indriids, *Daubentonia* and *L.* (*Pachylemur*), the deep auricular artery, which branches to the external auditory meatus and the tympanic membrane, crosses the wall of the bulla below the meatus; in *Megaladapis*, it lies in a long canal across the external wall. Two vessels penetrate inside the tympanic cavity: the anterior tympanic artery, which utilizes the petrotympanic groove and opens in the Glaserian fissure at variable levels, usually midway (cheirogaleines, indriids, *Daubentonia, Megaladapis*); and the internal carotid artery. This latter follows the carotid canal, which opens above the jugular foramen in *Phaner, Avahi, Archaeoindris,* and *Palaeopropithecus*, but which in all other genera lies in the neighborhood of the stylomastoid foramen. The tympanic cavity communicates directly with the exterior through the auditory tube, which opens near the apex of the pyramid.

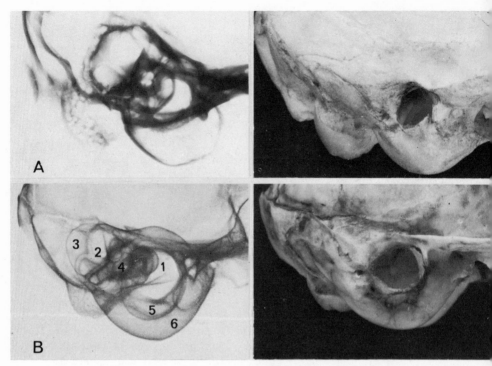

FIG. 11. Auditory region. Radiographs on the left, exocranial aspect on the right, lateral view. (A) *Lemur.* (B) *Microcebus.* (C) *Indri.* (D) *Daubentonia.* 1, Cochlea; 2, semicircular canals; 3, fossa subarcuata; 4, vestible; 5, tympanic ring; 6, auditory bulla

This tube consists of the petrosal and the alisphenoid, which form its roof, except in *Megaladapis* where the tympanic also enters into its composition.

In spite of the disappearance of the bulla in the large genera, the middle ear is represented by a vast tympanic cavity into which the ring and the cochlea project. In all genera the tympanic cavity is divided into three parts: the hypotympanic sinus, the region of the eardrum, and the epitympanic recess (Figs. 9–11). Lying below the cochlea, the first of these is always very spacious, but is less high where the bulla is lacking. Two septa connect it to the cochlea: the anterior, longitudinal septum, much more pronounced in the indriids and *Daubentonia,* joins the cochlea to the auditory tube (Pneumatic foramen) and follows the entocarotid groove; the posterior, transverse septum parallels the carotid canal. These septa delimit the three diverticula of the hypotympanic sinus: the anteroexternal (D1), beneath the auditory tube; the internal (D2), facing the eardrum; and the posteroexternal (D3), opposite

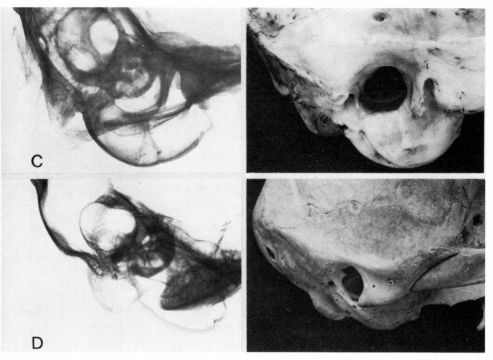

(hypotympanic sinus). Radiography and photography by Laboratoire d'Anatomie
Comparée du Muséum National d'Histoire Naturelle, Paris.

the cochlear fenestra. D1 is more voluminous among the indriids, and extends
laterally into the tympanic under the external auditory meatus in *Palaeopropithecus* and *Megaladapis*. D2 is always very large, while D3 is very small,
although it varies with the position of the carotid foramen. In *Palaeopropithecus* and *Megaladapis*, the region of the eardrum between the cochlea and
the tympanic ring is very narrow because of the conformation of the external
auditory meatus; in the latter, it is limited on its external side by the long anterior process of the tympanic. In *Hapalemur, Lepilemur,* and *Cheirogaleus,*
the alisphenoid participates in the area roofing the eardrum region; in all
other genera, this area is exclusively of petrosal formation and is crossed by
the groove for the tensor tympani muscle. The epitympanic recess inside the
petrosal receives the head of the malleus and the body of the incus, and opens
directly to the eardrum region. Its posterointernal wall is sculpted by the pro-
trusions of the horizontal and superior semicircular canals as well as, in *Av-*

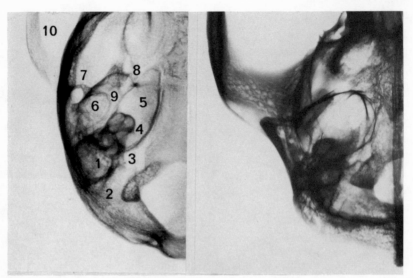

FIG. 12. Radiographs of the auditory bulla, ventral view. *Lemur* on the left, *Hadropithecus* on the right. 1, Horizontal semicircular canal; 2, fossa subarcuata; 3, vestibule; 4, cochlea; 5, auditory bulla (hypotympanic sinus, diverticulum D_2); 6, tympanic ring; 7, postglenoid foramen; 8, auditory tube; 9, eardrum region; 10, zygomatic process. Radiography by Laboratoire d'Anatomie Comparée du Muséum National d'Histoire Naturelle, Paris.

ahi, by the relief of the subarcuate fossa. Its orifice, bordered by the facial and stapedial canals, opens above the cochlea. Very large in *Propithecus, Indri*, and Archaeolemurinae, it is also higher, the ossicles occupying only its inferior part. Behind, it encroaches on the petromastoid (*Palaeopropithecus*), and sometimes an incomplete septum divides this part of the epitympanic recess (assimilated into the mastoid antrum) from that lodging the ossicles in *Megaladapis* and *Avahi*. Isolation of the mastoid antrum appears in *Cheirogaleus, and Megaladapis*; communication with the epitympanic recess is accomplished in these animals via the aditus and antrum, located at the level of the ampullae of the superior and horizontal semicircular canals.

Progression of Pneumatization

Although pneumatization of the temporal remains weak among the Lemuriformes as among all prosimians, a certain increase in cellular density can be noted among the indriids as compared to the lemurids. Pneumatic cells belong to the petromastoid, middle ear, and squama groups (Figs. 11 and 12).

These three cellular groups, although weak, are well delimited; they do not intercommunicate. With the exception of the cells of the squama, all pneumatization is concentrated around the bony labyrinth. The temporal squama presents, in all genera studied, a regular pneumatization of the zygomatic process. Pneumatization of the vertical portion, very weak in Lemuridae, tends to increase in Indriidae and *Daubentonia*, taking the form of very tiny cells (Fig. 11). The petromastoid group is represented among Lemuridae solely by the mastoid antrum, a large cavity opening into the epitympanic recess. Nevertheless, in *Lepilemur* and *Cheirogaleus*, the mastoid antrum shows the beginnings of partitioning as a result of the formation of fine bony divisions along its wall. This pneumatization is complicated in *Indri* and *Daubentonia* by the development of true periantral cells, while the tip of the mastoid process becomes very finely pneumatized in Archaeolemurinae (Fig. 12), *Palaeopropithecus*, and *Megaladapis*. The middle ear group includes caroticolabyrinthic and labyrinthic cells, the former represented by precochlear cells, small and sparse and distributed along the carotid canal, as well as by the cells of the apex of the pyramid, which, among Indriidae, are particularly well developed in *Hadropithecus* (Fig. 12). The latter are directly related to the bony labyrinth and are comprised of the supracochlear, postlabyrinthic, and petrosal crest cells. The supracochlear cells, small and few in number in Lemuridae and *Daubentonia* but more important in the indriids, are located beneath the endocranial cortex of the petrosal between the internal auditory meatus and the vestibule at the level of the superior ampulla. They invade the tentorium osseum when this enlarges, as in *Microcebus* and the indriids. The cells of the petrosal crest are large and grouped above the subarcuate fossa along the arch of the superior semicircular canal in Lemuridae and Indriidae. The retrolabyrinthic cells, small and numerous, follow the posterior semicircular canal and are located beneath the endocranial cortex at the level of the sigmoid groove and of the petro-occipital suture (Lemuridae, Indriidae, *Megaladapis*). They communicate with the petromastoid cells in *Lepilemur* and *Daubentonia* (Fig. 11). Translabyrinthic cells are never found among the lemurs.

The Auditory Ossicles

These auditory ossicles have been described for the living lemurs by Milne-Edwards and Grandidier (1875), Doran (1879), and Werner (1960), and for the extinct forms by Lamberton (1941). As among all primates, they are morphologically homogeneous. Ardouin (1935) gives particular importance to the functional axis around which the ossicular complex turns. This axis passes through the anterior processes of the malleus and the apex of the short crus of the incus. He also emphasizes the angle formed between this axis

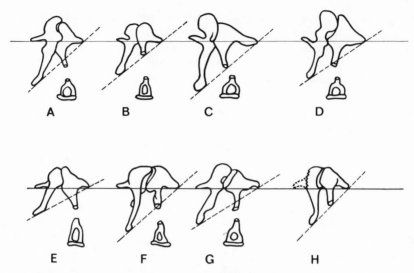

FIG. 13. The auditory ossicles of Primates showing the angle of the ossicular functional axis with the line of Helmholtz (dashed line). (A) *Lemur.* (B) *Propithecus.* (C) *Pan.* (D) *Man.* (E) *Archaeolemur.* (F) *Megaladapis edwardsi.* (G) *M. grandidieri* and *M. madagascariensis.* (H) *Palaeopropithecus.* (B) is from Milne-Edwards and Grandidier (1875), (C) from Ardouin (1935), (E–H) from Lamberton (1941).

and the long lever arm constituted by the line joining the end of the manubrium of the malleus to the end of the short crus of the incus. This angle remains constant within a species. However, the length of the long crus of the incus is extremely variable, allowing the lenticular process to lie along the line of the long lever arm (the three points of Helmholtz) or under it. Figure 13 compares several ossicles oriented along the functional axis. According to this scheme, living Lemuridae and Indriidae (*Daubentonia* shows the indriid condition) appear to exhibit differences in the alignment of the points of Helmholtz, the lenticular process lying on the line in the latter and away from it in the former. Among the subfossil forms (the ossicles are not known in *Archaeoindris* and *Hadropithecus*), *Archaeolemur* corresponds to the lemurid condition and *Palaeopropithecus* to the indriid type. On the other hand, in *Megaladapis* the end of the long crus of the incus, which bears the marks of the chorda tympani, tends to fall outside the line (a tendency accentuated in *M. madagascariensis* and *M. grandidieri*). *Archaeolemur* is distinguished by its very low angle (25°) from all other lemurs (35–40°). The stapes, always gracile, is shorter and wider among living lemurids than among the indriids.

In *Archaeolemur, Hadropithecus,* and the living lemurids excepting *Phaner,* a large space for the passage of the stapedial artery exists between the anterior and posterior crura of the stapes; this space is reduced among the indriines. On the other hand, the stapes of *Megaladapis* is more massive and the perforation still smaller as a consequence of the absence of the stapedial ramus.

THE INNER EAR

The internal ear is represented by a complex of intercommunicating but morphologically different cavities. The structure of these cavities is remarkably uniform throughout Lemuriformes. This complex is comprised of the bony labyrinth, formed, from front to back, by the cochlea, the vestibule, and the three semicircular canals, and easily recognizable in the radiograph of *Microcebus* (Fig. 11B). Each of these formations possesses specific topographical relationships.

The cochlea is represented by a canal rolled back on itself. The first turn of the spiral, in apposition to the fundus of the internal auditory meatus (Fig. 6), starts from the fenestra cochlea, and, reestablishing contact with the vestibule, connects with the first portion of the facial canal. The lower part of this whorl provides the floor of the internal auditory meatus, whereas the upper part lies in relationship with the cerebral surface of the petrosal. The next one and one-half turns of the spiral intrude into the region of the eardrum and the hypotympanic sinus, the cupola making contact with the entocarotid canal.

The superior wall of the vestibule, which encloses the ampullar orifice of the superior semicircular canal and the orifice of the common branch of the superior and posterior canals, provides the floor of the subarcuate fossa, which is limited by the horizontal semicircular canal (Fig. 11). The inferior wall, in relation with the eardrum region, is followed by the carotid canal and its prolongation, the entocarotid canal (Figs. 9 and 10). The external wall, where the stapedial canal runs (except in *Megaladapis* and *Phaner*), includes the oval (vestibular) fenestra and the ampullar orifice of the horizontal semicircular canal. This wall, in the area of the cochlear fenestra, bears the imprint of the ramifications of the tympanic nerve. The internal wall corresponds to the fundus of the internal auditory meatus (Fig. 2). The anterior wall, adjacent to the first turn of the cochlear spiral, comes into contact at the top with the first portion of the facial canal. The posterior wall includes the posterior ampulla.

The three semicircular canals form the superstructure of the subarcuate fossa, the endocranial opening of which is bordered by the superior canal (Figs. 2 and 11). In the three large subfossil genera, where there is no subarcuate fossa, the canals are enclosed in the compact bone through which the petromastoid canal still runs. Additionally, *Megaladapis* is characterized by the small diameter of its semicircular canals and by the shortness of the non-crossing horizontal and posterior canals. In all cases the horizontal ampulla is in relationship with the second portion of the facial canal, and with the stapedial canal where it exists. The posterior part of the common branch of the superior and posterior canals is in relationship with the vestibular aqueduct.

The small variations of the bony labyrinth can be brought out by comparing the tyridian angles and the degree by which the oval fenestra is raised (Saban, 1952), as well as by using the ampullar angle of Beauvieux (1934), which expresses the depression of the posterior canal. The thyridian angle, small (15–30°) in Indriinae, *Megaladapis*, and *Daubentonia* (20°), is larger in the living lemurids (35–37°) and Archaeolemurinae (38°); it is 60° in man. The angle of elevation of the oval fenestra, zero in man, is large in Archaeolemurinae (45°), *Daubentonia*, and *Megaladapis* (50°), but is much reduced in the indriines and lemurines (20–30°). On the other hand, the ampullar angle always remains small among the lemurs. In *Megaladapis* it measures around 5°, in *Archaeolemur* and *Daubentonia*, 15–17°, and 20° in all other living forms; in man it varies between 55° and 80°, according to cranial type.

THE BONY CANALS

Three kinds of canals, nervous, vascular, and muscular, cross the ear region.

Nervous Canals

The chorda tympani runs through two canals from one side to the other of the region of the eardrum. The posterior canal, in part set into the posterior petrotympanic suture, brings the chorda tympani into the tympanic cavity. It arises in front of the stylomastoid foramen and enters the region of the eardrum behind the posterior horn of the tympanic ring between the incus and the stapes. Shorter in *Megaladapis*, it does not extend beyond the edge of the ring. The anterior canal, formed by the petrotympanic groove inside the Glaserian fissure where the anterior tympanic artery also runs, opens in the anteroexternal roof of the eardrum region and ends at the level of the glenoid fossa.

In *Megaladapis* and the indriids, the tympanic canal, adjacent to the carotid canal, conducts the tympanic nerve on to the cochlea in front of the cochlear fenestra. In the lemurids there is no special canal, the nerve emerging from the carotid canal at the bifurcation of the stapedial and entocarotid arteries.

The small canal for the auricular ramus of the vagus nerve connects the jugular fossa (ostium introitum) with the facial canal. It enters within the petrosal, but is visible in the posterior part of the roof of the hypotympanic sinus in *Lepilemur* and the indriids (except *Avahi*).

Most important of all, the facial canal shelters the facial nerve and the stylomastoid vessels. It constitutes a large bony canal with three portions which traverse the petrosal from the fundus of the internal auditory meatus to the stylomastoid foramen. The first transverse portion, very short, travels from the fundus to the genu of the facial canal, occupying the space between the vestibule and the cochlea. The second portion, located perpendicular to the former, runs backward and downward. It appears in the eardrum region at the opening of the epitympanic recess and follows the oval fenestra beneath the protrusion of the horizontal semicircular canal. In *Palaeopropithecus* and *Microcebus*, the canal stops at the level of the oval fenestra. The third portion forms an angle of 120° with the second. It covers the pyramid, except in *Microcebus*, and ends at the stylomastoid foramen, through which the facial nerve exits from the cranium. This third portion, very long in *Palaeopropithecus*, crosses the tympanic wall (Fig. 10).

Vascular Canals

The canal of the deep auricular artery, which crosses the tympanic wall behind the external auditory meatus, runs along the floor of the meatus and opens in front of the tympanic ring in the three large subfossils. The carotid canal consists of a first section common to all lemurs, but which is only a simple groove in *Microcebus* and *Phaner*. This canal, visible in the posterior part of the hypotympanic sinus, carries the internal carotid artery on to the promontory. In *Cheirogaleus* and *Phaner*, its length increases due to the displacement of the carotid foramen. In all lemurs except *Phaner* and *Megaladapis*, the canal bifurcates at the level of the promontory into the stapedial and entocarotid canals. The stapedial canal penetrates the oval fenestra between the crura of the stapes; in *Lepilemur* and *Cheirogaleus*, it ceases at this level. It then intersects with the facial canal, skirts the opening of the epitympanic recess, crosses the tegmen tympani, and penetrates into the endocranium. In *Palaeopropithecus*, the stapedial canal, in the form of a small groove, travels around the oval fenestra and reaches into the endocranium through the roof of the eardrum region (this could correspond to the path of the intraorbital branch). It seems, however, that a minuscule twig branches off and crosses the oval fenestra. The entocarotid canal loops around the cochlea and rapidly becomes a groove along the longitudinal septum, then crosses the tegmen tympani near the auditory tube (in Lemurinae, Indriidae, and *Megaladapis*) or utilizes this tube itself, emerging opposite the middle lacerate foramen (Cheirogaleinae). In *Palaeopropithecus*, the groove forks at its origin into three parallel branches which follow the septum, then the auditory tube, and finally penetrate the endocranium either through the basisphenoid or through the petrobasisphenoid suture.

Among the venous canals, the aqueduct of Verga, which follows the petrosquamous suture, is of variable length (Fig. 2), and is in direct relationship with the roof of the tympanic cavity. It holds the petrosquamous sinus which reaches the exterior through the postglenoid foramen and thus debouches into the external jugular vein. This conduit possesses numerous exocranial tributaries (suprasquamous foramina) in *Megaladapis*. After the postglenoid foramen, the postglenoid canal remains simple in the lemurids, but already possesses a lateral branch in *Phaner*. Such ramifications become more important in the indriids (except *Avahi*) and *Daubentonia*, but attain their greatest complexity in the large subfossils (Saban, 1956, 1963).

Muscular Canals

The two ossicular muscles, stapedius and tensor tympani, are protected by a bony wall which constitutes the pyramid in the case of the first, and the tensor tympani canal in the case of the latter. In all genera, the pyramid represents a small elongated pocket attached to the third portion of the facial canal, with which it communicates by a small orifice for the passage of a twig of the facial nerve. Its anterior extremity gives off, in the interior of the region of the eardrum, the tendon of the muscle which attaches to the head of the stapes. At least among the lemurids, the bottom of the canal communicates with the exterior by a narrow orifice through which the nerves supplying the muscle pass.

The tensor tympani canal represents, among the lemurids and the indriids, a groove which widens posteriorly and which is limited by a raised border in front of the oval fenestra. Normally in the roof of the eardrum region, it is displaced on to the cochlea and the longitudinal septum in *Daubentonia* (Fig. 6). In *Megaladapis*, the canal comprises a thin, curved, bony lamina which, in the roof of the region of the eardrum, reaches the opening of the epitympanic recess. This lamina can be seen in Fig. 9 between the tympanic ring and the cochlea. The petrotympanic groove contains, besides the vasculonervous elements already described, the anterior ligament of the malleus and the anterior process to which it attaches.

DISCUSSION

As the foregoing considerations suggest, there is much uniformity throughout Lemuriformes in the structure of the bony ear.

The topography of the internal carotid artery, as it appears in the tympanic cavity, is constant except in *Phaner* and *Megaladapis*. The carotid canal divides on the cochlea into two grooves of about the same size: the stapedial and entocarotid canals, both penetrating into the endocranium. Among the lemurs, the irrigation of the brain is effected principally by the vertebral arteries, as among primitive mammals. The stapedial canal, still well developed, retains its superior (supraorbital) branch, which supplies the dura mater and the extrabulbar part of the orbit (Bugge, 1972). The inferior (infraorbital) branch of the stapedial artery, characteristic of insectivorous, may also have existed in *Palaeopropithecus*. The entocarotid artery, which joins the arterial circle of the brain near the hypophyseal fossa, deviates from its normal path in Cheirogaleinae, to anastomose with the "anterior" carotid; this large extratemporal artery leaves the common carotid to penetrate the cranium through the middle lacerate foramen en route to the orbit, as in Lorisiformes (Fig. 14A).

The venous drainage of the brain, at the level of the bony ear, is divided among living lemurs between the internal and external jugular systems. The former, found among all mammals, drains the blood of the lateral sinus of the jugular vein through the jugular foramen. The latter receives much of the cerebral blood through the petrosquamous sinus (aqueduct of Verga) and the postglenoid vein. This system is dominant in Cheirogaleinae, whose members are distinguished by the presence of emissary veins of the cavernous sinus, which joins the external jugular via the middle lacerate foramen (Fig. 14B). However, it is among the subfossils in which a reconstruction of the venous circulation has been attempted (Saban 1956, 1963) that a marked predominance of the external jugular system is seen; this is particularly marked in the three large genera.

Pneumatization, despite some tendency among Indriidae to increase, remains very little developed among the lemurs, and emphasizes the structural uniformity of the group.

Alongside these relatively homogeneous characteristics there exist, particularly in the large subfossil forms, certain traits which might in isolation be considered "advanced," since they are found among higher primates. Nonetheless, this would seem a false conclusion; they are better viewed as the results of parallelism. The origin of such parallelisms may lie in the evolution of the lemurs in the isolation of Madagascar, protected from competition and predation. As Millot (1972) has pointed out, the isolation of the Malagasy fauna "has effectively encouraged both palaeoendemism in Madagascar

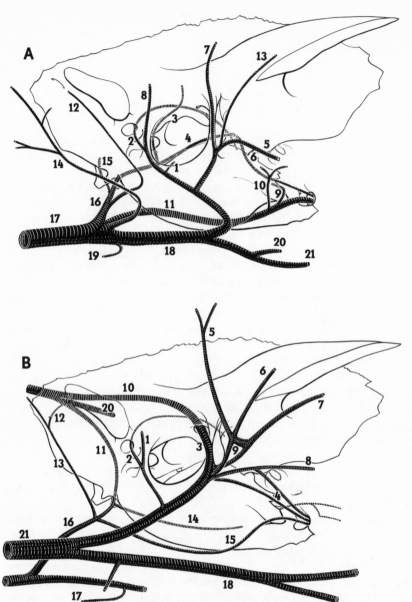

FIG. 14. *Cheirogaleus*. Scheme of the vascular connection of temporal bone. (A) Arteries. 1, Deep auricular a.; 2, stylomastoid a.; 3, stapedial a.; 4, entocarotid a.; 5, internal maxillary a.; 6, anterior tympanic a.; 7, anterior auricular a.; 8, posterior auricular a.; 9, tubal branch; 10, pterygoid branch; 11, anterior carotid a.; 12, digastric branch; 13, superficial temporal a.; 14, occipital a.; 15, posterior meningeal

through the refuge offered to the archaic species threatened with extinction, and neo-endemism by limiting, through reduction of competition, the elimination of mutations" (p. 751). Thus these evolutionary radiations, in which we find both archaic and modern forms, have resulted in the specific giantism which characterizes the terminal branches and which may in part have been responsible for their extinction (Saban, 1964a). This giantism, most marked in *Archaeoindris*, *Palaeopropithecus*, and *Megaladapis*, shows, as do cases of pathological giantism in man, a hyperdevelopment of various parts of the temporal, of which the components seem to have undergone a drawing-downward. This phenomenon is accompanied in man, at the level of the squama, by the formation of a shelf above the external auditory foramen; at the level of the tympanic by strong development of the vaginal process and of the external auditory meatus, whose axis is strongly directed downward; and at the petromastoid level by the great elongation of the mastoid process. These consequences are evident, and yet further accentuated, in the three large subfossils. It seems, for example, that the exaggerated development of the vaginal process of the tympanic may have created the lateral wall. Moreover, such results of acromegaly are reflected, as in man, in the circulatory system, where one finds an increase in the diameter of the principal arterial and venous trunks of the brain, an increase in the volume of the endocranial venous sinuses, and the persistence of the petrosquamous sinus, opening through the subglenoid and suprasquamous foramina. In the same way, a particularly well-developed vascular system must have existed among the large subfossils, as suggested by the number and size of the vascular orifices and canals. It should be noted again that at the level of the bony ear there is a strong similarity between the results of specific and pathological giantism. Otherwise, the absence of the auditory bulla in the large subfossils is no more than superficial, because the hypotympanic sinus remains voluminous and retains the structure seen in other lemurs. Pneumatization in these animals, a little more developed in the petromastoid region, cannot be compared with that of the apes; it results from the increased size of the skull, and in particular of the mastoid process, but does not reach the stage seen in either catarrhine or platyrrhine higher primates. The absence of the subarcuate fossa and the stapedial artery is merely a parallel development with the higher pri-

---

a.; 16, internal carotid a.; 17, common carotid a.; 18, external carotid a.; 19, cranial thyroid a.; 20, lingual a.; 21, facial a. (B) Veins. 1, Posterior auricular v.; 2, stylomastoid v.; 3, postglenoid v.; 4, cavernous sinus emissary v.; 5, superficial temporal v.; 6, zygomatic v.; 7, deep temporal v.; 8, internal maxillary v.; 9, pterygoid plexus; 10, petrosquamous sinus; 11, sigmoid sinus; 12, mastoid emissary v.; 13, deep occipital v.; 14, ventral petrosal sinus; 15, cavernous sinus emissary v. (petro-occipital sinus); 16, internal jugular v.; 17, cranial thyroid v.; 18, facial v.; 19, lingual v.; 20, dorsal petrosal sinus; 21, external jugular v.

mates. In fact, the subarcuate fossa is equally absent in certain edentates and in most ungulates (Artiodactyla, Perissodactyla). The disappearance of the stapedial artery in *Phaner* and *Megaladapis* does not have the same significance as in the higher primates, since the internal carotid system does not have the same importance in these two genera as in the latter; the entocarotid remains weak and similar to those of other prosimians.

The only feature which may possibly be regarded as evolved lies in the exteriorization of the tympanic and the formation of an external auditory meatus in the three large subfossils, whereas it is apparently muscular action which has influenced the formation of the tympanal wall and the development of the muscular process, a strong specialization in these forms.

As concerns morphological specialization of the bony ear among lemurs, we are largely restricted to the gradual development of the anterior process of the tympanic ring; weakly indicated in *Lepilemur, L. (Pachylemur)*, and *Indri*, it elongates and penetrates the auditory tube in *Hadropithecus, Daubentonia*, and *Megaladapis*. Unfortunately, the functional significance of such a process remains obscure. On the other hand, the specialization of the glenoid fossa in *Daubentonia* clearly results from the radical change in its maxillo-mandibular apparatus, which is adapted to mastication of rodent type.

REFERENCES

Ardouin, P., 1935, Considérations anatomiques sur les osselets de l'oreille chez certains Singes anthropomorphes, *Bull. Mém. Soc. Anthropol. Paris.* **5**:20–47.

Beauvieux, J., 1934, *Recherches Anatomiques sur les Canaux Semi-circulaires des Vertébrés*, 104 pp., Brusau Fr., Bordeaux.

Bugge, J., 1972, The cephalic arterial system in the Insectivores and the Primates with special reference to the Macroscelidoidea and Tupaioidea and the Insectivore–Primate boundary. *Z. Anat. Entwickl.-Gesch.* **135**:279–300.

Doran, A. H. G., 1879, Morphology of the mammalian ossicula auditus, *Trans. Linn. Soc. Lond. Zool.* **1**:371–497.

Grandidier, G., 1902, Observations sur les Lémuriens disparus de Madagascar, *Bull. Mus. Hist. Nat.* **8**:497–505, 587–591.

Lamberton, C., 1941, Contribution à la connaissance de la faune subfossile de Madagascar: Oreille osseuse des Lémuriens, *Mém. Acad. Malgache* **35**:1–134.

Millot, J., 1972, In conclusion. in: *Biogeography and Ecology in Madagascar.* (R. Battistini and G. Richard-Vindard, eds.), pp. 741–756, W. Junk, The Hague. Publ. :741–756.

Milne-Edwards, A., and Grandidier, A., 1875, *Histoire Physique, Naturelle et Politique de Madagascar,* Vol. 6: *Histoire Naturelle des Mammifères,* 396 pp. plus atlas, Impr. Nationale, Paris.

Saban, R., 1952, Fixité du canal semi-circulaire externe et variations de l'angle thyridien, *Mamalia.* **16**:77–92.

Saban, R., 1956, L'os temporal et ses rapports chez les Lémuriens subfossiles de Madagascar. I. Types à molaires quadrituberculées: Formes archaïques, *Mém. Inst. Rech. Sci. Madagascar Sér. A.* **10**:251–297.

Saban, R., 1963, Contribution à l'étude de l'os temporal des Primates. Description chez l'Homme et les Prosimiens. Anatomie comparée et phylogénie, *Mém. Mus. Hist. Nat* **29**:1–378.

Saban, R., 1964*a*, Contribution à l'étude du crâne et en particulier de l'os temporal des géants acromégales, *Bull. Mem. Soc. Anthropol. Paris 11ème Sér.* **6**:279–303.

Saban, R., 1964*b*, Sur la pneumatisation de l'os temporal des Primates adultes et son développement ontogénique chez le genre *Alouatta* (Platyrhinien), *Morphol. Jahrb.* **106**:569–593.

Saban, R., 1971. Le cartilage annulaire de l'oreille externe chez les Mammifères, *Bull. Ass. Anat. 56ème Congr. Nantes.* **154**:1152–1163.

Schlaikjer, E. M., 1933, Contribution to the stratigraphy and paleontology of the Goshen Hole area, Wyoming, I. A detailed study of the structure and relationships of a new zalambdodont insectivore from the middle Oligocene, *Bull. Mus. Comp. Zool. Harvard.* **76**:1–27.

Schlaikjer, E. M., 1934, A new fossil zalambdodont insectivore, *Ann. Mus. Novit.* **698**:1–8.

Werner, C. F., 1960, Das Ohr. A. Mittel und Innerohr, in: *Handbuch der Primatenkunde*, Vol. 3(5): *Primatologia* (H. Hofer, A. H. Schultz, and D. Starck, eds.), pp. 1–40, Karger, Basel.

Notes on the Cranial Anatomy of the Subfossil Malagasy Lemurs

7

IAN TATTERSALL

At a meeting of the Royal Society held on June 15, 1893, C. I. Forsyth Major described the skull of an extinct Malagasy primate, the first to come to scientific attention (Major, 1894). Since that time, the subfossil remains of some six genera and 12 species of extinct lemuroids have been recovered in Madagascar, many of them represented by quite abundant material.

The sites from which these forms are known are widely distributed over the western, central, and southern portions of the island (Fig. 1), and encompass the corresponding ecological zones (see Chapter 2). The central plateau sites, located in areas which are today almost entirely denuded of forest, appear to correspond to a forest environment containing many of the species, both floral and faunal, which typify the eastern humid forests. All the sites are of relatively recent origin: available ^{14}C dates range from 2850 ± 200 to 1035 ± 50 years B.P. (Mahé, 1965; Tattersall, 1973c; Mahé and Sourdat, in press). The subfossil lemurs, whose extinction was almost certainly due to the activities of man (Walker, 1967a), must therefore be regarded as belonging to the modern lemur fauna (Standing, 1908), and as shedding no direct light on lemuriform phylogeny.

IAN TATTERSALL   Department of Anthropology, American Museum of Natural History, New York, New York 10024

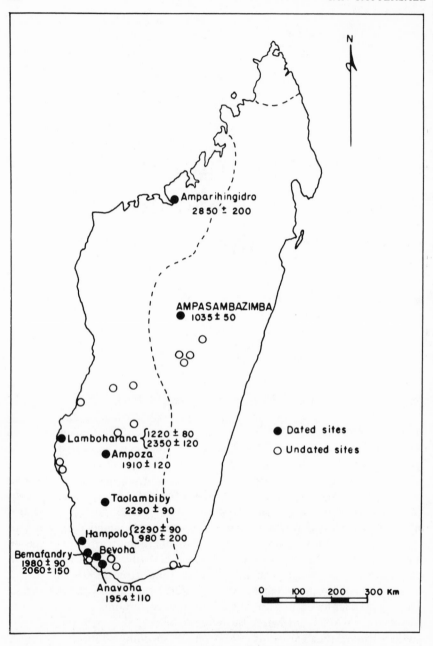

FIG. 1. Sites from which subfossil lemurs have been recovered, with dating where available. The dotted line indicates the western limit of the damp eastern flora. Dates from Mahé (1965), Mahé and Sourdat (in press), and Tattersall (1973c).

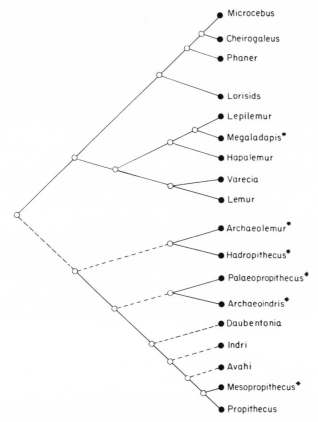

FIG. 2. Provisional theory of relationships of the Malagasy lemurs. Branching-points have no temporal or morphological significance. Extinct forms are denoted by asterisks.

The relationships of the subfossil forms to the living lemurs are relatively clear, and a tentative phylogeny of the Malagasy primates is given in Fig. 2. The scheme is presented here without justification; this may be found elsewhere (Tattersall and Schwartz, 1974). As Gingerich points out (Chapter 5), their considerable dental (and, to a lesser extent, other) similarities with *Adapis* suggest strongly that the lepilemurines (*Lepilemur* and *Hapalemur*) are of adapid derivation. Adapids formed a significant proportion of both the Eurasian and North American prosimian faunas during the Eocene, and it is not unlikely that they also penetrated Africa (and Madagascar) during that epoch, if indeed they did not originate there.

At this point, it is necessary to define the proposition of an African origin (e.g., Charles-Dominique and Martin, 1970; Martin, 1972; Tattersall, 1973*a,b*), as opposed to an Asian/Indian one (Gingerich, Chapter 5) for the

Malagasy prosimians. The argument against an Indian derivation rests partly upon the improbability of the presence of adapids (or, for that matter, of any primates) in India during the time period in question. The details of the breakup of Gondwana are still obscure. But whatever these may be it is virtually certain that India was well separated from Africa by the Jurassic (Dietz and Holden, 1970), while Madagascar, although perhaps isolated from the mainland by shallow seas at that time (Kent, 1972), remained in relatively close proximity. By the close of the Mesozoic, India was further isolated from Africa but was still remote from Asia. Madagascar's movement in the interim was evidently minimal (Kent, 1972).

The order Primates evidently had its origin in the latter part of the Cretaceous, but the similarities of certain Malagasy prosimians to characteristically Eocene forms suggests the descent of the former from representatives of the second major primate radiation, which originated and flourished during the time when India was perhaps at its farthest remove from any major land mass. It is thus far less likely that the adapids, a widespread Holarctic group, penetrated India during the latest Paleocene or earliest Eocene, than that they were in Africa, and subsequently Madagascar, at this time. The presence of an apparent adapid survivor in India during the Late Miocene (Tattersall, 1969) provides no evidence of the presence there of adapids during the subcontinent's period of isolation, since by then contact with the Asian mainland had been established and faunal interchange via the Afghan corridor would have been possible. Thus an invasion of Madagascar directly from the Asian mainland and an introduction from Asia by way of India seem equally unlikely in view of the vast distances involved.

Perhaps a more important consideration is that the Paleocene Eurasian fauna contains no primates which may plausibly be regarded as sister groups of any of the Recent Malagasy lemurs besides Lemuridae. If multiple invasions of Madagascar by prosimian primates are accepted, as increasing evidence suggests they should be, we should expect to find forms related to the invading groups in the parent fauna. However, as we have already seen, these are signally lacking in the Eurasian early Tertiary record in all cases except one. It seems more reasonable to suggest that the unknown African primate fauna of the appropriate age may have contained such forms, including (less specialized) adapids, than to posit that Eurasia was the ultimate source of the modern Malagasy prosimian fauna.

RECENTLY EXTINCT LEMUROIDS

The three extinct lemur subfamilies include at least eight species grouped into five genera. In addition, there are two known extinct indriine, one (or possibly two) extinct lemurine, and one extinct *Daubentonia* species. Most extinct genera are known from sites throughout the sampled area of the island; however, there are, in most cases, specific differences between congeneric forms from the central plateau and from the coastal regions. There may also be such differences within some of the latter; thus, for instance, the single cranium of *Archaeolemur* from Amparihingidro, the sole northwestern site, appears intermediate in size and morphology between the southern and southwestern species, *A. majori*, and the plateau species, *A. edwardsi*. It is difficult to make any concrete judgment on this, however, because the differences between the two species are of degree rather than of kind, and only a single specimen is available from the northwest.

Megaladapinae

Megaladapinae, a lemurid subfamily, contains the single genus *Megaladapis* (Fig. 3), represented by three species (*M. madagascariensis*, mean cranial length 240.5 mm, $n = 3$; *M. grandidieri*, mean cranial length 288.5 mm, $n = 3$; *M. edwardsi*, mean cranial length 296.0 mm, $n = 10$). As Major (1894) noted in describing the type of *Megaladapis*, "a *superficial* examination of the skull will certainly not suggest its classification among the Lemuroidea." In its highly elongated cranium, its peculiarly downturned nasal bones, its specialized ear region, and in many other characteristics, the skull of *Megaladapis* appears highly atypical for any lemur group. Yet an understanding of the functional anatomy of the skull together with attention to the details of dental and cranial morphology reveals the close relationship between this animal and the lepilemurines.

At the level of detail with which this chapter is concerned, the crania of the three species are sufficiently similar to be considered together, and all lend themselves admirably to analysis in terms of a model of the operation of the masticatory apparatus elaborated at some length elsewhere (Tattersall, 1973*a*, 1974; Roberts and Tattersall, 1974). Briefly, this model views the elevation of the mammalian jaw as being accomplished by a couple action between the two major groups of jaw-closing muscles, the masseter–internal pterygoid and the temporalis. In this system, the more anterior the point at which the bite force is exerted, the greater the effort required of the posterior (horizontal) portion of the temporalis muscle. Conversely, if biting takes place further posteriorly, the importance of the anterior (vertical) component of temporalis is emphasized.

According to this model, many of the unusual features of the cranium of *Megaladapis* may be traced to the animal's possession of a greatly elongated face, reflecting an anterior concentration of dental activity. Most obvious among these features are extreme hypertrophy and posterior displacement of the posterior moiety of the temporalis, together with the reduction of the an-

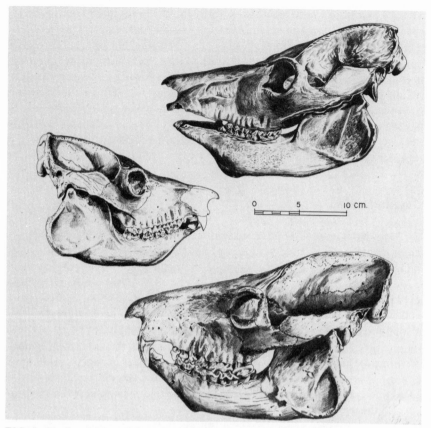

FIG. 3. Skulls of *Megaladapis* in lateral view. Above right: *M. grandidieri*; left: *M. madagascariensis*; below right: *M. edwardsi*.

terior portion of the muscle. Since relative to size, *Megaladapis* is extremely small brained, the posterior hypertrophy of the temporalis shows itself in the development of pronounced (posterior) sagittal and nuchal cresting. Further, the small size of the brain coupled with the need to shift the temporalis backward necessitated considerable spacing between the tiny neurocranium proper and the facial region. The development of truly enormous frontal sinuses correlates directly with this need. The failure of the brain itself to intrude anteriorly into the immediate postorbital area of the postfacial skeleton is reflected in the lack, noted by Radinsky (1970), of orbital impressions in the endocast, and in the inordinate length of the olfactory tracts.

The attenuation, for mechanical reasons, of the postfacial skeleton of *Megaladapis* also impinges on the structure of the cranial base, which exhibits a similar drawing-out. Beyond this, the atypical morphology of the cranial base can be attributed to two further major factors: first, the loss of the inflated auditory bulla, with the consequent rearrangement of the foramina of the cranial base (Saban, Chapter 6); and second, the development of the paroccipital process into a large, elongate, protruding structure. This expresses the enlargement of the digastric muscle, both bellies of which were hypertrophied in the medium- and large-sized subfossil lemurs and which reached the zenith of their development in *Megaladapis*. Another somewhat atypical character of the cranial base in this animal is the highly robust postglenoid process, the broad anterior surface of which articulates with a correspondingly extensive surface extending down from the posterior aspect of the mandibular condyle.

Modification of the masticatory apparatus in *Megaladapis*, largely in response to the great elongation of the face, thus accounts for many of its more striking differences in cranial construction from other members of Lemuridae. Why, then, this highly significant development of the face? As early as 1894, Major noted that *Megaladapis* possessed certain cranial features in common with the phalangerid marsupial *Phascolarctos cinereus*. Among such features are the facial elongation referred to above (although this is far more marked in *Megaladapis* than in *Phascolarctos*), the retroflexion of the facial skeleton relative to the plane of the cranial base, and the completely backward-facing foramen magnum, which is associated with an orientation of the occipital condyles perpendicular to the basicranial axis. Moreover, Walker (1967*b*) has recently shown that the locomotion of *Megaladapis* probably resembled that of *Phascolarctos*, the koala, which, although rather slow and clumsy, represents a form of vertical clinging and leaping.

Those characteristics of cranial construction common to both *Megaladapis* and the koala most likely represent adaptations to similar feeding patterns, since their effect is to transform the entire head into a long extension of the neck. *Phascolarctos* feeds, and *Megaladapis* presumably fed, by cropping leaves from branches pulled manually to within reach of the anterior teeth. The adaptations under discussion serve to maximize the radius within which feeding can take place from a single clinging position; this is advantageous for the relatively unagile koala, and would presumably have been yet more so for the vastly bulkier *Megaladapis*. The suggestion of a cropping habit for *Megaladapis* is reinforced by the loss in the adult of the upper incisor teeth and their presumed replacement by a horny pad, an adaptation characteristic also of certain ruminants. Further, the possession of a mobile snout by *Megaladapis* may be indicated by the long nasal bones, downwardly flexed at their anterior extremeties and overlapping the nasal aperture, together with the highly vascularized nature of the bone covering the nasal region, which hints at a thick tissue covering.

Archaeolemurinae

Archaeolemurinae is an indriid subfamily containing the three species *Archaeolemur majori* (mean cranial length 130.6 mm, *n* = 17), *A. edwardsi* (mean cranial length 147.0 mm, *n* = 17), and *Hadropithecus stenognathus* (two crania known, lengths 128.2 and 141.8 mm) (Fig. 4). Distinguished from the indriines both by morphological specialization of the dentition and by the possession of an additional premolar in each jaw, they nonetheless quite closely resemble them in cranial structure (Tattersall, 1973a,b). Indeed, one might establish a morphological sequence of skull form: *P. verreauxi → A. majori → A. edwardsi → H. stenognathus*, which largely expresses increasing specialization among the archaeolemurines (Tattersall, 1973a).

It is primarily from early studies of the archaeolemurines (e.g., Major, 1896; Lorenz von Liburnau, 1899; Standing, 1908) that the notion of "advanced" lemuroids has stemmed. More recent work, however, has shown that such views are untenable; the archaeolemurines are in no meaningful sense

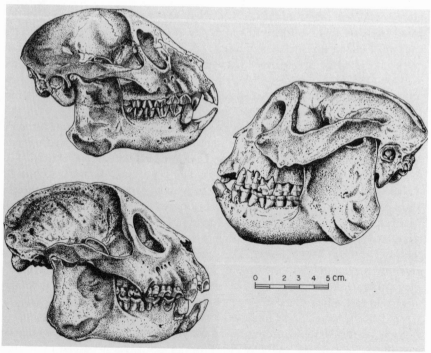

FIG. 4. Archaeolemurine skulls in lateral view. Above left: *Archaeolemur majori*; right: *Hadropithecus stenognathus*; below left: *Archaeolemur edwardsi*.

"advanced" over the indriines in any gradal characteristic (Lamberton, 1937; Piveteau, 1948; Tattersall, 1973a).

The most striking departures of the archaeolemurines from the ancestral indriid condition, most closely approximated today by *Propithecus*, are an increase in size and specialization of the masticatory apparatus. Dentally, the two species of *Archaeolemur* are extraordinarily similar; there are no consistent diagnostic differences which separate them, and there is even overlap in the size ranges of individual teeth. Both, then, are characterized by premolar rows modified into continuous shearing blades and by the possession of bilophodont molars. Such molar teeth are easily derivable from the quadricuspid indriid condition but are clearly distinct from it. If analogy with the bilophodont cercopithecids is permissible, the relatively low molar cusps in *Archaeolemur* suggest a predominantly frugivorous rather than folivorous diet. The central upper incisor is expanded, while the lateral, closely approximated to it, is small. The lower incisors are stout and somewhat forwardly inclined (possibly indicating derivation from a procumbent condition), the lateral exceeding the central teeth in size.

In *Hadropithecus*, the molar teeth are broad and high-crowned, with complex enamel folds replacing the simple cusp-and-loph pattern of *Archaeolemur*. Unlike the molars of the latter, these teeth wear rapidly. The posterior premolars are molarized, although the anterior ones, at least in their unworn state, retain a somewhat bladelike conformation. The incisors, canine and caniniform, are very greatly reduced; this is the primary, if not the only, reason for the remarkable orthognathy evinced by *Hadropithecus*, which in this respect contrasts markedly with the somewhat prognathous *Archaeolemur*.

The dental dissimilarities between *Hadropithecus* and *Archaeolemur* correspond quite closely to those described by Jolly (1970a) between *Theropithecus* and *Papio* (Jolly, 1970b; Tattersall, 1973a, 1974), and may quite plausibly be attributed to related selective pressures. In each case, *Papio* and *Archaeolemur*, respectively, may be viewed as approximating a primitive, dietarily more generalized, condition. A diet reminiscent of that of *Theropithecus*, i.e., of grass blades or tough, gritty morsels gathered with the fingers at ground level and placed directly between the cheek teeth, would satisfactorily explain the anterior dental reduction of *Hadropithecus*, together with the enormous expansion and complication of ,ts posterior teeth. In turn, these opposite trends in the anterior and posterior dentitions are correlated with major modifications in cranial architecture.

The development of a powerful posterior grinding apparatus is intimately linked to that of the expanded masticatory musculature reflected in the sagittal and nuchal cresting characteristic of *Hadropithecus* and, to a slightly lesser extent, of *A. edwardsi*. Further, according to the model of masticatory mechanics mentioned earlier, the posterior concentration of masticatory activity in *Hadropithecus* is the cause of the facial deepening and the forward shifting of the muscles of mastication in this animal relative to the other archaeolemurines.

Also related to mastication is perhaps the most striking qualitative distinction between the archaeolemurines and the indriines: the fusion of the mandibular symphysis in the former, with which is correlated a considerable increase in mandibular robusticity. These allegedly "monkey-like" adaptations correspond merely to the adoption of a more powerful mode of chewing by the archaeolemurines: one in which, in particular, lateral mandibular movements could have been aided, if not primarily effected, by the contralateral muscles.

Palaeopropithecinae

Frequently considered inseparable from Indriinae, the two large-sized species *Archaeoindris fontoynonti* (length of single known cranium 269.0 mm) and *Palaeopropithecus ingens* (here taken to include *P. maximus*; mean cranial length 194.5 mm, $n = 8$) are in fact more divergent cranially from the indriines than are the archaeolemurines (Fig. 5). Dentally, however, their size differences notwithstanding, the indriines and palaeopropithecines are remarkably close, although the lower anterior teeth of the latter are nowhere near fully procumbent.

The primary characteristic distinguishing the palaeopropithecines from the indriids lies in the construction of the middle ear (Saban, Chapter 6). Instead of possessing the "lemuroid" condition, with the tympanic ring free inside an inflated bulla, the palaeopropithecines lack the latter and the ectotympanic communicates with the exterior via a bony tube. I have noted elsewhere (Tattersall, 1973b) that the occurrence of a similar conformation in the remotely related *Megaladapis* suggests that the condition is size-related in lemuroids possessing large crania and relatively small brains.

The cranial anatomy of the palaeopropithecines has been described in some detail by Standing (1908) and Lamberton (1934), and in this brief review it is probably sufficient to note that the cranial differences between these animals and the indriines have been overstated. *Palaeopropithecus*, with its relatively long and low cranial contour, is not unreminiscent of *Indri* (although in certain features of its dentition, notably in the reduction of M^3 and in the disposition of the tooth rows it more closely resembles *Propithecus*). The relatively smaller brain and orbits of the subfossil form, possibly allometric manifestations, account for many of the major apparent distinctions between the two genera in overall cranial appearance. Thus, for instance, the more depressed profile of *Palaeopropithecus* in the region of the junction of the orbital and nasal capsules is probably to be traced to the relative reduc-

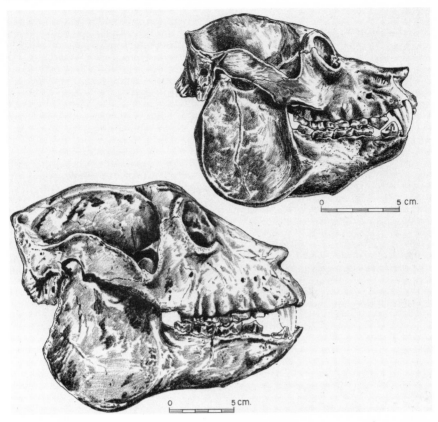

FIG. 5. Palaeopropithecine skulls in lateral view. Above: *Palaeopropithecus ingens* (= *P. maximus*); below: *Archaeoindris fontoynonti.*

tion of its orbits. In *Indri*, conversely, the interorbital area, because it is flanked by the relatively large orbits, is filled out by the development of considerable frontal sinuses.

Propithecus and *Archaeoindris* both possess crania which are relatively abbreviated anteroposteriorly. Structurally, *Propithecus* is extraordinarily similar to *Indri*; virtually all the dissimilarities between the crania of the two genera are directly due to their differences in facial length. The flatter face of *Propithecus* is functionally linked to the shortening of its neurocranium and to the forward shifting of the origin of temporalis. Thus, although relative to skull length, the cranium of *Propithecus* is rather high when compared with that of *Indri*, this is not a real difference in functional terms. In *Archaeoindris*, however, deepening of the face and neurocranium appears to be biologically significant, although the relative longitudinal proportions of the splanchnocranium and neurocranium are determined by the requirements of the masticatory apparatus in a pattern very similar to that evinced by *Propithecus*. It seems plausible that the genuine facial deepening seen in *Archaeoindris* reflects the need to resolve large occlusal forces. The fusion of the mandibular symphysis in both palaeopropithecines, and particularly its foreshortening in *Archaeoindris*, would appear to support this suggestion.

Other Subfossil Forms

Subfossil indriines are limited to the genus *Mesopropithecus* (Fig. 6), with a central plateau species, *M. pithecoides*, and a southern and southwestern one, *M. globiceps* (Tattersall, 1971). Both species bear close resemblances to *Propithecus* in their crania, although *M. pithecoides*, in particular, is very much more robust, possessing, for example, both sagittal and nuchal crests. Whether or not these forms will ultimately prove congeneric with *Propithecus* depends partially on the acquisition of more knowledge of their

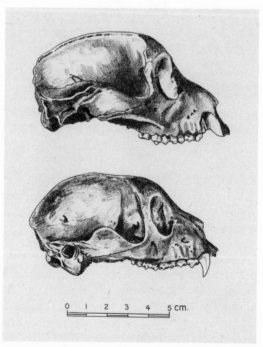

FIG. 6. Crania of the extinct indriines. Above: *Mesopropithecus Pithecoides*; below: *M. globiceps*.

postcrania; at present, despite their cranial similarities, *Propithecus* and *Mesopropithecus* are best regarded as generically distinct.

The two subfossil lemurines, *Varecia insignis* and *V. jullyi*, are close in cranial and dental morphology to their extant congener, although they are somewhat larger in size. *V. jullyi*, known from the plateau site of Ampasambazimba, possesses, on average, a slightly longer cranium than that of the southern *V. insignis*, but in general the differences between the two forms are no greater than those which, for instance, separate the larger eastern subspecies of *Avahi*, *A. laniger laniger*, from the smaller western subspecies, *A. l. occidentalis*. I have therefore suggested (Tattersall, 1973*b*) that the two subfossil forms be synonymized under *Varecia insignis*.

The remaining extinct lemur, *Daubentonia robusta*, differs in known parts (postcranial bones and a few teeth) from its living congener in size alone.

REFERENCES

Charles-Dominique, P., and Martin, R. D., 1970, Evolution of lemurs and lorises, *Nature* **227**:257–260.

Dietz, R. S., and Holden, J. C., 1970, The breakup of Pangaea, *Sci. Am.* **223**(4):30–41.

Jolly, C. J., 1970*a*, The seed-eaters: A new model of hominid differentiation based on a baboon analogy, *Man (n.s.)* **5**:5–26.

Jolly, C. J., 1970*b*, *Hadropithecus*, a lemuroid small-object feeder, *Man (n.s.)* **5**:525–529.

Kent, P. E., 1972, Mesozoic history of the east coast of Africa, *Nature* **238**:147–148.

Lamberton, C., 1934, Contribution à la connaissance de la faune subfossile de Madagascar: Lémuriens et ratites, *Mém. Acad. Malgache* **17**:19–22.

Lamberton, C., 1937, Contribution à la connaissance de la faune subfossile de Madagascar. Note 3: Les Hadropithèques. *Bull. Acad. Malgache* **20**:1–44.

Lorenz von Liburnau, L. R., 1899, Einen fossilen Anthropoiden von Madagaskar, *Anz. K. Acad. Wiss. Wien* **19**:255–257.

Mahé, J., 1965, *Les Subfossiles Malgaches*, 11pp., Imprimerie National, Tananarive.

Mahé, J., and Sourdat, M., 1974, Sur l'extinction des vertébrés subfossiles et l'ardification du climat dans le sud-ouest de Madagascar, *Bull. Soc. Géol. France.*

Major, C. I. F., 1894, On *Megaladapis madagascariensis*, an extinct gigantic lemuroid from Madagascar, *Phil. Trans. Roy. Soc. Lond. Ser. B* **185**:15–38.

Martin, R. D., 1972, Adaptive radiation and behaviour of the Malagasy lemurs, *Phil. Trans. Roy. Soc. Lond. Ser. B* **264**:295–352.

Piveteau, J., 1948, Recherches anatomiques sur les lémuriens disparus: le genre *Archaeolemur*, *Ann. Paleontol.* **34**:127–171.

Radinsky, L. B., 1970, The fossil evidence of prosimian brain evolution, in: *The Primate Brain.* (C. R. Noback, and W. Montagna, eds.), pp. 209–224, Appleton-Century-Crofts, New York.

Roberts, D., and Tattersall, I., 1974, Skull form and the mechanics of mandibular elevation in mammals, *Am. Mus. Novit.* **2536**:1–9.

Standing, H., 1908, On recently discovered subfossil primates from Madagascar, *Trans. Zool. Soc. Lond.* **18**:69–112.

Tattersall, I., 1969, More on the ecology of north Indian *Ramapithecus*, *Nature* **224**:821–822.
Tattersall, I., 1971, Revision of the subfossil Indriinae, *Folia Primatol.* **16**:257–269.
Tattersall, I., 1973a, Cranial anatomy of Archaeolemurinae, *Anthropol. Pap. Am. Mus. Nat. Hist.* **52**(1):1–110.
Tattersall, I., 1973b, Subfossil lemuroids and the "adaptive radiation" of the Malagasy lemurs, *Trans. N.Y. Acad. Sci.* **35**:314–324.
Tattersall, I., 1973c, A note on the age of the subfossil site of Ampasambazimba, Miarinarivo Province, Malagasy Republic, *Am. Mus. Novit.* **2520**:1–6.
Tattersall, I., 1974; Facial structure and mandibular mechanics in *Archaeolemur*, in: *Prosimian Biology* (R. D. Martin, G. A. Doyle, and A. C. Walker, eds.), Duckworth, London.
Tattersall, I. and Schwartz, J. H., 1974, Craniodental morphology and the systematics of the Malagasy lemurs (Prosimii, Primates). *Anthropol. Pap. Am. Mus. Nat. Hist.* **52**(3):141–192.
Walker, A., 1967a, Patterns of extinction among the subfossil Madagascan lemuroids, in: *Pleistocene Extinctions*. (P. S. Martin, and H. E. Wright, Jr., eds.), pp. 425–432, Yale University Press, New Haven.
Walker, A., 1967b, Locomotor adaptation in recent and subfossil Madagascan lemurs, Ph.D. thesis, University of London.

The Lemur Scapula 8

DAVID ROBERTS AND ISOBELLE DAVIDSON

In this chapter we seek to examine the causal relationships existing between the various modes of locomotion adopted by prosimian primates, particularly the lemurs, and the development of the scapula, the main structural unit of the pectoral girdle.

To paleoprimatologists, prosimians are of particular interest because in some respects, for instance the dentition and cranial osteology, they represent the closest existing approximation to the evolutionary stage reached by the primates of the Eocene radiation.

This is probably less true, however, in terms of their locomotor behavior. It seems likely that the earliest primates were scurrying, rodent-like animals similar to the living tree shrews (*Tupaia*), with a marked proclivity toward an orthograde posture that has persisted in most subsequent primate species (Simons, 1972). Some species of living prosimians have utilized the orthograde posture in developing a form of locomotion unique to this suborder, which has been defined by Napier and Walker (1967) as "vertical clinging and leaping." The remaining prosimian species are all arboreal quadrupeds, although only the mouse lemur (*Microcebus*), which is described by Napier and Napier (1967) as having a "scurrying gait . . . like a rodent," possibly resembles the ancestral forms in locomotor habit. Quadrupedal prosimians from Asia and the mainland of Africa have adopted the gait described as "slow-climbing," whereas the Malagasy lemurs include species which locomote in a

DAVID ROBERTS   School of Veterinary Medicine, University of Pennsylvania, Philadelphia, Pennsylvania 19174. ISOBELLE DAVIDSON   Department of Anthropology, University of Pennsylvania, Philadelphia, Pennsylvania 19148

manner resembling that of some quadrupedal monkeys, including horizontal leaping from branch to branch and arm-swinging (Buettner-Janusch, 1967). Prosimian locomotion will be discussed more fully later, but it may be observed at this point that the spectrum of locomotor types presented by the species of the prosimian suborder can be grouped into several categories along the lines indicated above.

The scapulae of species within each prosimian category possess distinctive sets of morphological characteristics which suggest that the morphology of the bone is solely determined by locomotor functions. Previous studies have indicated that there is a form–function relationship active in determining the morphology of all parts of the skeleton. To investigate this relationship in the scapula, its biomechanics must be examined.

BIOMECHANICAL SYSTEMS AND FORM–FUNCTION RELATIONSHIPS

Living organisms are essentially composed of many functional systems which are complex integrations of fundamental chemical, electrical, and mechanical components. The basic components of such a system interact not only to bring about the functions for which the system as a whole is responsible, but also to maintain the system at the optimal level of efficiency required to perform those functions. The vertebrate musculo–skeletal system provides an excellent example. The function of this system is to transfer loads in order to carry out movements essential for the behavior of the organism. This is achieved mechanically through muscle contractions, which are themselves the result of complex electrochemical reactions within the muscle sarcomeres. Contraction of muscles causes the bony struts to which they are attached to move with respect to one another. This motion of bone on bone is governed by simple physical laws relating to mechanical efficiency. Thus if the load placed upon some particular part of the musculo–skeletal system is somewhat greater than that normally imposed on it, abnormal levels of stress are generated. These abnormal stresses produce electrical potentials on and within the bones that appear to trigger a complex series of chemical reactions which in turn stimulate the osteoblasts of the bone, and may ultimately result in remodeling taking place. Within certain practical limits, continual application of abnormal stress can produce a general reorganization of the bony morphology such that the form and function of the system are again in a state of dynamic equilibrium.

Knowledge of functional systems and their contained feedback subsystems has led many biologists to attempt to correlate skeletal morphology and behavior. It is obvious that the establishment of such form–function relationships provides a valuable tool for use in interpreting functional and perhaps even behavioral patterns for fossil species of which only skeletal evidence remains.

BIOMECHANICS OF THE SCAPULA

Although there are fewer bony elements in the mammalian pectoral girdle than in that of a reptile or amphibian, the locomotor functions of the forelimb are quite complex. Moreover, the forelimb is frequently used to perform a wide range of functions other than those relating to locomotion. The bony elements of the mammalian pectoral girdle are the clavicle and scapula. In some mammals the clavicle is lost, but it is always present in primates. To compensate for the loss of pectoral bones that are present in animals more primitive in this respect, the mammalian scapula has acquired several "new" morphological features. These include the development of a scapular spine, which may be prolonged into an acromial process, and of a supraspinous fossa.

The form–function relationships of the scapula have attracted much attention in the past. Smith and Savage (1956) pointed out that the morphology of the inferior angle (Fig. 1) and the development of a postscapular fossa ap-

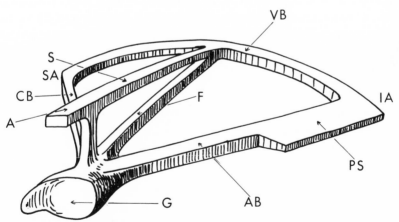

FIG. 1. Schematic representation of the mammalian scapula, looking obliquely onto the suprascapular surface. The bone is resolved into its component struts which form a three-dimensional framework, imparting great strength. A, Acromion; AB, axillary border; CB, coracoid border; F, line of fusion between the scapular spine and the blade; G, glenoid articular surface; IA, inferior angle; PS, postscapular fossa; SA, superior angle; VB, vertebral border. CB, VB, and F enclose the supraspinous fossa; AB, VB, and F enclose the infraspinous fossa.

peared to be determined by the need for a long lever arm for the teres major muscle. Müller (1967) attempted to show that the form of the vertebral border of the scapula was related to load bearing (body weight), and Oxnard (1967) suggested that the shape of the primate scapular blade may be determined by the nature of the predominant stresses (compressional or torsional) that it is called on to withstand during locomotion. Ashton *et al.* (1971) concluded, as the result of a statistical analysis of scapular dimensions, that the morphology of the primate scapula correlates in some definite way with the behavior of the species concerned. Roberts (1973) examined various aspects of scapular morphology, in part through experimental biomechanical studies, and reached similar conclusions to those of Oxnard (1967) and Ashton *et al.* (1971), but disagreed with Müller (1967) in finding that body weight alone probably does not account for the extremely long vertebral border observed in the scapulae of some primate species. Some of the current ideas concerning scapular form–function relationships are summarized below.

SCAPULAR STRUCTURE AND FUNCTION

To understand the form–function relationships of the scapula, it is first necessary to appreciate some aspects of its physical structure. The scapula is composed essentially of two flat triangular plates of bone: the blade and the spine. The spine is fused to the blade in a plane perpendicular to it along a line referred to here as "the base of the scapular spine." The suprascapular surface of the blade is thus divided into two areas: the supraspinous and infraspinous fossae (Fig. 1). This general arrangement of two plates of bone fused together, one perpendicular to the other, imparts great mechanical strength to the scapula. Further structural strength is gained by thickening the free margins of the spine and blade to form bony struts. These peripheral struts and the thickening along the base of the scapular spine where the two scapular plates are fused form a three-dimensional framework (Fig. 1). Roentgenograms of scapulae show that trabecular structures tend to parallel the struts forming the frameworks of the bones. This microstructural orientation is an indication that stresses acting on scapulae are distributed along their framework struts, and comparison of the frame structures of different scapulae provides some indication of the differential ranges and sizes of the stresses to which the bones are subjected.

The suprascapular fossae support muscles forming part of the rotator cuff musculature of the shoulder. The function of these muscles, the supraspinatus and infraspinatus, together with the subscapularis muscle—which occupies the whole of the subscapular surface—is twofold: first, to secure the shoulder joint from dislocation and, second, to initiate and provide power in forelimb motions (Inman *et al.*, 1944). The size of the fossae is determined by the size of the muscles they contain. Therefore, in an animal whose forelimb functions place a premium on the ability to secure the shoulder joint (as, for example, in a hanger and swinger), and where powerful motions are required, the fossae of the scapula will be well developed. Development of the infraspinatus muscle and infraspinous fossa is associated with flexion (retraction) of the forelimb, and development of the supraspinatus muscle and supraspinous fossa is associated with extension (protraction) of the limb. Powerful flexion of the forelimb is a characteristic locomotor function in most mammals whether they are cursors or climbers; accordingly, the infraspinous fossa is usually fairly well developed and strengthened by peripheral thickening. Extension of the forelimb appears to play a less important part in many quadrupedal mammals (probably because gravity assists in the action, which is not a power stroke) and in such forms the supraspinous fossa may be poorly developed and lack peripheral thickening.

In order to demonstrate the relative development of the scapular fossae, Roberts (1973) has suggested the use of supraspinous and infraspinous indices. Photographs of examined scapulae were made such that the length of the base of the scapular spine was brought to a constant value. The index for fossa was calculated from the formula

$$\frac{\text{area of the fossa} \times 100}{\text{length of the base of the scapular spine}}$$

The length of the base of the scapular spine was used to rationalize the indices because experimental work by Miller (1932) and others has suggested that this dimension, which represents the "length" of the scapula, is not affected by external influences that can result in abnormal development of other aspects of scapular morphology.

Figure 2 shows a bivariate plot of scapular indices of various mammals, permitting comparison of the development of the scapular fossae and indicating how this may relate to locomotor function.

The scapular framework is subject not only to forces generated by the contraction of muscles used to stabilize the shoulder joint and to move the forelimb, but also to the transmission of part of the body weight (all of it if the animal is hanging by the forelimbs). Müller (1967) suggested that the arched form of the vertebral border in carnivores was developed because this was the optimal structural form for supporting the weight of the body; it was assumed, probably correctly, that the body weight of the animal was transmitted from the vertebral border to the forelimb via the struts formed by the posterior border of the blade and the base of the scapular spine. This argument is weakened, however, when the shape of the vertebral border is considered in other mammals. In the horse, where the weight of the body relative to the size of the forelimb is high, the vertebral border is almost straight. In

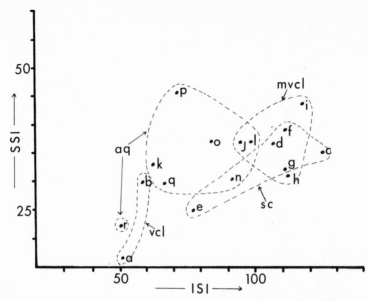

FIG. 2. Bivariate plot of the supraspinous and infraspinous indices given in Table I. The broken lines indicate recognized locomotor categories. Note that there is a tendency for the infraspinous fossa to be larger in the slow climbers than in the arboreal quadrupeds. The tendency to possess a large infraspinous fossa in the group designated mvcl suggests that these animals may also use their forelimbs to support body weight in climbing, hanging, or swinging. The small fossa in the squirrel-like *Microcebus* is typical of cursors. a, *Tarsius spectrum*; b, *Galago* sp.; c, *Perodicticus potto*; d, *Arctocebus calabarensis*; e, *Nycticebus coucang*; f, *Loris tardigradus*; g, *Indri indri*; h, *Avahi laniger*; i, *Propithecus diadema*; j, *Propithecus verreauxi*; k, *Lemur catta*; l, *Lemur macaco*; n, *Lemur variegatus*; o, *Lemur mongoz*; p, *Hapalemur griseus*; q, *Lepilemur mustelinus*; r, *Microcebus murinus*. aq, Arboreal quadrupeds; vcl, vertical clingers and leapers; mvcl, Malagasy vertical clingers and leapers; SSI, supraspinous index; ISI, infraspinous index.

some primates, for example *Propithecus*, the border tends to be markedly straight, particularly where it forms the boundary of the infraspinous fossa. Furthermore, since the muscles reponsible for attaching the scapula to the thorax insert along the vertebral border of the scapula, it may be assumed that the length of the border should reflect the power of those muscles. In a carnivore such as the cat, the length of the vertebral border is about the same as the length of the base of the scapular spine. In the relatively, as well as absolutely, heavier-bodied horse, the length of the vertebral border is only about half the length of the base of the scapular spine. This suggests that the length of the vertebral border is not directly related to the body weight of the animal.

It may, however, relate to the dynamic forces generated by the body weight when the forelimb makes contact with the substrate at the end of a leap or stride (Roberts, 1973). This would explain the long border observed in the cat, since it is a leaping animal. The curvature of the border would thus relate to the need to provide a long insertion for powerful serratus and rhomboid muscles in a scapula with relatively narrow fossae. In a climbing animal, the required long vertebral border may not be curved because the fossae are wide to accommodate the large rotator cuff muscles used to secure the shoulder joint. In order to facilitate comparisons between prosimian scapulae, a vertebral border index may be defined:

$$\frac{\text{length of vertebral border} \times 100}{\text{length of base of scapular spine}}$$

FUNCTIONAL DIMENSIONS OF THE SCAPULA

Frey (1923) suggested that the range of limb circumduction of which an animal is capable is determined by the width of the scapular fossae; that is, the wider the fossae the greater the range of potential circumduction. The scapula is, however, extremely mobile on the dorsum. In running quadrupeds, it rocks backward and forward as the limbs are flexed and extended. In climbing animals, it moves dorsomedially toward the backbone when the forelimb is abducted, and ventrolaterally when it is adducted. Thus it is possible to obtain forelimb circumduction without much scapulohumeral movement. In fast-running animals such as the horse and some dogs, the scapula is narrow and seems to function as a lever that can be pulled rapidly across the relatively broad dorsum. It is possible that this kind of motion is more efficient for high-speed forelimb motions than would be the case if the scapula was held relatively immobile and a large amount of scapulohumeral motion took place instead. Climbing animals such as primates, however, require reasonably broad fossae to contain the large rotator cuff muscles used to secure the shoulder joint against dislocation when it is subjected to potentially tensile stress. However, a relatively broad scapula appears to pose a dynamic problem inasmuch that it cannot be moved rapidly through a large arc across the dorsum.

To overcome this difficulty in primates that require rapid movement of the forelimb during locomotion, there is a tendency for reorganization of the outline of the scapular blade that results in a bone which is effectively narrow but which still contains broad fossae. The best examples of such "obliquely reorganized" scapulae are seen in the brachiating gibbon and in the pygmy chimpanzee (which has also been observed to brachiate; D. R. Pilbeam, personal communication). Here, the coracoid (cranial) border of the blade is shortened and the vertebral border (which is long and straight) is rotated so that it makes an acute angle with the spine; at the same time the glenoid articular facet is turned cranially (Fig. 3). The bone is now said to be *functionally narrow*.

"Length" and "breadth" of the fossae, blade dimensions used by some authors in the past (e.g., Inman *et al.*, 1944), are misleading in the case of a functionally narrow bone. For this reason, new variables or "functional dimensions" have been defined. Functional dimensions are related to a line drawn tangential to the glenoid articular facet (ignoring the ventrally turned "lip" of the glenoid) in the plane of the blade. The functional breadth of the blade is found by drawing perpendiculars to the line tangential to the coracoid and axillary borders of the blade. The functional length is found by

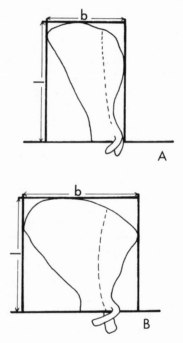

FIG. 3. To show the construction of the "functional parameters" of the scapula in (A) a functionally narrow scapula (index: 152), and (B) a functionally broad scapula (index: 97). l, Length; b, breadth.

drawing a tangent to the distal margin (usually some part of the vertebral border) perpendicular to the lines used to define the breadth of the blade and parallel to the datum line drawn tangential to the glenoid. This construction can be easily understood by reference to Fig. 3. By this method, functionally narrow scapulae can be distinguished from scapulae with narrow fossae because the former have a high vertebral border index and high supraspinous and infraspinous indices.

The functional dimensions described have been used to define a "blade index":

$$\frac{\text{functional length of blade} \times 100}{\text{functional breadth of blade}}$$

SHAPE OF THE SUPERIOR ANGLE OF THE SCAPULA

The superior angle of the scapula varies in form among mammal species. In some it is acutely angular, and in others it is smoothly rounded. Roberts (1973) concluded that the shape of the superior angle relates to the size of the scapula with respect to the size of the dorsum. Smoothly rounding the superior angle permits a relatively large scapula to slide forward beneath the integument of the neck without causing undue disruption of the tissues.

PROSIMIAN LOCOMOTOR CATEGORIES

There is a large range of variability in the locomotor behavior of different prosimian species. As with any group of primates, it is difficult to draw sharp boundaries between some of the locomotor categories represented, although individual species display distinctive locomotor behavior. Prosimians are usually grouped into slow climbers, vertical clingers and leapers, and quadrupeds with or without vertical clinging and leaping tendencies. The following locomotor categorization follows Napier and Napier (1967):

1. Slow climbers:
   Loris spp.*
   Nycticebus spp.*
   Perodicticus potto*
   Arctocebus calabarensis*
2. Vertical clingers and leapers:
   Tarsius spp.*
   Galago spp.*
   Propithecus spp.

   Avahi laniger
   Indri indri
   Lepilemur mustelinus
3. Arboreal quadrupeds:
   Lemur spp.
   Hapalemur spp.
   Microcebus spp.
   Phaner furcifer
   Cheirogaleus spp.

The locomotor behavior of Asian and African species (denoted by aster-isks above) appears to be unambiguous; but certain of the Malagasy species represented appear to have variable locomotor habits. Napier and Napier (1967) note, for instance, that the gait of some primarily quadrupedal lemurs contains elements approaching those of vertical clinging and leaping. There is biomechanical evidence to suggest that the indriines utilize their forelimbs in some form of climbing. This latter conclusion is supported to some extent by field observations made by I. Tattersall (personal communication), who has observed *Propithecus* and *Indri* climbing tree trunks with a "shinning mo-tion" and descending hand-over-hand. Petter (1962) has also recorded arm-swinging in *P. verreauxi.*

PROSIMIAN SCAPULAE: METHODS AND MATERIALS

Scapulae of a wide variety of prosimian species were studied. Attention was paid to general proportions, surface areas of the supraspinous and in-fraspinous fossae, length of the vertebral border, and functional dimensions. Surface areas of the scapular fossae were calculated as follows: The bones to be compared were photographed and were reduced in the printing process so that the length of the base of the scapular spine was uniform. Millimeter squared paper was placed over the photograph and the squares in each fossa were counted. A value was thus obtained for each fossa which could be used to calculate supraspinous and infraspinous index ratios. The indices are plotted in Fig. 2. The dotted lines represent known locomotor categories.

Functional length and functional width were measured on a photograph of each scapula as described. The relationships of functional dimensions (blade index) among prosimian primates are indicated in Table I. The lowest index value represents the widest scapula, and the highest value represents a long, narrow blade.

The vertebral border is defined as the length of the serratus muscle in-sertion. This border was measured using a precision odometer graduated in millimeters, and its length used to calculate the vertebral border index for each bone studied.

DESCRIPTIONS OF SCAPULAE

The following descriptions are not meant to be formal and complete, but are intended to illustrate morphological features that are useful in inter-preting the functions of the bones.

TABLE I. Relationships of Functional Dimensions among Prosimian Primates[a]

Species	Vert. bord. index	Funct. dim. index	Inf. sp. index	Sup. sp. index
Loris tardigradus	145	112	111.6	44.1
Arctocebus calabarensis	143	104	105	37
Nycticebus coucang	121	117	77.4	24.4
Perodicticus potto	129	107	124	34.7
Indri indri	133	108	115	30.4
Avahi laniger	148	124	112.8	30.6
Propithecus diadema	150	107	117.5	43.2
Propithecus verreauxi	125	136	93.6	37.5
Lemur catta	91	161	63.6	33
Lemur macaco	118	158	98.8	37
Lemur variegatus	108	137	92.2	30.1
Lemur mongoz	124	142	83.7	36.8
Hapalemur spp.	125	150	72.4	45.2
Lepilemur mustelinus	108	170	68.9	29.2
Microcebus spp.	62	206	51.3	22.2
Tarsius spp.	71	195	52.2	16.4
Galago spp.	53	180	59.3	29.9

[a] Vert. bord., vertebral border; Funct. dim., functional dimensions (the index is functional length divided by functional breadth); Inf. sp., infraspinous; Sup. sp., Supraspinous. The indices represent average values for numbers of individuals ranging from two to nine. The small sample sizes were dictated by the available material.

Tarsius spectrum (Fig. 4). The blade of the scapula is long and narrow with a blade index of 195. The vertebral border is short (index 71), indicating that the bone is not functionally narrow. The supraspinous fossa is particularly narrow, and a narrow infraspinous fossa is extended slightly at the inferior angle to provide adequate leverage for the teres major muscle. An acromial process is present but is not particularly long.

Galago crassicaudatus (Fig. 5). The scapula of this species is very similar to that of *Tarsius spectrum*. It has a blade index of 180 and a vertebral border index of 53.

Loris tardigradus (Fig. 6). The scapula of this species is short and broad with a blade index of 112. Most of the width is made up by the very well developed infraspinous fossa, but the supraspinous fossa is also fairly well developed. The vertebral border is fairly long with an index of 145, and is either straight or gently convex. There is a sharp superior angle and a moderately well-developed acromial process.

Nycticebus coucang (Fig. 6). The scapula of this species is similar to that of *Loris tardigradus*. It has a blade index of 117 and a vertebral border index of 121.

Perodicticus potto (Fig. 6). The scapula of this species is very broad, with

FIG. 4. The scapula of *Tarsius spectrum*.

FIG. 5. The scapula of *Galago crassicaudatus*.

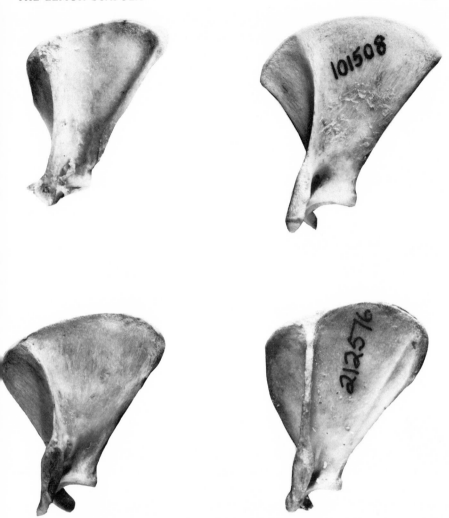

FIG. 6. Lorisine scapulae. Above left: *Loris tardigradus*; above right: *Nycticebus coucang*; below left: *Perodicticus potto*; below right: *Arctocebus calabarensis*.

a blade index of 107. Both the suprascapular fossae are well developed, but the infraspinous fossa accounts for more than two-thirds of the blade surface. The vertebral border is long in keeping with the wide blade with an index of 129. It is, however, straight or only gently curved. The superior angle is acute and the acromial process moderately well developed.

　　Arctocebus calabarensis (Fig. 6). It has been suggested (Buettner-Janusch, 1967) that this species should be included in the genus *Perodicticus*. The scapula of *Arctocebus* is certainly very similar to that of the potto. It has a blade index of 104 and a vertebral border index of 143.

Microcebus sp. (Fig. 7). The scapula of this species is long and narrow, having a blade index of 206. The vertebral border is straight and therefore short, with an index of 62. The superior angle is rounded and the acromial process is not large. There is, however, a metacromial process developed on the inferior border of the scapular spine, a feature not generally characteristic of primates.

Propithecus diadema (Fig. 8). The scapula of this species is broad with a blade index of 107. The vertebral border is long, having an index of 150, and is straight or gently concave above the infraspinous fossa. This fossa is relatively large and is extended backward by the development of a postscapular fossa, a characteristic associated with a powerful teres major muscle and powerful limb flexion. The superior angle is rounded and the supraspinous fossa is moderately well developed. The acromial process is only moderately well developed.

Propithecus verreauxi (Fig. 8). The scapula of this species is similar to that of *P. diadema* but is slightly narrower. The blade index is 136 and the vertebral border index 125.

Indri indri (Fig. 8). The scapula of this species is intermediate in form between that of *Propithecus* spp. and *Lemur* spp. The blade index is 108, indicating that the blade is as broad as that of *Propithecus diadema* and even broader than that of *Propithecus verreauxi* which has a higher index. The vertebral border is long and tends to be straight; its index is 133. The superior angle is smoothly rounded and there is a well-developed acromial process.

FIG. 7. The scapula of *Microcebus murinus*.

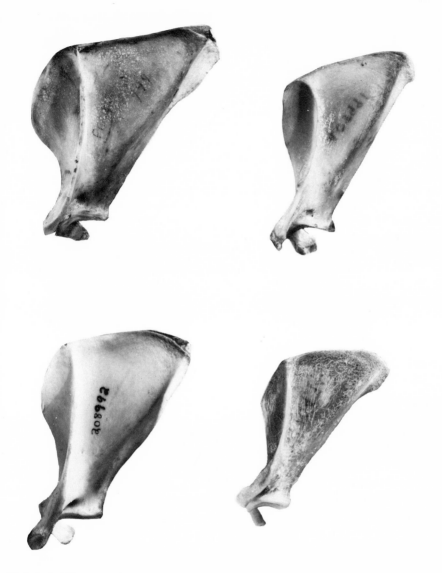

FIG. 8. Indriine scapulae. Above left: *Propithecus diadema*; above right: *Propithecus verreauxi*; below left: *Indri indri*; below right: *Avahi laniger*.

Avahi laniger (Fig. 8). The scapula of this species resembles that of *Indri indri* but is narrower and has a distinct postscapular fossa.

Lemur catta (Fig. 9). The scapula of this species has a blade index of 161, indicating a fairly narrow bone. The vertebral index is 91. The vertebral bor-

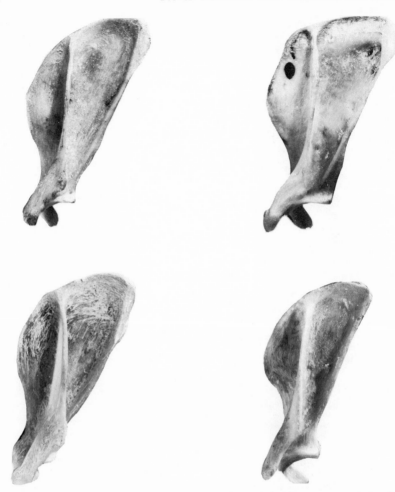

FIG. 9. *Lemur* scapulae. Above left: *L. catta*; above right: *L. variegatus*; below left: *L. macaco*; below right: *L. mongoz*.

der is proportionally longer than that of *Tarsius spectrum*, *Galago crassi-caudatus*, and *Microcebus* spp. when compared to the width of the blade. This proportional increase, which is in the region of 20%, appears to be taken up by a slight oblique reorganization of the blade profile. The superior angle is not marked, being a smooth transition between the vertebral and coracoid borders. The acromion is moderately well developed.

    Lemur variegatus (Fig. 9). The scapula of this species of *Lemur* is similar in general form to that of *L. catta*, but it is slightly broader and has a relatively longer vertebral border. The blade and vertebral border indices are 137 and 108, respectively.

Lemur macaco (Fig. 9). The blade index of this species of *Lemur* is 158, which is intermediate between that of *L. catta* and *L. variegatus*. The vertebral border index, however, is 118, indicating that the vertebral border is proportionally about as long as that of *L. variegatus*.

Lemur mongoz (Fig. 9). The blade of the scapula of *Lemur mongoz* is slightly broader than that of the other lemur species discussed, with the exception of *L. variegatus*. The blade index is 142. The vertebral border index, however, is 124, indicating that this species possesses relatively the longest vertebral border of any of the lemur species. As in these other species, the long vertebral border is accommodated by slight oblique reorganization of the scapular blade.

Hapalemur sp. (Fig. 10). The general appearance of the scapula of this species is similar to that of *Lemur catta*. But with a blade index of 150, the vertebral border index is 125, indicating a proportionally very long vertebral border. There is a distinctly developed postscapular fossa and the superior angle differs from that seen in species of *Lemur* by being acute rather than rounded. The general appearance of being obliquely reorganized is greater than in the *Lemur* species described.

Lepilemur mustelinus (Fig. 10). The scapula of this species is narrower than those of *Lemur* and *Hapalemur* species with a blade index of 170. The vertebral border index is 108. The scapula has a hooklike development of the post-scapular fossa and a rounded superior angle.

Figure 11 shows the scapulae described above arranged on the basis of blade morphology. The morphological grouping thus obtained corresponds very closely to the locomotor categories noted on page 61—clearly demonstrating that a form–function relationship exists.

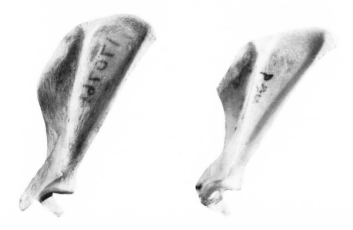

FIG. 10. The scapulae of *Hapalemur griseus* (left), and *Lepilemur mustelinus* (right).

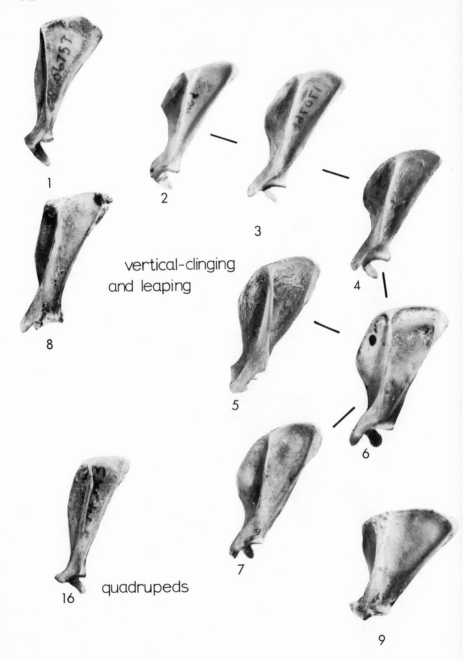

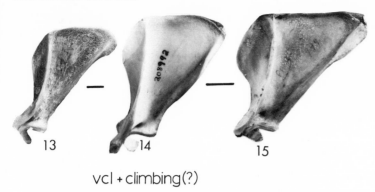

vcl + climbing(?)

FIG. 11. Arrangement of Prosimian Scapulae based on morphological characteristics. Grouping based on morphological characteristics corresponds closely to locomotor categories. 1, *Tarsius spectrum*; 2, *Lepilemur mustelinus*; 3, *Hapelemur griseus*; 4, *Lemur macaco*; 5, *Lemur mongoz*; 6, *Lemur variegatus*; 7, *Lemur catta*; 8, *Galago crassicaudatus*; 9, *Loris tardigradus*; 10, *Nycticebus coucang*; 11, *Arctocebus calabarensis*; 12, *Perodicticus potto*; 13, *Avahi laniger*; 14, *Indri indri*; 15, *Propithecus diadema*; 16, *Microcebus spp.*

slow - climbing

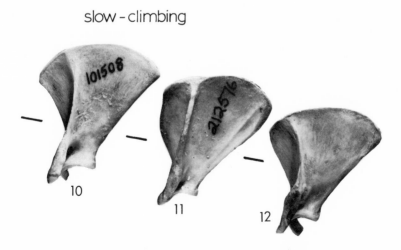

LOCOMOTOR CHARACTERISTICS OF PROSIMIAN SPECIES AND THEIR RELATIONSHIP TO SCAPULAR MORPHOLOGY

Slow-climbing species may suspend themselves beneath a branch in a slothlike fashion, or may climb with an orthograde posture from one support to another, raising the body weight by means of the forelimbs. This behavior suggests the necessity for strong rotator cuff musculature to secure the shoulder joint against dislocation, and implies that the scapular fossae will be broad. Table I and descriptions given previously show that in the slow-climbing prosimians the scapula is relatively broader than in most other prosimian species, and particularly so with respect to those species that are unequivocally vertical clingers and leapers or quadrupeds. In the slow climbers, the use of the forelimb in an extended position requires a relatively well-developed supraspinatus muscle and supraspinous fossa. The development of a broad scapular blade that is not obliquely reorganized suggests that there is little need for rapid movements of the bone across the dorsum. Presumably, the very rapid motions of the forelimbs made in capturing prey are brought about by intrinsic scapulohumeral and humeroradioulnar extensions. Obviously, the long vertebral border determined by the broad fossae is sufficient to provide an adequate area of insertion for the scapulovertebral muscles, and need not be increased by convex curvature.

Prosimian species that are unequivocally vertical clingers and leapers belong to the genera *Tarsius* and *Galago*. In these species, the forelimbs are not directly utilized in their preferred mode of locomotion. The need for strong rotator cuff musculature is thus considerably reduced and the scapular fossae are correspondingly narrow. As no body weight is transmitted through the forelimb, the vertebral border of the scapula can be very short, the scapula-vertebral musculature having merely to support the weight of the limb. Very narrow scapulae with short vertebral borders are not typical, however, of the Malagasy prosimian species that are reportedly vertical clingers and leapers. This implies that these latter species may in fact utilize a broader range of locomotor behavior than those of the genera *Tarsius* and *Galago*.

The scapulae of the Malagasy species belonging to the subfamily Indriinae are characterized by fairly broad fossae, especially the infraspinous fossae. The superior angles are rounded and the vertebral borders are long and straight. The relatively great development of the scapular fossae suggests powerful rotator cuff musculature, a condition which, as mentioned, is not necessary in animals that are purely vertical clingers and leapers. The biomechanical implications of the broad fossae are that these relatively heavy-bodied animals support or lift either all or a good part of their body weight on occasion. As the indriines are described by most students of prosimians as being essentially orthograde animals, they may have a tendency to climb or descend vertical supports in a bearlike manner. Such behavior contrasts with that of *Galago* or *Tarsius* in which even vertical travel is accomplished by leaping that utilizes the hindlimbs alone. Bearlike climbing of vertical supports would certainly explain the apparent ability of *Propithecus* and *Indri* to provide powerful forelimb flexion as suggested by the well-developed infraspinatus muscle.

Broad scapular fossae also provide sufficient shoulder joint stabilization to permit arm-swinging should this mode of locomotion be incipient in the general locomotor behavior of the animal concerned. Tattersall (personal communication) reports having observed arm-swinging behavior in *Indri* and further suggests that in *Indri* and *Propithecus* the forelimb may function during vertical clinging and leaping locomotion to support part of the body weight immediately after initial contact has been made with the substrate by the hindlimb at the end of a leaping phase. The support of part or all of the body weight in *Indri* and *Propithecus* by the forelimb does not pose any problem as far as scapulovertebral musculature is concerned because the broad fossae already provide an adequately long vertebral border for the insertion of powerful serratus and rhomboid muscles.

Quadrupeds do not, in general, need such well-developed rotator cuff musculature as do climbing and hanging animals. This is because the body weight of the animal acts compressively across the scapulohumeral joint and insures its integrity. The scapulae of quadrupeds such as dogs, horses, and antelopes thus tend to be long, narrow bones. Arboreal quadrupeds, however, must contend with a discontinuous substrate calling for frequent use of the forelimb in an abducted position. Shoulder joint integrity through the compressive action of body weight is less constant under these circumstances and the rotator cuff musculature and scapular fossae therefore tend to be better developed in arboreal quadrupeds than in terrestrial cursors.

The functional dimensions of the scapulae of quadrupedal *Lemur* species indicate that these are relatively narrow bones, especially if they are compared with those of the slow-climbing prosimians. It should be noted, however, that the fossae of *Lemur* scapulae are relatively broader than those of *Microcebus* species. These latter, although arboreal, are small animals that run along branches and up and down the trunks of trees. That is, because of their small size, the discontinuity of their environmental substrate is effectively reduced.

Although the scapulae of *Lemur* species tend to be narrow, the vertebral borders tend to be moderately well developed, a trait achieved by a reorganization of the profile of the blade. The vertebral border thus provides adequate insertion area for scapulovertebral musculature that can support the body weight and absorb the shock of footfall at the end of a quadrupedal leap. With respect to the vertebral border, therefore, lemurines differ markedly from the vertical clinging and leaping *Galago* and *Tarsius*. However, examination of the scapular morphology of the lemurines indicates that they are probably not a homogeneous group in terms of locomotor behavior. *Lemur* species appear to form a triradiate series that has as end members forms approximating the Asian and African vertical clingers and leapers, the Malagasy "climbing" vertical clingers and leapers, and arboreal quadrupeds such as cercopithecine monkeys. On the basis of this apparant triradiate adaptation, it could be suggested that these lemurs were in the past more generalized "climbing and vertical leaping" animals than they are today. Adaptive radiation has produced the behavioral extremes represented by *Lepilemur mustelinus* (vertical clinging and leaping) on the one hand and *Lemur catta* (semiterrestrial and arboreal quadruped) on the other, with *Lemur variegatus* possibly representing most closely the ancestral behavioral pattern. This speculation can be supported by observing the manner in which the scapulae of *Lemur catta* and *Lepilemur mustelinus* can be derived from a morphology similar to that of *Lemur variegatus* through a process of profile reorganization.

CONCLUSIONS

This chapter has contained a very cursory review of scapular biomechanics and an attempt to correlate them with the scapulae of prosimian primates in a manner that sheds light on their locomotor behavior. The relationships thus observed between scapular morphology and locomotor categorizations are clear and not unexpected. Two interesting conclusions may be drawn from the study; first, the species belonging to the subfamily Indriinae may include climbing and arm-swinging as a normal part of their locomotor behavior, and, second, the lemurines form a complex behavioral spectrum whose elements may possibly represent the course of the past adaptive radiation of the group.

ACKNOWLEDGMENTS

The research reported here was supported in part by a grant (No. 2735) from the Wenner-Gren Foundation for Anthropological Research, Inc., and by Grant No. RRO 5337-11 from the National Institutes of Health to the School of Dental Medicine, University of Pennsylvania.
The specimens used in this study were made available by the Department

of Mammalogy, the American Museum of Natural History, the Department of Mammalogy of the U.S. National Museum, Washington, D.C., and the Division of Vertebrate Paleontology, Peabody Museum, Yale University.

REFERENCES

Ashton, E. H., Flinn, R. M., Oxnard, C. E., and Spence, T. F., 1971, The functional and combined classificatory significance of combined metrical features of the primate shoulder girdle, *J. Zool.* **163**:319–350.

Buettner-Janusch, J., 1967, *Origins of Man*, Wiley, New York.

Frey, H., 1923, Untersuchungen über den Scapula, speziell über äussere Form und Abhangigkeit von Funktion, *Z. Gesamt. Anat.* **68**:276–324.

Inman, V. T., Saunders, M., Abbott, A., and Leroy, C., 1944, Observations on the function of the shoulder joint, *J. Bone Joint Surg.* **26**(1):1–30.

Miller, R. A., 1932. Evolution of the pectoral girdle and forelimb in the primates, *Am. J. Phys. Anthropol.* **18**:1–56.

Müller, H. J., 1967, Form und Funktion der Scapula: Vergleichend-analytische Studien bei Carnivoren und Ungulaten, *A. Anat. Entwicklungsgesch.* **126**:205–263.

Napier, J. R., and Napier, P. H., 1967, *A Handbook of Living Primates*, Academic Press, London.

Napier, J. R., and Walker, A. C., 1967, Vertical clinging and leaping: A newly recognised category of locomotor behavior among primates, *Folia Primatol.* **6**:180–203.

Oxnard, C. E., 1967, Aspects of the mechanical efficiency of the scapula in some primates, *Anat. Rec.* **157**:296.

Petter, J. J., 1962, Recherches sur l'écologie et l'éthologie des lémuriens malgaches, *Mém. Mus. Natl. Hist. Nat. Sér. A.* **27**:1–146.

Roberts, D., 1973, Structure and function of the primate scapula, in: *Primate Locomotion* (F. A. Jenkins, ed.), Academic Press, New York.

Simons, E. L., 1972, *Primate Evolution: An Introduction to Man's Place in Nature*, Macmillan, New York.

Smith, J. M., and Savage, R. J. G., 1956, Some locomotory adaptations in mammals, *J. Linn. Soc. (Zool.)* **42**:603–622.

Osteology and Myology of   9
the Lemuriform Postcranial
Skeleton

FRANÇOISE K. JOUFFROY

The study of limb morphology is no longer, as it has been until recently, simply a complement to the study of external characters and the skull, traditionally regarded by systematists and paleontologists as more significant. The concerns of zoologists today are much wider, extending from pure systematics to ecoethology, and the study of the movements involved in locomotion and prehension requires a profound understanding of limb morphology.

The extant Malagasy lemurs are all more or less well adapted to arboreal jumping, with numerous variants which run from quadrupedal (Cheirogaleinae) to bipedal (Indriidae and *Lepilemur*) leaping. Certainly, there are other primates which are equally good or even better jumpers, but this propensity is so general among the Lemuriformes that it can be considered as characteristic of the group as a whole. This finds expression in the fact that the hindlimb shows many more specialized characters than does the forelimb. Since it is not possible within the limitations of this chapter to review all the characters of the skeleton and musculature, I shall briefly call attention to the most important traits of the limbs, and dwell at greater length upon the ex-

FRANÇOISE K. JOUFFROY   Laboratoire d'Anatomie Comparée, Muséum National d'Histoire Naturelle, 55 Rue de Buffon, 75 Paris Ve, France

tremities, the morphology of which is perhaps most closely related to locomotion and prehension.

LIMB PROPORTIONS

The order Primates exhibits a wide diversity in limb proportions; this diversity correlates closely with locomotor behavior. The intermembral index

$$\frac{\text{humerus length} + \text{radius length} \times 100}{\text{femur length} + \text{tibia length}}$$

is well over 100 in brachiators (140 in *Hylobates*, 150 in *Pongo*), around 100 in quadrupeds (e.g., *Papio*, *Mandrillus*), and well under 100 in jumpers (down to 55 in *Tarsius*) and in man. All the Malagasy lemurs (Table I) possess indices below 76. Only *Tarsius* and *Galago* among the prosimians, and *Aotus* among the higher primates, possess comparable indices. Among the lemurs, two groups can be distinguished: the indriids and *Lepilemur*, with indices between 56 and 65, and the remaining lemurs, with indices of 65–76. These two groups are distinct in both their jumping adaptations and their posture, the former group being almost exclusively bipedal and vertical, the latter group quadrupedal and horizontal in these traits.

The femur is always longer than the humerus, and the tibia is longer than the radius. The femorohumeral index

$$\frac{\text{femur length} \times 100}{\text{humerus length}}$$

varies from 136 in *Cheirogaleus* to 209 in *Avahi*; the tibioradial index

$$\frac{\text{tibia length} \times 100}{\text{radius length}}$$

varies from 124 in *Varecia* to 154 in *Propithecus*.

THE FORELIMB

Proportions (Table I)

The forelimb, compared to the trunk (precaudal vertebral column), is *very short in Lemurinae and Cheirogaleinae* (75%–87%). Such shortness is exceptional among monkeys (Lessertisseur, 1970), but does characterize

TABLE I. Intermembral Index and Intersegmental Indices of the Forelimb and Hindlimb (Averages, in Percent)[a]

Genus	$\frac{H+R}{F+T}$	$\frac{F}{H}$	$\frac{T}{R}$	$\frac{P}{M}$	$\frac{FL}{S}$	$\frac{H}{FL}$	$\frac{R}{FL}$	$\frac{R}{H}$	$\frac{HL}{S}$	$\frac{F}{HL}$	$\frac{T}{HL}$	$\frac{T}{F}$
Lemur	72	149	134	146	86	35	38	108	122	37	35	96
Varecia	76	138	124	127	83	33	34	103	104	36	34	93
Hapalemur	68	150	140	154	87	34	36	103	130	35	34	96
Lepilemur	60	179	153	148	87	32	35	109	141	36	34	94
Cheirogaleus	63	136	139	172	79	34	34	100	117	33	33	101
Microcebus	71	140	139	172	75	34	37	106	108	32	34	106
Propithecus	58	191	154	126	98	32	34	108	153	39	34	86
Indri	63	188	136	116	114	30	36	120	165	38	34	87
Avahi	57	209	149	129	103	29	36	120	165	39	34	87
Daubentonia	68	145	138	97	111	29	30	101	135	34	33	96

[a]F, Femur; FL, entire forelimb; H, humerus; HL, entire hindlimb; M, manus; P, pes; R, radius; S, precaudal spine; T, tibia.

some other prosimians, including jumpers such as *Galago* or slow climbers such as *Perodicticus*. With few exceptions (*Microcebus, Cheirogaleus, Daubentonia*, and *Varecia*), where the radius is sometimes shorter than or equal in length to the humerus, the antebrachium is longer than the brachium. It should be noted that in this characteristic *Varecia* differs strongly from the other lemurs. The length of the antebrachium compared to the brachium is maximal in the indriines and *Lepilemur*; the brachial indices

$$\frac{\text{radius length} \times 100}{\text{humerus length}}$$

of these forms are on the order of 110–120, values equaled elsewhere among the primates only by *Hylobates*, *Loris*, and *Galago*, and surpassed only by *Tarsius* at 128.

Skeleton (Figs. 1, 2, and 3)

I do not intend to discuss the scapula here, since this bone is covered in Chapter 8. However, this element is compared among various lemurs in Fig. 1.

The clavicle is relatively long, and is enlarged at its acromial extremity.

The humerus is robust; the index of robusticity (minimum circumference as a percentage of maximum length) is highest among all living primates in *Lemur* and *Lepilemur* (28–30). The diaphysis is rectilinear; the head, directed backward, is ellipsoidal. The deltoid crest is prominent and overhangs the bicipital groove. There is always an entepicondylar foramen.

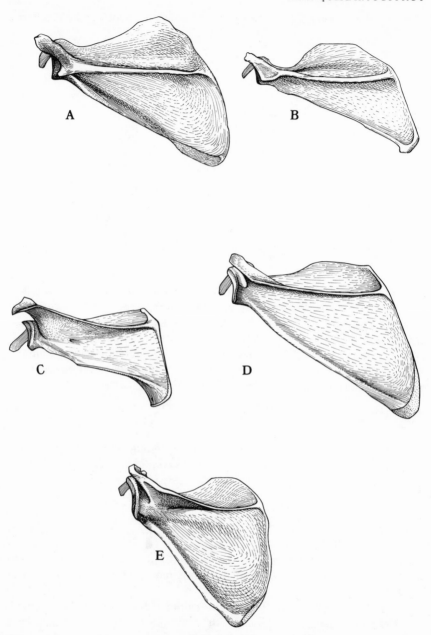

FIG. 1. Left scapulae from the dorsal aspect. (A) *Lemur*, (B) *Microcebus*, (C) *Daubentonia*, (D) *Propithecus*, (E) *Indri*. Not to scale.

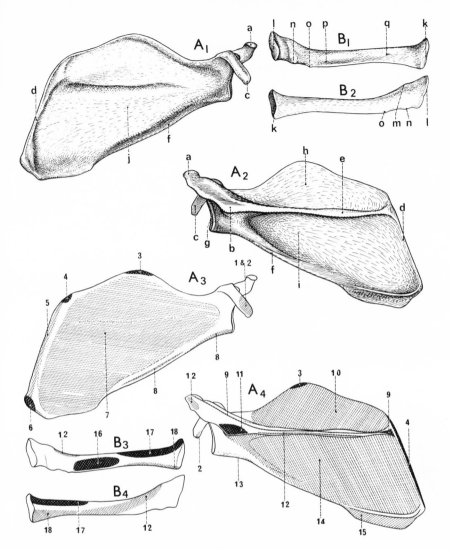

FIG. 2. *Varecia* (= *Lemur*) *variegatus*. Left scapula from the ventral aspect (A_1, A_3) and from the dorsal aspect (A_2, A_4); left clavicle from the caudal aspect (B_1, B_3) and from the cranial aspect (B_2, B_4). Relief (A_1, A_2, B_1, B_2) and muscular attachments (A_3, A_4, B_3, B_4,). Light, origins; dark, insertions. a, Acromion; b, metacromion; c, processus coracoideus; d, margo vertebralis; e, spina scapulae; f, margo axillaris; g, cavitas glenoidalis; h, fossa supraspinata; i, fossa infraspinata; j, fossa subscapularis; k, extremitas sternalis; l, extremitas acromialis; m, tuberculum deltoideum; n, tuberculum coracoideum; o, margo anterior; p, sulcus subclavius; q, foramen nutricium. 1, Biceps; 2, coracobrachialis; 3, omohyoideus; 4, rhomboideus capitis; 5, levator scapulae dorsalis; 6, serratus anterior; 7, subscapularis; 8, triceps, caput longum; 9, trapezius; 10, supraspinatus; 11, levator scapulae ventralis; 12, deltoideus; 13, teres minor; 14, supraspinatus; 15, teres major; 16, subclavius; 17, sternocleidomastoideus; 18, pectoralis major.

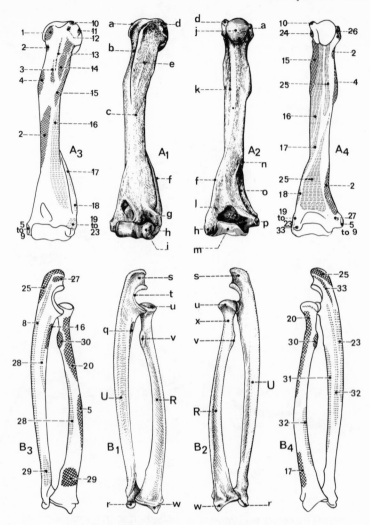

FIG. 3. *Varecia* (=*Lemur*) *variegatus*. Left humerus from the anterior aspect (A₁, A₃) and from the posterior aspect (A₂, A₄); left ulna and radius from the anteromedial aspect (B₁, B₃) and from the posterolateral aspect (B₂, B₄). Relief (A₁, A₂, B₁, B₂) and muscular insertions (A₃, A₄, B₃, B₄). Light, origins; dark, insertions. a, Tuberculum minus; b, sulcus inter-tubercularis; c, tuberositas deltoidea; d, tuberculum majus; e, "V" del-toideum; f, crista epicondylia lateralis; g, fossa radialis; h, epicondylus lateralis; i, condylus lateralis; j, caput humeri; k, linea m. tricipitis; l, fossa olecrani; m, trochlea humeri; n, foramen nutricium; o, foramem supra-trochleare; p, epicondylus medialis; q, fossa subcoronoidea; R, radius; r, processus styloideus ulnae; s, olecranon; t, incisura trochlearis; U, ulna; u,

The ulna possesses a very high olecranon process, compressed laterally and bearing a plane surface for triceps insertion. The diaphysis is straight.

The radius possesses a strong external curvature. The bicipital tuberosity is very prominent and is marked by a groove running along the diaphysial axis. The radial head forms a circular depression. The neck is long (15–20% of the total length of the diaphysis between the distal part of the bicipital tuberosity and the head).

Musculature (Figs. 4–8; Table II)

The musculature of the pectoral limb is in complete conformity with the general scheme among primates. Within this region, the muscles of the pectoral girdle are innervated by the accessory nerve (XI) and by cervical vertebral branches. The muscles of the limb proper are innervated by the brachial plexus (both by collateral branches and by the four principal terminal branches, nn. radialis and axillaris in the case of the extensors, nn. musculocutaneus and medianus in the case of the flexors) (Fig. 4). Figures 2, 3 and 20 show the principal muscular attachments in the forelimb of *Lemur*.

Since it is impossible to note all of the details of the musculature here (see Jouffroy, 1962; also Figs. 5–8), I shall discuss a general aspect of forelimb muscular morphology among the lemurs, the very oblique disposition of numerous muscles attached far on either side of the articulation, which they hold together strongly. In the shoulder and in the elbow there are also musculocutaneous membranes which give the forelimb a peculiar appearance, sometimes reminiscent (in *Microcebus* especially) of the patagium of certain gliding mammals (Milne-Edwards and Grandidier, 1875; Anthony and Bortnowsky, 1914). At shoulder level, the muscles comprising the membrane are primarily the dorsoepitrochlearis (or accessory of m. latissimus dorsi),

---

caput radii; v, tuberositas radii; w, processus styloideus radii, x, collum radii. 1, Subscapularis; 2, coracobrachialis; 3, latissimus dorsi; 4, teres major; 5, pronator teres; 6, flexor carpi radialis; 7, palmaris longus; 8, flexor carpi ulnaris; 9, flexor digitorum superficialis; 10, supraspinatus; 11, infraspinatus; 12, pectoralis minor; 13, pectoralis major; 14, deltoideus, pars clavicularis; 15, deltoideus, pars acromialis et pars scapularis; 16, brachialis; 17, brachioradialis; 18, extensor carpi radialis longus; 19, extensor carpi radialis brevis; 20, supinator; 21, extensor digitorum communis; 22, extensor digiti quinti; 23, extensor carpi ulnaris; 24, teres minor; 25, triceps; 26, subscapularis; 27, epitrochleoanconeus; 28, flexor digitorum profundus; 29, pronator quadratus; 30, biceps; 31, extensor digitorum profundus; 31, abductor pollicis longus; 33, epicondylocubitalis.

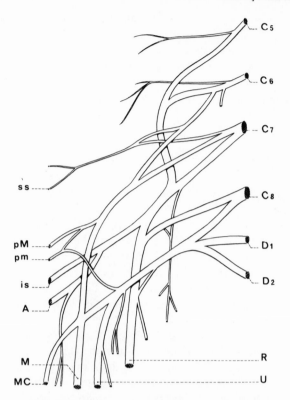

FIG. 4. *Propithecus*. Right brachial plexus. C_5-C_8
and D_1, D_2, Spinal nerves. A, n. axillaris; is, n.
infrascapularis (subscapularis); M, n. medianus; MC,
n. musculocutaneus; pM., n. to the pectoralis major
muscle; pm, n. to the pectoralis minor muscle; R, n.
radialis; ss, n. suprascapularis; U, n. ulnaris.

which unites the trunk and the forearm, and a superficial muscle derived
from the panniculus carnosus, the dorsohumeralis. This unites the trunk to
the arm and is more or less adherent either to m. pectoralis or (in *Propithecus* and *Lepilemur*) to latissimus dorsi. At elbow level, the major
muscles to note are brachioradialis and extensor carpi radialis longus
(Fig. 8), which among the indriines extend along the arm up to the insertion
of the deltoid.

Thus among the lemurs, to varying degrees and in a variety of ways, the
mobility of the articulations of the forelimb is limited, particularly in the case
of antebrachial rotation.

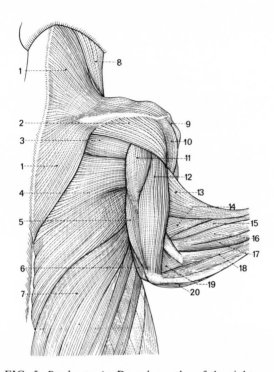

FIG. 5. *Daubentonia*. Dorsal muscles of the right shoulder and upper arm from the dorsal aspect, superficial sheet. 1, Trapezius; 2, deltoideus, pars scapularis; 3, teres major; 4, latissimus dorsi; 5, triceps, caput mediale; 6, dorsoepitrochlearis; 7, panniculus carnosus; 8, sternocleidomastoideus; 9, deltoideus, pars clavicularis; 10, deltoideus, pars acromialis; 11, triceps, caput longum; 12, triceps, caput laterale; 13, brachioradialis; 14, extensor carpi radialis longus; 15, extensor carpi radialis brevis; 16, extensor digitorum communis; 17, extensor digiti quinti; 18, extensor carpi ulnaris; 19, anconeus; 20, flexor carpi ulnaris.

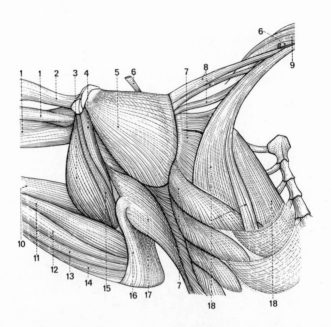

FIG. 6. *Varcecia* (= *Lemur*) *variegatus*. Muscles of the right shoulder and arm, from the ventral aspect (clavicle and shoulder blade removed laterally). 1, Pectoralis major; 2, pectoralis minor; 3, biceps, caput longum; 4, coracobrachialis; 5, subscapularis; 6, omohyoideus; 7, serratus anterior; 8, levator scapulae dorsalis; 9, splenius; 10, brachioradialis; 11, flexor carpi radialis; 12, flexor digitorum superficialis; 13, palmaris longus; 14, flexor carpi ulnaris; 15, biceps, caput breve; 16, dorsoepitrochlearis; 17, latissimus dorsi; 18, scaleni.

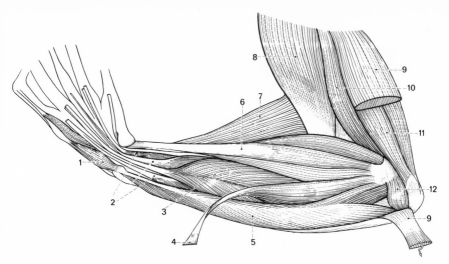

FIG. 7. *Daubentonia*. Muscles of the right forearm, from medial aspect. 1, Abductor digiti quinti brevis; 2, flexor digitorum profundus; 3, flexor digitorum superficialis; 4, palmaris longus; 5, flexor carpi ulnaris; 6, flexor carpi radialis; 7, brachioradialis; 8, biceps brachii; 9, dorsoepitrochlearis; 10, brachialis; 11, triceps; 12, epitrochleo-anconeus.

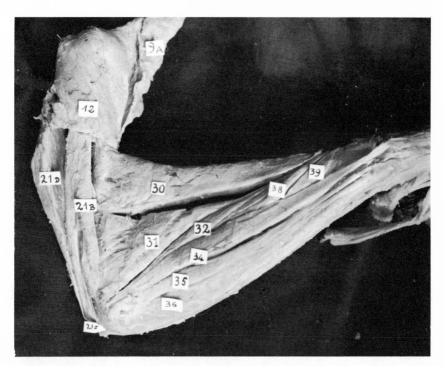

FIG. 8. *Propithecus*. Muscles of the right forearm from the lateral aspect. 9A, Pectoralis major; 12, deltoideus; 21B, triceps, caput laterale; 21D, triceps, caput longum; 21F, anconeus; 30, brachioradialis; 31, extensor carpi radialis longus; 32, extensor carpi radialis brevis; 34, extensor digitorum communis; 35, extensor lateralis; 36, extensor carpi ulnaris; 38, 39, abductor pollicis longus.

TABLE II. List of the Pectoral Limb Muscles, with Their Insertions and Innervation[a]

I. Branchiomeric muscles

Muscles	Insertions	Innervation
Sternocleidomastoideus	Cr — St + Cl 17	XI and cervical plexus
Trapezius	Vct[b] — Sc 9	

II. Myomeric muscles

Dorsal groups

Muscles	Insertions	Innervation
Rhomboideus	Cr + Vct – Sc 4	}
Levator scapulae ventralis	Vc – Sc 11	} cervicobrachial
Serratus anterior	Co – Sc 6	} plexus (C$_{3-8}$)
Levator scapulae dorsalis	V1 – Sc 5	}
Latissimus dorsi	Vtl – H 3	n. thoracodorsalis
Subscapularis	Sc 7 – H 1 and 26	} n. subscapularis
Teres major	Sc 15 – H 4	}
Deltoideus scapularis	Sc 12 – H 15	}
Deltoideus clavicularis	Cl 12 – H 14	} n. axillaris
Teres minor	Sc 13 – H 24	}
Triceps caput longum	Sc 8 – U 25	}
Triceps caput laterale + mediale	H 25 – U 25	} n. radialis
Dorsoepitrochlearis	m.l.d. – U with 25	}

Ventral groups

1. Shoulder and upperarm

Muscles	Insertions	Innervation
Subclavius	Co$_1$ – Cl 16	n. subclavius
Pectoralis major	St + Cl 18 – H 13	} nn. thoracales
Pectoralis minor	Co – H 12	} anteriores
Panniculus carnosus	m.p.M. – skin	} (n. pectoralis proprius)
Supraspinatus	Sc 10 – H 10	} n. subscapularis
Infraspinatus	Sc 14 – H 11	}
Coracobrachialis	Sc 2 – H 2	}
Biceps	Sc 1 – R 30	} n. musculocutaneus
Brachialis	H 16 – U 16	}

2. Forearm and hand

Muscle	Insertion	Nerve
Supinator	H 20 – R 20	n. radialis
Brachioradialis	H 17 – R 17	n. radialis
Extensor carpi radialis longus	H 18 – M 15	n. radialis
Extensor carpi radialis brevis	H 19 – M 16	n. radialis
Extensor digitorum communis	H 21 – M 13	n. radialis
Extensor digitorum lateralis	H 22 – M 12	n. radialis
Epicondylocubitalis (anconeus)	H 33 – U 33	n. radialis
Extensor carpi ulnaris	H 23 – M 14	n. radialis
Abductor pollicis longus	R + U 32 – M 9	n. radialis
Extensor digitorum profondus (= e. pollicis longus + indicis proprius)	U 31 – M 17	n. radialis

Muscle	Insertion	Nerve
Pronator teres	H 5 – R 5	n. medianus
Flexor carpi radialis	H 6 – M 8	n. medianus
Palmaris longus	H 7 – a.p.	n. medianus
Flexor digitorum superficialis	H 9 – M 2	n. medianus
Flexor digitorum profundus	R + U 28 – M 1	n. medianus
Pronator quadratus	U 29 – R 29	n. ulnaris
Epitrochleoanconeus	H 27 – U 27	n. ulnaris
Flexor carpi ulnaris	H 8 – M 11	n. ulnaris
Palmaris brevis	a.p. – a.p.	nn. ulnaris + medianus
Lumbricales	t.fl. – M 4	nn. ulnaris + medianus
Contrahentes digitorum	M 3 – M 3	nn. ulnaris + medianus
Interossei	M 5 – M 5	nn. ulnaris + medianus
Abductor pollicis brevis	M 6 – M 6	nn. ulnaris + medianus
Opponens pollicis	M 7 – M 7	nn. ulnaris + medianus
Abductor digiti V brevis	M 10 – M 10	nn. ulnaris + medianus

[a] The numbers in the insertions column are those mentioned on the bones, Fig. 2, 3, and 20. a.p., aponeurosis palmaris; Cl, clavicula; Co, costae; Cr, cranium; H, humerus; M, manus; m.l.d., m. latissimus dorsi; m.p.M., m. pectoralis major; R, radius; Sc, scapula; St, sternum; t.fl., tendines flexo-rium; U, ulna; Vc, vertebrae cervicales; Vl, vertebrae lumbales; Vt, vertebrae thoracicae; V_1, atlas.
[b] In *Daubentonia* also occipitalis.

THE HINDLIMB

Proportions (Table I)

We have already noted the great length of the hindlimb relative to the forelimb (130–155%) in reference to the intermembral index, among the lemurs. This high value is almost entirely due to the shortness of the forelimb, since, if it is compared with trunk length, the relative length of the hindlimb (femur plus tibia plus foot: from 104 in *Varecia* to 165 in *Indri* and *Avahi*). does not distinguish the lemurs from other primates (Lessertisseur, 1970). Indeed, compared to trunk length, the hindlimb of *Lemurids* is among the shortest in the order.

If one considers the relative lengths of the three principal segments of the limb (femur, tibia, and pes, Table I), two groups may be distinguished:

1. Those forms in which the three segments are subequal in length, the cheirogaleines and *Daubentonia*. It may be noted that *Microcebus* is the only lemur whose tibia is slightly longer than the femur (segmental formula: 32.4–34.5–33.1).

2. Those forms in which the femur is the longest segment, the foot the shortest. These are, in order of increasing strength of this character, Lemurinae (35–34–30); *Indri* and *Propithecus* (39–34–27); and, above all, *Avahi* (40–34–26).

The crural index

$$\frac{\text{tibia length} \times 100}{\text{femur length}}$$

reflects the preceding observations. A crural index of around 100 is found in the cheirogaleines (101 in *Cheirogaleus*, 105 in *Microcebus*), *Daubentonia* (96), and the lemurines (95); it is much lower among the indriines (84–87). Variations in this index among the lemurs encompass the same range as among Primates as a whole. It is exceptional to find within Primates a tibia longer than the femur (*Galago*, *Erythrocebus*, callithricids); similarly, it is unusual to find a femur relatively as long as that of *Avahi* (*Pan*, *Gorilla*).

Skeleton (Figs. 9–12)

The morphology of the pelvis varies widely among the lemurs. Very elongated and gracile in Cheirogaleinae and *Lepilemur*, it is thick and squat in the indriines and intermediate in the lemurines and *Daubentonia*. The ilium is long and rodlike in the cheirogaleines, short and bladelike in the in-

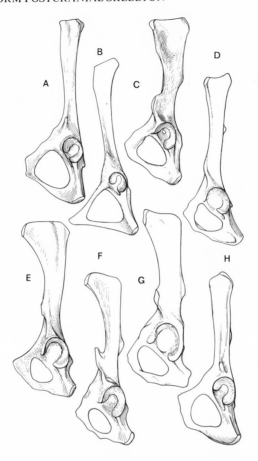

FIG. 9. Pelvis from the lateral left aspect. (A) *Cheirogaleus*, (B) *Microcebus*, (C) *Lemur*, (D) *Lepilemur*, (E) *Avahi*, (F) *Propithecus*, (G) *Indri*, (H) *Daubentonia*. Not to scale.

driines; the width of the iliac blade is 23% of the total length of the ilia in the former, 40% in the latter. The sacroiliac joint is much further removed from the acetabulum (three-quarters of the length of the ilium) in the cheirogaleines than in the indriines (at midlength of the ilium). Above the acetabular depression, the ilium is triangular in section among the indriines, and the anteroinferior spine is salient. The surface of the iliac blade (attachment area of mm. iliacus, gluteus medius, and sartorius) is enlarged, without increasing the weight of the bone, by a ligament stretching between the superior and inferior

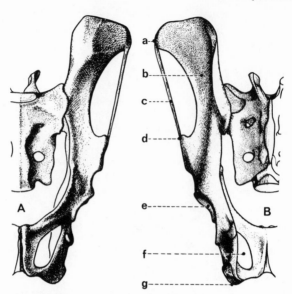

FIG. 10. *Propithecus*. Left hemipelvis, from the ventral aspect (A) and from the dorsal aspect (B). a, Spina iliaca anterior superior; b, ala ossis ilii; c, ligamentum interspinosum; d, spina iliaca anterior inferior; e, acetabulum; f, foramen obturatum; g, tuber ischiadicum.

anterior spines (*Propithecus*, Fig. 10). Finally, the symphysis is much shorter in the cheirogaleines, where it involves no more than a small region of the ascending pubic process, than it is in the lemurines, *Propithecus* and *Avahi*, where it reaches the ischiopubic extremity. In *Indri* and *Daubentonia*, even the ischium is involved in formation of the symphysis.

The femur possesses a very straight shaft, and in all living lemurs exhibits a third trochanter representing the terminal tubercle of the crest of insertion of gluteus maximus; this lies below the greater trochanter on the external border of the diaphysis. A third trochanter is not found elsewhere among the primates except in *Perodicticus*, *Galago*, and the callithricids. The shaft is subcircular in section; the index of flattening is close to 100.

The tibia is strongly compressed laterally (platycnemic index around 100), and the fibula is remarkably gracile.

Numerous sesamoids are present. Two, the fabellae, are found in the tendons of the origin of m. gastrocnemius, one, the cyamella, in the tendon of popliteus, and one, the superior patella, in the distal tendon of vastus intermedius. In distinction to the patella proper, which is a half-cartilaginous, half-bony sesamoid, the superior patella is a vesiculofibrous thickening in the

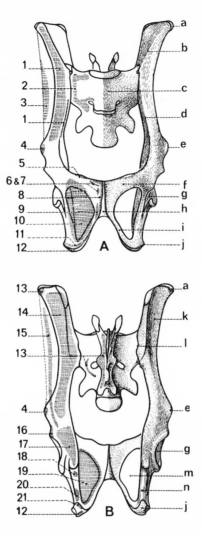

FIG. 11. *Varecia* (= *Lemur*) *variegatus*. Pelvic bone from (A) ventral aspect and (B) dorsal aspect. Right: relief; left: muscular insertions. Light, origin; dark, distal insertion. a, Spina iliaca anterior et superior; b, linea limitans; c, sacrum; d, linea terminalis; e, spina iliaca anterior et inferior; f, ramus superior ossis pubis; g, acetabulum; h, symphysis pubica; i, ramus inferior ossis pubis; j, tuber ischiadica; k, facies glutea; l, articulato sacroiliaca; m, foramen obturatum; n, incisura ischiadica minor. 1, Iliopsoas; 2, piriformis; 3, sartorius; 4, rectus femoris; 5, pectineus; 6, adductor longus; 7, adductor brevis; 8, gracilis; 9, adductor magnus; 10, obturator externus; 11, quadratus femoris; 12, semimembranosus; 13, gluteus maximus; 14, gluteus medius; 15, scansorius; 16, articularis coxae; 19, obturator internus; 20, gemellus inferior; 21, biceps plus semitendinosus.

166 FRANÇOISE K. JOUFFROY

FIG. 12. *Varecia* (*=Lemur*) *variegatus*. Left femur from the anterior aspect (A_1, A_2) and from the posterior aspect (A_3, A_4); left tibia and fibula from the anterior aspect (B_1, B_3) and from the posterior aspect (B_2, B_4). Relief (A_1, A_3, B_1, B_2) and muscular insertions (A_2, A_4, B_3, B_4). Light, origins; dark, insertions. a, Caput femoris; b, trochanter minor; c, margo interna; d, fossa supratrochlearis; e, facies patellaris; F, fibula; f, trochanter major; g, trochanter tertius; h, margo externa; i, condylus lateralis femoris; j, condylus medialis femoris; k, condylus medialis tibiae; l, tuberositas tibiae; m, tuberositas inferior; n, membrana interossea; o, malleolus medialis; p, condylus lateralis tibiae; q, malleolus lateralis; r, sulcus malleolaris; T, tibia. 1, Vastus internus; 2, iliopsoas; 3, articularis coxae; 4, vastus intermedius; 5, gluteus minimus plus scansorius; 6, vastus lateralis; 7, popliteus; 8, gluteus medius plus piriformis; 9, quadratus femoris; 10, gluteus maximus; 11, caudofemoralis; 12, femorococcygeus; 13, gastrocnemius lateralis; 14, obturator internus plus gemelli; 15, obturator externus; 16, pectineus; 17, adductor brevis; 18, adductor longus; 19, adductor magnus; 20, gastrocnemius medialis; 21, semimembranosus; 22, quadriceps femoris (=vasti plus rectus femoris); 23, pes anseris (semitendinosus, gracilis, sartorius); 24, tibialis anterior plus abductor hallucis longus; 25, biceps femoris; 26, extensor digitorum longus; 27, peroneus longus; 28, peronei digiti quarti et digiti quinti; 29, extensor hallucis longus; 30, peroneus brevis; 31, soleus; 32, tibialis posterior; 33, flexor fibularis; 34, peroneotibialis; 35, flexor tibialis.

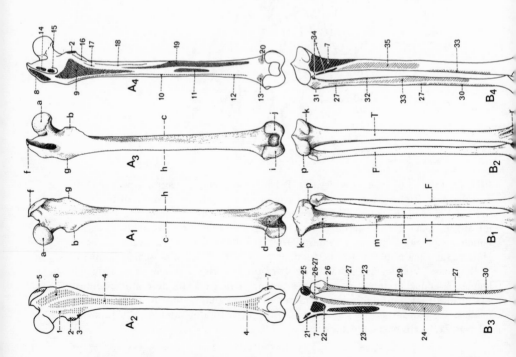

posterior aspect of the tendon of vastus intermedius. The presence of this sesamoid is an attribute of primates adapted to jumping, including the lemurs, *Tarsius*, galagines, callithricids, and *Aotus* (Vallois, 1914). It is absent among the lorisines.

Musculature (Figs. 13–18; Table III)

The muscles of the hindlimb are innervated by branches of the lumbosacral plexus (Fig. 13). The muscles of the hip and thigh receive innervation from collateral branches of the sacral plexus as well as from the femoral and obturator nerves; the muscles of the lower leg and feet are innervated by nn. ischiadicus dorsalis (peroneus communis) and ischiadicus ventralis (tibialis).

The general arrangement of muscles in the leg follows the standard primate pattern. Figures 11 and 12 show the principal muscle attachments in *Lemur*. A variety of points of interest may be noted (Jouffroy, 1962; Figs. 14 and 17):

1. The capsularis is always present.
2. The origin of sartorius varies according to the form of the ilium. Where the latter is rodlike, the muscle attaches on the anterior border of the iliac blade; it is very short and holds the hip and knee joints together strongly, as do the hamstrings. Among the indriines, with a bladelike ilium, sartorius takes origin from the ligament between the superior and inferior anterior iliac spines; here, it is broad and very long and does not limit mobility at the joint. The former condition thus holds for quadrupedal jumpers, which utilize the anklejoint, while the latter is found in vertical clingers and leapers, which depend on the hip and knee joints.
3. The quadriceps femoris shows clear adaptations to jumping both in the strong development of vastus externus and in the limitation of the origins of the two vasti in the region of the superior femoral epiphysis, as well as in the presence of the "superior patella."
4. A caudofemoralis is present only in Lemuridae. This muscle, present in many mammals, is exceptional among the primates; in the lemurines and callithricids it possesses a typical coccygeal origin, but in *Tarsius*, *Galago*, and cheirogaleines it has an ischial origin, perhaps as an adaptation to a specific kind of jumping.
5. The peronei quarti digiti and quinti digiti coexist with the short extensors of digits IV and V (Fig. 26) as in *Tarsius*, whereas only peroneus quinti digiti is found in the lorisiforms and simians.
6. The flexor digitorum fibularis is larger than the flexor digitorum tibialis; both contribute to the formation of the perforating tendons of each toe.

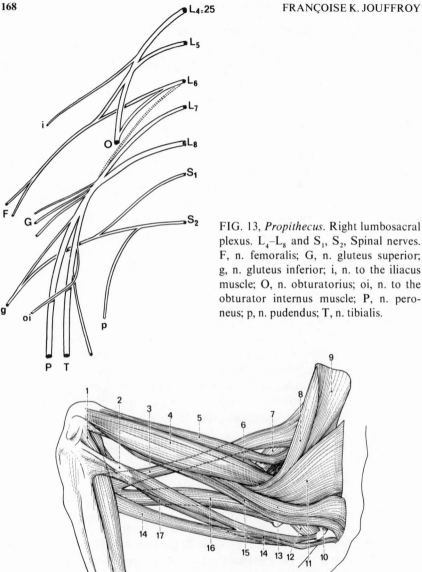

FIG. 13, *Propithecus*. Right lumbosacral plexus. L_4–L_8 and S_1, S_2, Spinal nerves. F, n. femoralis; G, n. gluteus superior; g, n. gluteus inferior; i, n. to the iliacus muscle; O, n. obturatorius; oi, n. to the obturator internus muscle; P, n. peroneus; p, n. pudendus; T, n. tibialis.

FIG. 14. *Varecia* (=*Lemur*) *variegatus*. Muscles of the left hip and thigh from the lateral aspect, superficial sheet. 1, Gastrocnemius lateralis; 2, biceps femoris; 3, sartorius; 4, vastus lateralis; 5, rectus femoris; 6, vastus medialis; 7, iliopsoas; 8, tensor fasciae latae; 9, gluteus medius; 10, gemellus inferior; 11, gluteus maximus; 12, quadratus femoris; 13, femorococcygeus; 14, semitendinosus; 15, caudofemoralis; 16, gracilis; 17, semimembranosus.

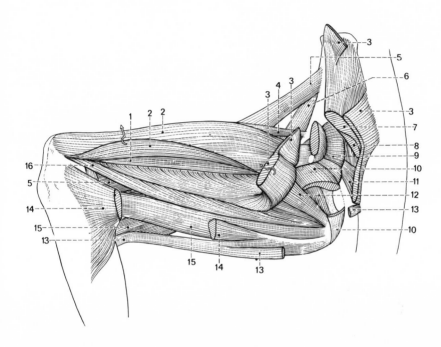

FIG. 15. *Daubentonia*. Muscles of the left hip and thigh, from the lateral aspect, middle sheet. 1, Vastus intermedius; 2, vastus lateralis; 3, gluteus maximus; 4, rectus femoris; 5, sartorius; 6, iliopsoas; 7, gluteus medius; 8, piriformis; 9, gluteus minimus; 10, gemellus inferior; 11, ischiocaudalis; 12, quadratus femoris; 13, semitendinosus; 14, biceps femoris; 15, semimembranosus; 16, adductor magnus.

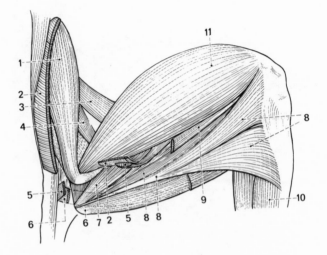

FIG. 16. *Microcebus*. Muscles of the right hip and thigh,
from the lateral aspect, middle sheet. 1, Gluteus medius;
2, gluteus maximus; 3, sartorius; 4, iliopsoas; 5, femorococcygeus; 6, semitendinosus; 7, quadratus femoris; 8,
biceps femoris; 9, semimembranosus; 10, gastrocnemius
lateralis.

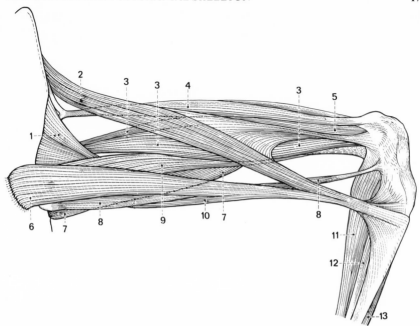

FIG. 17. *Propithecus*. Muscles of the left hip and thigh from the medial aspect. 1, Iliopsoas; 2, sartorius; 3, vastus medialis; 4, rectus femoris; 5, vastus lateralis; 6, gracilis; 7, femorococcygeus; 8, semimembranosus; 9, adductor magnus; 10, semitendinosus; 11, gastrocnemius medialis; 12, soleus; 13, tibialis anterior.

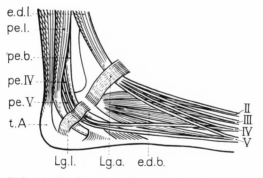

FIG. 18. *Daubentonia*. Semischematic right foot, from the lateral aspect. e.d.b., Extensor digitorum brevis; e.d.l., extensor digitorum longus; Lg.a., ligamentum annulare; Lg.l., ligamentum laterale; pe.b., peroneus brevis; pe.l., peroneus longus; pe.IV, peroneus digiti quarti; pe.V, peroneus digiti quinti; t.A, tendo Achillis. II–V, toes.

TABLE III. List of the Pelvic Limb Muscles, with Their Insertions and Innervation[a]

	Dorsal groups		Ventral groups		
Muscles	Insertions	Innervation	Muscles	Insertions	Innervation
1. Hip and thigh					
Lumbar part					
Psoas	Vl – Fe 2	⎫	Adductor longus	Pu 6 – Fe 18	⎫
Iliacus	Il 1 – Fe 2	⎬ n. femoralis	Adductor brevis	Pu 7 – Fe 17	⎬ n. obturatorius
Articularis coxae	Il 16 – Fe 3	⎪	Adductor magnus	Pu 9 – Fe 19	⎪
Pectineus	Pu 5 – Fe 16	⎭			⎭
			Obturator externus	o.m. 10 – Fe 15	
Sartorius	Il 3 – T 23	⎫			
Rectus femoris	Il 4 – Pa and T 22	⎪	Gracilis	Pu 8 – T 23	
Vastus lateralis	Fe 6 – Pa and T 22	⎬ n. femoralis			
Vastus intermedius	Fe 4 – Pa and T 22	⎪			
Vastus medialis	Fe 1 – Pa and T 22	⎭			
Sacral part					
Femorococcygeus	Vca – Fe 12	⎫	Quadratus femoris	Is 11 – Fe 9	⎫
Gluteus maximus (+ tensor fasciae latae)	Il 13 – Fe 10	⎪	Gemellus inferior	Is 18 – Fe 14	⎪
		⎬ nn. gluteus inferior, gluteus superior, ischiadicus dorsalis	Gemellus superior	Is 20 – Fe 14	⎬ Collateral branches of sacral plexus and n. ischiadicus ventralis
Gluteus medius	Il 14 – Fe 8	⎪	Obturator internus	o.m. 19 – Fe 14	⎪
Piriformis	Sa 2 – Fe 8	⎪	Biceps femoris	Is 21 – Ti 25	⎪
		⎪	Semimembranosus	Is 12 – Ti 21	⎪
Gluteus minimus	Il 17 – Fe 5	⎪	Semitendinosus, caput ventrale	Pu 8 – Ti 23	⎪
Scansorius[b]	Il 15 – Fe 5	⎭	Semitendinosus, caput dorsale	Vca – Ti 23	⎪
			Caudofemoralis[c]	Vca – Fe 11	⎭

2. Leg and foot

Anterior muscles

Tibialis anterior	T 24 – P 6	} n. peroneus profundus
Abductor hallucis longus	T 24 – P 18	
Extensor digitorum longus	T + Fi 26 – P 13	
Extensor hallucis longus	Fi 29 – P 17	
Extensores digitorum breves[d]	P 14 – P 14	

Lateral muscles

Peroneus longus	T + Fi 27 – P 5	} n. peroneus superficialis
Peroneus brevis	Fi 30 – P 16	
Peronei digitorum IV and V	Fi 28 – P 15	

Posterior muscles

Gastrocnemius medialis	Fe 20 – P 12	
Gastrocnemius lateralis	Fe 13 – P 12	
Soleus	Fi 31 – P 12	
Plantaris	Fe 13 – a.p.	
Tibialis posterior	Fi 32 – P 7	
Flexor fibularis	T + Fi 33 – P 1	
Flexor tibialis[e]	T 35 – P 8	
Popliteus	Fe 7 – T 7	} n. tibialis
Peroneotibialis	Fi 34 – T 34	
Flexor digitorum brevis, caput superficiale	a.p. + P 19 – P 9	
Flexor digitorum brevis, caput profundum	t.f.l. – P 9	
Lumbricales	t.f.l. – d.a.	
Contrahentes	P 2 – P 2	
Interossei	P 3 – P 3	
Abductor hallucis brevis	P 4 – P 4	
Abductores V and M V	P 10 – P 10	
Opponens V	P 11 – P 11	

[a] The numbers in the insertions column are those mentioned on the bones, Fig. 11, 12, and 27. a.p., Aponeurosis plantaris; d.a., dorsal toes-aponeurosis; Fe, femur; Fi, fibula, Il, ilium; Is, ischium; o.m., obturating membrane; P, pes; Pa, patella; Pu, pubis; Sa, sacrum; T, tibia; t.fl., tendines flexorium; Vca, vertebrae caudales; Vl, vertebrae lumbales.

[b] Wanting in the Cheirogaleinae.
[c] Wanting in *Propithecus*.
[d] Extensor hallucis brevis normally wanting.
[e] Quadratus plantae wanting.

THE HAND (Figs. 19–24; Table IV)

The hands of the Malagasy lemurs (Jouffroy and Lessertisseur, 1959a), even when they seem to display only weak specialization as among the lemurids, are in fact far removed from the primitive mammalian hand, and differ fundamentally from those of the higher primates. Such differences lie not only in external characters and general proportions, but above all in the morphological axis (axony) of the hand. Among the lemurs this passes through the fourth digit (ectaxony), and not through the third (mesaxony) as in the higher primates and man.

External Characters

The tactile pads are arranged (as generally among primates) in three sets; five apical, at the enlarged, spatulate digital extremities; three interdigital, at the base of the digits (not in *Daubentonia*); and two at the base of the palm. These pads are covered with papillary crests, parallel or in loops (Hill, 1953; Midlo and Cummins, 1942; Cummins and Midlo, 1961; Biegert, 1961; Rakotosamimanana and Rumpler, 1970). The palmar pads are very prominent in the cheirogaleines, whereas in the indriines, their presence is marked only by the pattern of papillary crests. Among the lemurids (Fig. 19), Indriidae (except *Indri*), and *Daubentonia*, the center of the palm supports a granular zone (insulae lenticulares). The area of this zone is variable; in *Daubentonia*, it occupies the whole palm except for the small area of the inter-

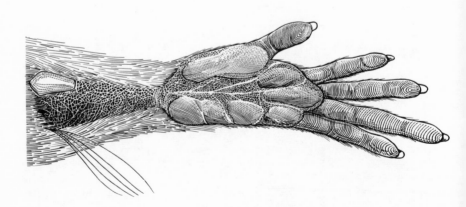

FIG. 19. *Hapalemur*, left hand from the volar aspect.

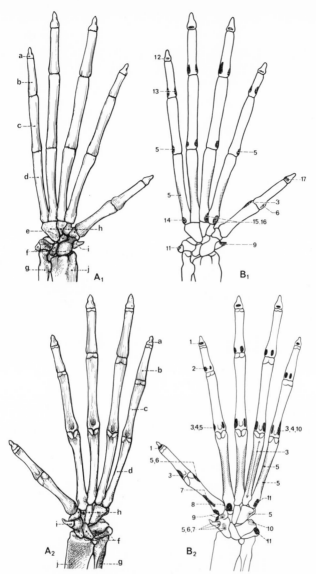

FIG. 20. *Varecia* (= *Lemur*) *variegatus*. Skeleton of the left manus from the dorsal aspect (A₁, B₁) and from the volar aspect (A₂, B₂). Relief (A₁, A₂) and muscular insertions (B₁, B₂). Light, origins; dark, insertions. a, b, c, Third, second, and first phalanges; d, metacarpus; e, os centrale; f, proximal row of the carpus; g, cubitus; h, distal row of the carpus; i, praepollex; j, radius. 1, flexor digitorum profundus; 2, flexor digitorum superficialis; 3, contrahentes digitorum; 4, lumbricales; 5, interossei; 6, abductor pollicis brevis; 7, opponens pollicis; 8, flexor carpi radialis; 9, abductor pollicis longus; 10, abductor digiti quinti brevis; 11, flexor carpi ulnaris; 12, extensor digiti quinti (= extensor lateralis); 13, extensor digitorum communis; 14, extensor carpi ulnaris; 15, extensor carpi radialis longus; 16, extensor carpi radialis brevis; 17, extensor digitorum profundus.

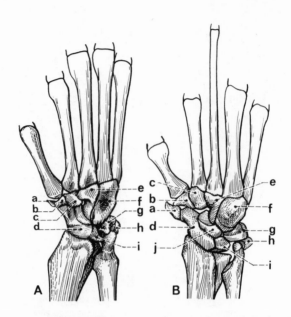

FIG. 21. Right carpus of (A) *Propithecus* and (B)
Daubentonia from the dorsal aspect. a, Praepollex; b,
trapezium; c, trapezoideum; d, scaphoideum; e, capi-
tatum; f, hamatum; g, triquetrum; h, pisiforme; i,
lunatum; j, os centrale.

digital pads, while in *Propithecus* and *Avahi*, it is reduced to a thin axial rib-
bon. In *Indri*, it is lacking; instead, the palm is totally covered with papillary
crests.

Among the indriines, an interdigital web extends between digits III, IV,
and V as far as the distal ends of the first phalanges (and even beyond in *Indri*). The digits terminate in more or less carinated nails, with the sole exception of *Daubentonia*, all of whose fingers possess claws (atypical but characterized by their double matrix; cf. Le Gros Clark, 1936). Thus, contrary to
the condition observed in the higher primates where only the most primitive
forms (Callithricidae) possess claws, among the lemurs claws characterize the
form whose hand is the most highly specialized.

Only in the lemurids do vibrissae, glands, and callosities exist in the carpal region. In males of *Hapalemur* (Fig. 19) and of *Lemur catta*, the glandular region extends widely over the forearm, constituting a secondary sexual
character.

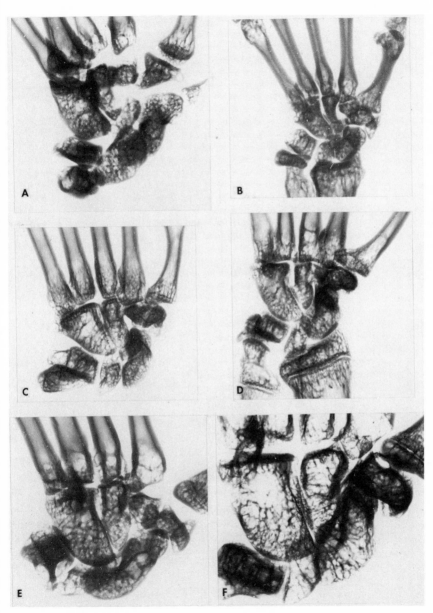

FIG. 22. Radiographs of carpus. (A) *Daubentonia*, (B) *Microcebus*, (C) *Lemur*, (D) *Lepilemur*, (E) *Propithecus*, (F) *Indri*. Free os centrale: A, B, C, E.

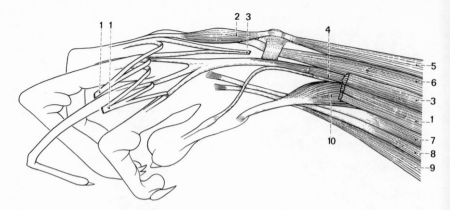

FIG. 23. *Daubentonia*. Muscles of the right hand from the dorsal aspect, middle sheet. 1, Extensor digitorum communis; 2, abductor digiti quinti brevis; 3, extensor digiti quinti (extensor lateralis); 4, extensor digitorum profundus; 5, flexor carpi ulnaris; 6, extensor carpi ulnaris; 7, brachioradialis; 8, extensor carpi radialis brevis; 9, extensor carpi radialis longus; 10, abductor pollicis longus.

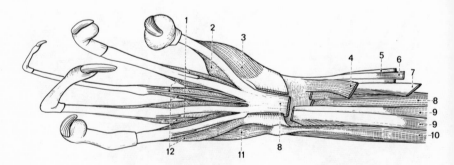

FIG. 24. *Daubentonia*. Muscles of the right hand from the volar aspect, deep sheet. 1, Lumbricales; 2, contrahens (=adductor brevis) pollicis; 3, abductor pollicis brevis; 4, flexor carpi radialis; 5, brachioradialis; 6, abductor pollicis longus; 7, pronator teres; 8, flexor digitorum profundus; 9, flexor digitorum superficialis; 10, flexor carpi ulnaris; 11, abductor digiti quinti brevis; 12, interossei.

Form and Proportions (Table IV)

Among the lemurids and indriids, the hand is of medium size; its length represents between 20% and 40% of trunk length, as is usual among non-brachiating primates. In this group the lemurines and cheirogaleines have relatively the shortest hands, while those of the indriines are the longest. Only *Daubentonia* is characterized by an extreme relative hand length (45% of trunk length), approached only by Hylobates and equaled by *Tarsius* alone.

In all lemurs, but especially in *Avahi*, *Indri*, and *Daubentonia*, the hand is greatly elongate in form, much more than in the light arm-swingers (*Ateles*, *Hylobates*, *Symphalangus*; Jouffroy and Lessertisseur, 1960). Hand length is on the order of six times the width of the palm (as compared with four times in the gibbon).

The Palm and Digits II–V

The hands of the lemurs, as of all prosimians except *Tarsius*, are ectaxonic, i.e., the morphological axis of the palm passes through the fourth digit; this digit is the longest. In correlation, digit II is very short, although without showing a degree of atrophy comparable to that seen in *Perodicticus* and *Arctocebus*.

Ectaxony is less accentuated among the lemurines and cheirogaleines than in the indriines and *Daubentonia*. The digital formula of the former is IV ≥ III > II > V; that of the latter is IV > III > V > II.

TABLE IV. Relative Length of the Manus and Intrinsic Proportions of Its Segments (Averages, in Percent)[a]

Genus	$\dfrac{M}{S}$	$\dfrac{M}{FL}$	$\dfrac{M}{R}$	$\dfrac{c}{M}$	$\dfrac{m_3}{M}$	$\dfrac{d_4}{M}$	$\dfrac{d_2}{d_3}$	$\dfrac{d_4}{d_3}$	$\dfrac{d_1}{M}$	$\dfrac{m_1 + d_1}{M}$
Lemur	22	27	73	16	32	56	85	107	27	46
Varecia	27	31	92	15	30	57	83	109	26	44
Hapalemur	26	33	84	18	29	57	79	107	28	47
Lepilemur	28	33	94	15	27	59	79	110	25	41
Cheirogaleus	23	31	89	18	29	56	88	104	28	47
Microcebus	22	29	78	15	29	60	86	113	29	49
Propithecus	32	33	99	13	30	58	76	114	26	44
Indri	39	34	97	11	33	55	80	109	27	47
Avahi	36	34	95	10	31	58	76	114	20	34
Daubentonia	45	41	139	11	29	72	92	139	24	36

[a]c, Carpus; d_1–d_4, digits; FL, entire forelimb; M, manus; m_1–m_4, metacarpals; R, radius; S, precaudal spine.

Daubentonia conforms with the general pattern of ectaxony, but in a unique manner. Here, the fourth digit is not only the longest but also the most robust, while the third is very gracile and attenuated. However, ectaxony, very pronounced at the level of the digital extremities since digit IV is almost half as long again as the subequal II and V, only shows itself beyond the extremity of the second phalanx. In fact, the length of metacarpal 3 and that of the first phalanx of digit III insure that the latter greatly surpasses the first phalanx of digit IV. In compensation, the second and third phalanges of digit II are very short.

In the carpus (Figs. 21 and 22), the most notable characteristic is the large size of the hamate (unciform) and of the scaphoid. The hamate is larger than the capitate (os magnum), and may in *Indri* form entirely the mediocarpal condyle. The scaphoid forms almost the entire radiocarpal joint. Beyond this, the os centrale, free in *Lemur, Microcebus, Cheirogaleus, Propithecus*, and *Daubentonia*, is fused with the scaphoid in adults of *Hapalemur, Lepilemur, Indri*, and *Avahi*. It articulates with the hamate; this is sometimes considered a primitive character, but is more probably an adaptive character correlated with the large size of the hamate.

In the mesaxonic primitive mammalian hand (e.g., *Tupaia*), the largest carpal bone is the capitate. Among the Malagasy lemurs as among the Lorisiformes, it is those elements corresponding to the thumb (scaphoid) and to digit IV (hamate) which become largest. Concomitant carpal reduction affects the trapezoid (multangular minus), correlating with reduction of digit II. The more specialized the hand, the more this tendency is accentuated; in this regard, Lemurinae and Cheirogaleinae are less specialized than Indriinae and *Daubentonia*.

Except in *Daubentonia*, ectaxony extends to the level of the metacarpals; M1 is always the longest. The terminal phalanges are always very short.

The digital musculature of the lemurs conforms to the general primate pattern (Table II; Fig. 20); the long extensors and flexors reach all digits. No specializations are evident; not even the predominance of digit IV or the brevity of digit II is accompanied by any special muscular disposition. The great power and large number of extensor tendons in digit IV of *Daubentonia* may be noted (Fig. 23), together with the gracility of the nonperforated flexor tendon of III.

In compensation, ectaxony reveals itself in certain short muscles, the contrahentes and interossei. As Jouffroy and Lessertisseur (1959*b*) noted, the primitive mesaxonic disposition is accompanied by the absence of contrahens (C3), thus underlining the axial role of digit III. In the case of the ectaxonic lemurs, one may observe various arrangements:

1. The absence of C4.

2. The existence of a fasciculus ("distrahens") symmetrical with C4, i.e., on the ulnar border of digit IV.
3. The presence of C3.

At the level of the interossei (Lessertisseur, 1958), ectaxony shows itself in the normal presence of two dorsal interossei at digit IV, and of only one at digit III (Jouffroy and Lessertisseur, 1959a), while in mesaxony digit III is flanked by two dorsal interossei.

The Pollex

The pollex is long and originates high on the palm of the hand; the pollicial index (length of thumb, minus metacarpal, as a percentage of hand length), generally below 25 in the higher primates, is almost constant and always above this figure in the lemurs (except *Avahi*). Beyond this, in *Indri*, as in the gibbon, the free length of the thumb corresponds not only to the phalanges, but also to a part of the metacarpus.

The length of the thumb, as well as the arrangement of the trapezium, permits its "pseudo-opposability," and is characteristic of the lemurs (Jouffroy and Lessertisseur, 1959a; Napier, 1961; Petter, 1962; Bishop, 1964). When the hand grasps a relatively small object, the five digits flex simultaneously. The object is held between the extremities of digits II–V and the palmar interdigital pads, the hand thus forming a sort of hook. The thumb, also flexed, has its metacarpal widely abducted and divergent from the other metacarpals so that its tip becomes braced on the side of the object.

The divergence of the first metacarpal and its inclination relative to the plane of the other metacarpals is due to the highly external position of the trapezium, which articulates with a prominent tubercle of the scaphoid. The saddle shape of the trapezometacarpal joint limits the movement of the metacarpal. Relative to the plane of the other metacarpals, the axis of this articulation is very oblique toward the palm.

Beside the trapezium, a radial sesamoid or "prepollex" (Fig. 21) has been observed in *Lemur, Lepilemur, Propithecus*, and *Daubentonia* (Milne-Edwards and Grandidier, 1875; Jouffroy, 1962). The presence of such a bone is highly exceptional among the higher primates.

THE FOOT (Figs. 25–27; Table V)

The foot is always more specialized than the hand. The large, wide hallux is opposed to the other toes. The foot is semi-digitigrade, the tarsus raised during locomotion. Weight is borne by both tarsometatarsal and last interphalangeal joints.

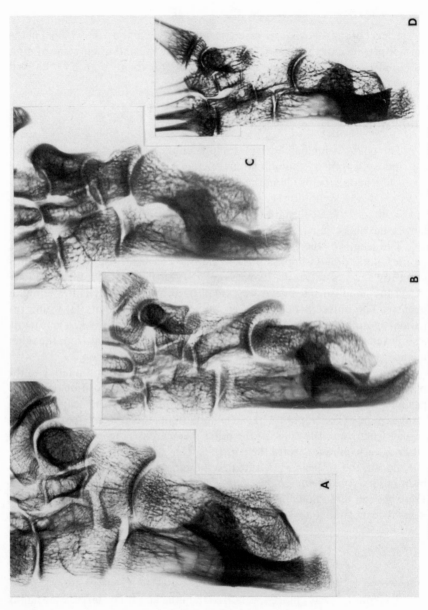

FIG. 25. Radiographs of tarsus. (A) *Indri,* (B) *Daubentonia,* (C) *Lepilemur,* (D) *Hapalemur.*

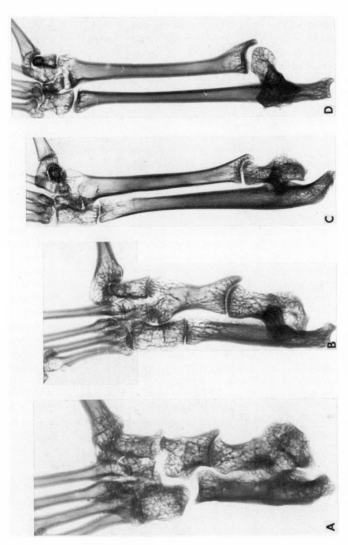

FIG. 26. Radiographs of tarsus. (A) *Cheirogaleus*, (B) *Microcebus*, (C) *Galago*, (D) *Tarsius*.

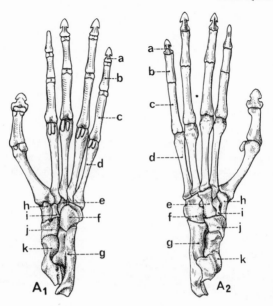

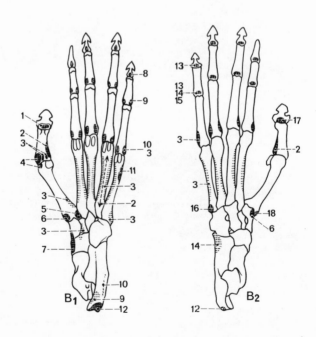

External Characters

The presence of papillary crests on the sole of the primate foot is considered to be an evolved character. The extent of such crests is extremely variable among the lemurs (Hill, 1953; Midlo and Cummins, 1942; Cummins and Midlo, 1961; Biegert, 1961; Rakotosamimanana and Rumpler, 1970). Crests exist on all four interdigital pads and apical pads at the extremities of the digits. In *Daubentonia*, these are the only regions bearing papillary crests, while in the cheirogaleines there exist also more proximal crest-bearing pads, the thenar and hypothenar. In *Lemur* and *Hapalemur*, these proximal pads are missing, but in their place exists a granular zone with small islands of relief; these latter bear crests. Among the others, the entire surface of the sole may be covered with crests as in *Indri*, or only a small central area may lack them, as in *Lepilemur*, *Avahi*, and *Propithecus*.

Among all lemurs except *Daubentonia*, each toe bears a nail except for the second, which bears instead a toilet claw. In *Daubentonia*, the hallux alone possesses a nail; the other toes are clawed. In the indriines, an interdigital web extends between digits III–V to the distal extremity of the first phalanx.

Form and Proportions (Table V)

The foot, compared to trunk length, has an average value of 30–46%. There exists among the primates a certain correlation between hand and foot, and for this reason one finds long hands and feet in jumping forms (lemurs, galagines, *Tarsius*) as well as in arm-swingers (*Hylobates*, *Ateles*, *Pongo*).

FIG. 27. *Varecia* (= *Lemur*) *variegatus*. Skeleton of the left pes from the plantar aspect (A_1, B_1) and from the dorsal aspect (A_2, B_2). Relief (A_1, A_2) and muscular insertions (B_1, B_2). Light, origins; dark, insertions. a, Third phalanx; b, second one; c, first one; d, metatarsal; e, third cuneiform; f, cuboid; g, calcaneus; h, first cuneiform; i, second one; j, navicular; k, talus. 1, Flexor hallucis longus; 2, contrahens (= adductor brevis) hallucis; 3, interossei; 4, abductor hallucis brevis; 5, peroneus longus; 6, tibialis anterior; 7, tibialis posterior; 8, flexor digitorum longus; 9, flexor digitorum brevis; 10, abductor digiti quinti; 11, opponens digiti quinti; 12, tendo achillis (gastrocnemii plus soleus); 13, extensor digitorum longus; 14, extensor digitorum brevis; 15, peroneus digiti quinti; 16, peroneus brevis; 17, extensor hallucis brevis; 18, abductor hallucis longus.

TABLE V. Relative Length of the Pes and Intrinsic Proportions of Its Segments (Averages, in Percent)[a]

Genus	$\dfrac{P}{S}$	$\dfrac{P}{HL}$	$\dfrac{t}{P}$	$\dfrac{m_3}{P}$	$\dfrac{d_4}{P}$	$\dfrac{d_2}{d_3}$	$\dfrac{d_4}{d_3}$	$\dfrac{d_1}{P}$	$\dfrac{m_1 + d_1}{P}$
Lemur	33	28	38	27	39	90	108	24	50
Varecia	30	29	36	26	40	89	110	21	44
Hapalemur	41	31	37	26	38	87	108	25	53
Lepilemur	42	34	39	24	40	91	119	24	47
Cheirogaleus	40	34	40	23	40	90	113	23	47
Microcebus	36	33	45	20	40	93	119	20	41
Propithecus	42	27	32	31	41	92	108	29	60
Indri	46	28	29	31	42	93	108	29	59
Avahi	46	28	31	29	43	96	115	24	52
Daubentonia	45	32	29	25	47	86	109	20	44

[a]d_1–d_4, Digits; HL, entire hindlimb; m_1–m_4, metatarsals; P, pes; S, precaudal spine; t, tarsus.

When foot length is expressed as a percentage of hindlimb length (Lessertisseur, 1970; Lessertisseur and Jouffroy, 1974), it emerges that the lemurs exhibit two types which are extreme for Primates:

1. Short-footed forms (26–28%): the indriines.
2. Long-footed forms (33–35%): *Lepilemur* and the cheirogaleines.

Lemur, Hapalemur, and *Daubentonia* are intermediate.

No primate except man has so short a foot compared to the hindlimb as do the indriines. But numerous primates have a foot as long (*Tarsius, Galago,* callithricids) or even longer (*Pongo*) than that of the cheirogaleines and *Lepilemur.*

Two further types can be distinguished through comparison of the three principal segments of the foot (tarsus, metatarsus, and phalanges: Lessertisseur and Jouffroy, 1974):

1. Short tarsals (29–33% of total foot length) and long metatarsals (25–31%): indriids and *Daubentonia.*
2. Long tarsals (37–45%) and short metatarsals (19–26%): *Lepilemur* and the cheirogaleines.

Again, *Lemur* and *Hapalemur* represent an intermediate type (tarsals, 37%; metatarsals, 26.5%). The length of the toes is hardly of any significance here since it is relatively constant among the lemurs (38.5–42%). In comparison with other primates, however, the toes of lemurs are very long; only *Pongo* and *Lagothrix* achieve such proportions (42%).

Among Primates as a whole, the lemur tarsus is long; only the short type, and that of *Indri* in particular, approaches monkeylike proportions (*Ateles,*

Lagothrix). The great length of the tarsus results from the elongation of the navicular and calcaneum. In *Microcebus* (45%; Fig. 26), these bones are rodlike and comparable to those seen in *Tarsius* (46%) and *Galago* (55%).

The Sole and Digits II–V

The lemur foot is ectaxonic to an even greater extent than the hand: the morphological axis of the sole passes through the fourth toe, the longest. The difference in length between digits III and IV is largest in *Lepilemur* and *Microcebus* (IV/III:119). Among Prosimians as a whole, only *Perodicticus* (124) and *Tarsius* (138) possess a relatively longer fourth toe (Lessertisseur and Jouffroy, 1974).

Contrary to the hand, digit II of the lemur foot is as developed as that of the monkeys (II/III:90–93). Among the prosimians, the last (toilet-claw-bearing) phalanx of this digit is relatively long, and an extreme shortness of the two first phalanges is necessary to prevent this from affecting the length of the entire digit. Here again, *Daubentonia* is distinguished from the other lemurs by the relative brevity of its second toe (86%).

The musculature of the toes conforms to general primate morphology (Table III, Fig. 27). The long extensors and flexors reach all toes. The ectaxony of the foot hardly expresses itself at all except at the level of the intrinsic muscles, contrahentes and interossei (Lessertisseur, 1958; Jouffroy and Lessertisseur, 1959b; Jouffroy, 1962). It should be noted that in a typical mesaxonic disposition (higher primates), the contrahentes occupy the position and perform the functions of adductors and are therefore also known as "short adductors."

Among the lemurs, the disposition of the contrahentes is slightly modified in correlation with the displacement of the axis of the foot to digit IV. There is generally found a contractor of III, inserted on the aspect of the toe facing IV, and a fasciculus on IV. The latter is symmetrical with the contrahens, and is known for this reason as a "distrahens." In *Daubentonia*, C4 is displaced onto the axis of the toe.

As far as the interossei ("short flexors") are concerned, one can only note that a dorsal interosseous is frequently present on either side of digit IV, while one only is associated with digits III, II, and V.

The Hallux

The hallux is long, 41–60% of the length of the foot. In the nonhominid higher primates, this value varies from 32% to 45%. In order of increasing hallucial length, the lemurs fall into the following sequence: *Daubentonia* and

Microcebus (40–43%); *Cheirogaleus* and *Lepilemur* (47%); *Lemur* and *Hapalemur* (50–52%); and finally, the indriids (53–60%), in which this figure is the highest in Primates. The relative length of the hallux is inversely proportional to the length of the foot; thus *Daubentonia* has the longest foot and the shortest hallux, and *Indri* the shortest foot and the longest hallux.

In all the lemurs except the cheirogaleines, an abductor hallucis longus is present among the hallucial muscles, formed from the splitting of a fasciculus of the tibialis anterior, which inserts at the base of the first metatarsal. This is auxiliary to extensor hallucis longus. Also present in numerous higher primates and in man, this fasciculus is unknown among the other prosimians.

In general there is in the lemur hallux neither a tendon from flexor digitorum brevis (although one is occasionally observed in *Daubentonia*), nor an extensor hallucis brevis (very infrequently seen in *Lemur*). Adductor hallucis brevis (contrahens 1) is strong; it may, as in *Propithecus*, form a layer without oblique or transverse heads, or, conversely, it may, as in *Daubentonia*, form several fasciculi which are widely separated and unite the hallux to metatarsal 4. Abductor hallucis brevis and flexor hallucis brevis (interossei sheet) complete the hallucial musculature.

CONCLUSIONS

The indriids, vertical clingers and leapers in the trees and largely bipedal on the ground, are the most specialized and evolved of the lemurs. The hind limb is very long, but the two traits most characteristic of the group, which differentiate them totally from other prosimian vertical clingers and leapers such as *Galago* and *Tarsius*, are the great length of the femur and the shortness of the tarsus. There is no spectacular tibiotarsal adaptation; propulsion is provided by all the joints of the hindlimb.

The indriid foot forms a strong prehensile grasping organ. The hand is strongly specialized in its elongated form and in its ectaxony (less, however, than in *Daubentonia*, and in a different way).

The hand is also evolved in possessing weak or nonexistent relief of the palmar pads and by its extensive dermatoglyphic covering.

The cheirogaleines, the most quadrupedal of the lemurs, are the least specialized. The hindlimb, if relatively longer compared to the forelimb than in quadrupedal higher primates and in the subfossil lemurs, is nonetheless much shorter than in the indriids. The three segments of the limb are subequal; the articulations, strongly held together, exhibit no play. The hands and feet, although typically lemuriform, are the least specialized among the group, and the least elongated and ectaxonic. The prominent pads and the sharp carinated nails permit the animals to climb like those primates possessing claws (callithricids). The hand is imperfectly adapted to precise manipulation. It should be noted that *Phaner* has not been studied for this report.

The most spectacular, if not the only, specialization lies in the elongation of the tarsus of *Microcebus*, a leaping adaptation which one finds, in slightly different forms, in *Lepilemur* and especially in *Galago* and *Tarsius*.

The anatomy of the limbs leads to the separation of *Lepilemur* from *Lemur* and *Hapalemur*, as do behavior and ecology (Petter, 1962). The latter two genera, quadrupedal runners as much as leapers in most of their characters, form a group intermediate between Cheirogaleinae and Indriinae. This holds for the intermembral index, proportions of the various segments of the hindlimb, length of the hand, ectaxony, and so forth. The hand of *Lemur* and *Hapalemur* is characterized above all by the presence of glands, carpal vibrissae, and callosities.

Within genus *Lemur*, there is a strong separation between the species *L. variegatus* (which in fact should constitute a genus of its own, *Varecia* Gray, 1863) and the others. The intermembral index of *Varecia* (76) is higher than in *Lemur* (72), and the brachial (103 vs. 108), crural (93 vs. 96), and humerofemoral (138 vs. 149) indices are lower.

Lepilemur presents a series of characters which not only distinguish it from the other lemurines, but which are reminiscent either of the indriids or cheirogaleines. The vertical position of the body at rest and during locomotion is comparable to that of the indriids. The low value of the intermembral index, reflecting the predominance of the posterior limb, is the most striking anatomical resemblance. However, in the morphology of its foot, *Lepilemur* exhibits characters close to those of the cheirogaleines: elongation of the calcaneum and navicular, short metatarsals, great length of digit IV, ectaxony, and moderate length of the hallux. One might almost say that *Lepilemur* leaps like *Propithecus* with the foot of *Cheirogaleus*!

Daubentonia is the strangest of the lemurs. There is no longer any question, as there was at the time of its discovery, that it should not be regarded as a rodent, but its systematic position remains controversial. It has been placed in its own family, superfamily and even infraorder. In the morphology of its limbs, the most striking characteristics are:

1. The presence of claws with double matrix (Le Gros Clark, 1936) on all fingers and toes except the hallux.

2. The great length of the hand and foot (each 45% of trunk length), so exceptional that among the primates similar elongation is found only in the most specialized arm-swingers (*Hylobates*, *Pongo*) and in *Tarsius*.

3. The special morphology of the third digit of the hand. This digit is extremely gracile, with a ball-and-socket joint at the highly distal metacarpophalangeal articulation. It is mobile on its metacarpal in all di-

rections and constitutes a specialized instrument for impaling insect larvae, emptying coconuts and eggs, and drinking (Petter and Peyrieras, 1970). This specialization reduces the efficiency of the hand in both prehension and locomotion.

Among other characters, one may note in comparison with the other lemurs that the forelimb is relatively less short (111% of trunk length) and that the hindlimb is among the longest (135%). The intermembral index is on the order of that of the cheirogaleines (70%), and, as in this group, the upper arm is longer than the forearm and the three segments of the hindlimb are subequal. The animal travels in a quadrupedal attitude and uses its claws in climbing. Like *Cheirogaleus*, *Daubentonia* is not an accomplished jumper.

The hand of this animal (contrary to the situation in Primates and even in mammals in general) is longer and more specialized than the foot. Besides the morphology of digit III, it is distinguished by

1. Its extraordinary length: among the higher primates, a hand as long as the foot is seen only in *Hylobates*, while a hand longer than the forearm is unknown.

2. Its elongated form (length more than six times width): more so than in *Indri*, and much more so than in the gibbon (length four times width).

3. Its extreme ectaxony: digit IV is almost 1.5 times the length of II or V.

None of these characters is found in its much more banal foot; the tarsus is short and the metatarsals long, as in the indriines and many higher primates. As in the cheirogaleines, the hallux is relatively short, as is digit II, since it lacks the toilet claw.

Finally, the palm, like the sole, is primitive in the weak development of papillary crests. These are confined, in the metacarpophalangeal region, to the ill-defined interdigital pads and elsewhere to the terminal digital pads. The remaining surfaces, including the palmar aspects of the fingers and toes, are covered with papilliform granulations.

If one adds that other characters such as the occipital insertion of trapezius and the position below this of levator scapulae ventralis (Jouffroy, 1962) are strongly simian and unknown in other prosimians, it is clear that this rapid examination of limb morphology points strongly to the dissimilarity of *Daubentonia* from the other lemurs. Today, these latter form in their postcranial anatomy a more homogeneous and restricted group than was the case among the lemurs until very recent times.

REFERENCES[1]

Anthony R., and Bortnowsky, I., 1914, Recherches sur un appareil aérien de type particulier chez un Lémurien, *Arch. Zool. Exp. Gen.* **53**:399–324.

Anthony, R., and Coupin, P., 1931, Tableau résumé d'une classification générique des Primates fossiles et actuels, *Bull. Mus. Hist. Nat. Paris Sér. 2* **3**:566–569.

Biegert, J., 1961, Volarhaut der Hände und Füsse, *Primatologia* **2(1)**:1–326.

Bishop, A., 1964, Use of the hand in lower primates, in: *Evolutionary and Genetic Biology of Primates*, Vol. 2 (J. Buettner-Janusch, ed.), pp. 133–225, Academic Press, New York and London.

Charles-Dominique, P., and Hladik, C. M., 1971, Le *Lepilemur* du sud de Madagascar: Ecologie, alimentation et vie sociale, *Terre Vie* **25**:3–66.

Cummins, H., and Midlo, C., 1961, *Fingerprints Palms and Soles, An Introduction to Dermatoglyphics*, 319 pp., Dover, New York.

Hill, W. C. O., 1953, Primates, *Comparative Anatomy and Taxonomy*, Vol. 1: *Strepsirhini*, 798 pp., Edinburgh University Press, Edinburgh.

Jolly, A., 1966, *Lemur Behavior*, 187 p. University of Chicago Press, Chicago.

Jouffroy, F. K., 1960*a*, Caractères adaptatifs dans les proportions des membres chez les Lémurs fossiles, *C. R. Acad. Sci.* **251**:2756–2757.

Jouffroy, F. K., 1960*b*, Le squelette des membres et ses rapports musculaires dans le genre *Lemur*. I. L'humerus, *Bull. Mus. Hist. Nat. Paris Sér. 2* **32**:259–268.

Jouffroy, F. K., 1962, La musculature des membres chez les Lémuriens de Madagascar: Etude descriptive et comparative, *Mammalia* **26(2)**:1–326.

Jouffroy, F. K., 1963, Contribution à la connaissance du genre *Archaeolemur, Ann. Paleontol.* **49**:3–29.

Jouffroy, F. K., and Lessertisseur, J., 1959*a*, La main des Lémuriens malgaches comparée à celle des autres Primates, *Mém. Inst. Sci. Madagascar Sér. A* **13**:195–219.

Jouffroy, F. K., and Lessertisseur, J., 1959*b*, Réflexions sur les muscles contracteurs des doigts et des orteils (contrahentes digitorum) chez les Primates, *Ann. Sci. Nat. Zool. Sér. 12* **1**:211–235.

Jouffroy, F. K., and Lessertisseur, J., 1960, Les spécialisations anatomiques de la main chez les Singes à progression suspendue, *Mammalia* **24**:93–151.

Lamberton, C., 1946, Les Pachylémurs, *Bull. Acad. Malgache n.s.* **27**:7–22.

Le Gros Clark, W. E., 1936, The problem of the claw in Primates, *Proc. Zool. Soc. (Lond.)*:1–24.

Lessertisseur, J., 1958, Doit-on distinguer deux plans de muscles intersseux à la main et au pied des Primates, *Ann. Sci. Nat. Zool.* **11**:77–103.

Lessertisseur, J., 1970, Les proportions du membre postérieur de l'Homme comparées à celles des autres Primates, *Bull. Mém. Soc. Anthropol. Paris Sér. 12* **6**:227–241.

Lessertisseur, J., and Jouffroy, F. K., 1974, Tendances locomotrices des Primates traduites par les proportions du pied, *Primatologia* **20**:125–160

Martin, R. D., 1972, Adaptive radiation and behaviour of the Malagasy lemurs, *Phil. Trans. Roy. Soc. Lond.* **264B(862)**:295–352.

Midlo, C., and Cummins, H., 1942, Palmar and plantar dermatoglyphics in the Primates, *Am. Anat. Mem.* **20**:1–198.

Milne-Edwards, A., and Grandidier, A., 1875–1890, Histoire naturelle des Mammifères, in: Grandidier, A. (ed.), *Histoire Physique, Naturelle et Politique de Madagascar*, vols. 6, 9, and 10, Imprimerie Nationale, Paris.

Napier, J. R., 1961, Prehensility and opposability in the hands of Primates, *Symp. Zool. Soc. Lond.* **5**:115–132.

---

[1]This section is a short bibliography; some references are not specifically cited in the text.

Petter, J. J., 1962. Recherches sur l'écologie et l'éthologie des Lémuriens malgaches, *Mém. Mus. Nat. Hist.* **A27**:1–146.

Petter, J. J., 1969, Speciation in Madagascar Lemurs, *Biol. J. Linn. Soc.* 77–80.

Petter, J. J., and Petter-Rousseaux, A., 1964, Première tentative d'estimation des densités de peuplement des Lémuriens malgaches, *Terre Vie* **4**:427–435.

Petter, J. J., and Peyrieras, A., 1970, Nouvelle contribution à l'étude d'un Lémurien malgache, l'Aye-aye (*Daubentonia madagascariensis* E. Geoffroy), *Mammalia* **34**(2):167–193.

Petter, J. J., Schilling, A., and Pariente, G., 1971, Observations éco-éthologiques sur deux lémuriens malgaches nocturnes: *Phaner furcifer* et *Microcebus coquereli*, *Terre Vie* **118**:287–327.

Pocock, R. I., 1918, On the external characters of the lemurs and of *Tarsius*, *Proc. Zool. Soc. Lond.* 19–53.

Rakotosamimanana, R. B., and Rumpler, Y., 1970, Etude des dermatoglyphes palmaires et plantaires de quelques Lémuriens malgaches, *C. R. Ass. Anat.* **55**:493–510.

Schultz, A. H., 1963, Relations between the lengths of the main parts of the foot skeleton in Primates, *Folia Primatol.* **1**:150–171.

Schwarz, E., 1931, A revision of the genera and species of Madagascar Lemuridae, *Proc. Zool. Soc. Lond.* 399–428.

Seligsohn, D., and Szalay, F. S., 1974, Dental occlusion and the masticatory apparatus in *Lemur* and *Varecia*: Their bearing on the systematics of living and fossil primates, in: *Prosimian Biology* (R. D. Martin, G. A. Doyle, and A. C. Walker, eds.), Duckworth, London.

Starmühlner, F., 1960, Beobachtungen an Mausmaki (*Microcebus murinus*), *Natur Volk* **90**:194–204.

Vallois, H. V., 1914, *Etude Anatomique de l'Articulation de Genou chez les Primates*, 467 pp., Montpellier.

Vallois, H. V., 1955, Primates, in: Grassé, P. R., *Traité de Zoologie* **17**(2):1854–2206.

Walker, A., 1967, Locomotor adaptation in recent and fossil Madagascan lemurs, 535 pp., thesis, London University.

Body Temperatures and Behavior of Captive Cheirogaleids

10

ROBERT JAY RUSSELL

The behavior of dwarf and mouse lemurs is of particular interest to prima-
tologists since some species are thought to resemble the earliest prosimians
in terms of their small body size and bush habitat (Charles-Dominque and
Martin, 1970; Cartmill, 1972). Species of both dwarf and mouse lemurs
are said to exhibit lethargy (von Weidholz, 1932; Petter, 1962), but no study
of the thermoregulatory physiology of these prosimians has been reported.
I report here the preliminary results of a laboratory study conducted in an
effort to determine possible daily or annual patterns of change in the body
temperatures or behavior of the lesser mouse lemur, *Microcebus murinus*,
and the fat-tailed dwarf lemur, *Cheirogaleus medius*.

METHODS

Nine *Microcebus murinus* (gray variety) and three *Cheirogaleus medius*
housed at the Duke University Primate Facility were used in this study.
Three male *M. murinus*, referred to here as "young males," were born in the

ROBERT JAY RUSSELL Department of Anatomy, Duke University, Durham, North
Carolina 27706

colony 18 months prior to the observations. All other animals were caught wild at least 30 months before the observations began.

The animals were divided into the following groups: one male and two female *C. medius*; three young male *M. murinus*; two female *M. murinus*; and two separate pairs of *M. murinus*. These groups were each housed separately in glass-fronted cages (115 by 115 by 28 cm or larger) for at least 8 months before the first observations on body temperatures were made. All animals were given an *ad lib* diet. The pair of females received a high-protein diet of chopped beef heart and beef liver; all other animals received a standard mixed diet of chopped beef, beef liver, apples, oranges, raisins, lettuce, Purina Monkey Chow, and a vitamin-mineral supplement. All animals were provided with water.

During the observations and for at least 8 months beforehand, all animals received 12 h of white incandescent and fluorescent light (light phase approximately 700 lms/m^2) and 12 h of red incandescent light (dark phase approximately 350 lms/m^2) each day. The light phase (L-phase) began at 06:00 h, local time; the dark phase (D-phase) began at 18:00 h. The ambient temperature (T_A) was maintained at 21°C, although some sporadic variation occurred. The maximum T_A recorded during the study was 24°C; the minimum T_A was 18°C. This temperature range is similar to the mean maximum and minimum daily temperatures reported for Fort Dauphin, Madagascar, during the months of June, July, and August (Griffiths and Ranaivoson, 1972). The relative humidity varied noncyclically from 40 to 60% during the study. Light intensity was estimated using a Gossen Luna Pro light meter. A Pacific Transducer 24-h recording thermometer was used to monitor the ambient room temperature. Relative humidity was calculated at selected intervals using a Taylor Mason's hygrometer.

Body temperature (T_B) was recorded to the nearest 0.2°C using a Yellow Springs Instrument Company model 44 telethermometer equipped with thermistor probes. The probes were inserted into the rectum to depths of 4.5 cm (*M. murinus*) or 6.0 cm *(C. medius)*. Through observations of rectal probe placement in cadavers, I determined that in both cases the tips of the probes came to rest approximately 2.0 cm below the diaphragm, at the left colic flexure. This deep insertion apparently did not disturb the animals; they could defecate freely past the probe and appeared to notice the attending wire only when it inhibited their movements.

Two data-gathering techniques were employed to determine T_B: capture temperature readings and continuous observations of single animals. Capture temperature readings were conducted by capturing an animal and recording its T_B within 1 min. The animal was observed several minutes before its capture and its activity state ("active," "awake," or at "rest," cf. Russell, in preparation) was noted. No animal was captured more than once in any 12-h period. Body temperatures recorded in this manner probably do not vary more than $\pm 0.3°$C from the "actual" T_B prior to capture, since T_B observations of cheirogaleids taken during prolonged handling of inactive animals revealed a maximum T_B increase of $3.0°$C in the first 12 min (T_A = 20°C). During the months of March and April, capture temperature observations were recorded during the L-phase between 1100 and 1300 h and during the D-phase between 1900 and 2100 h. The T_B of *M. murinus* was recorded 32 times during the L-phase and 19 times during the D-phase. The T_B of *C. medius* was recorded nine times during the L-phase and nine times during the D-phase in March and April. Considerably more capture temperature observations were made during the months of September through December. Capture temperature observations of all the animals were recorded during the following hours, local time: 3, 5, 6, 8, 10–14, 16–24. From September through December, the capture T_B of *M. murinus* was recorded 76 times during the L-phase and 73 times during the D-phase; the capture T_B of *C. medius* was recorded 27 times during the L-phase and 30 times during the D-phase.

Observations of the T_B of a single animal for periods of from 13 to 15 h were conducted during September and October. First, an animal was captured just before the onset of the L-phase and a rectal probe was inserted. The wire leading to the thermistor probe was taped securely to the animal's tail to prevent the probe from becoming displaced. The animal was then placed in a wire cage (21 by 11 by 13 cm), which allowed the animal considerable movement. This cage was placed in turn within a sound-proof chamber (1.25 by 0.75 by 0.75 m) constructed from Gold Bond Deciban sound-deadening panels lined with Owens-Corning 2 1/4-inch Fiberglas insulation. Sound attenuation from outside to inside this chamber was measured at 95.4% of a 76-db 1000-hz signal. A baffled exit fan and entry vent provided slow circulation of fresh air. The chamber was provided with double-walled Plexiglas ports, one for viewing the animal within the cage and one for transmission of room light. The only obvious time clue presented to the animal was the illumination (12-h L-phase/12-h D-phase). From a position outside the chamber, I recorded the activity state and T_B of the animal and the T_A and relative humidity of the chamber at intervals of 60 min or less. Four *M. murinus* and two *C. medius* were observed in this manner.

The animals used in this study were captured once each week and the sexual state (approximate testis size and estrus or anestrus as per Petter-Rousseaux, 1964), body weight (± 0.5 g using an Ohaus animal balance), tail length (from anal aperture to the skin covering the last caudal vertebra), and tail displacement (to ± 0.5 ml of water) of each animal were recorded. The latter measurements were used to construct an index of absolute tail fatness, the "tail index" (T.I.), equal to the displacement of the animal's tail in milliliters divided by the length of its tail in centimeters.

RESULTS

Annual Patterns of Activity

All *M. murinus* and *C. medius* observed in this study displayed correlated annual changes in body weight, tail fat storage, and sexual activity
even when maintained under conditions of constant ambient temperature
(T_A 18–24°C), day length (12-h L-phase/12 h D-phase), and diet for more
than 1 yr (Fig. 1). Successful pregnancies have been recorded in animals
maintained under these conditions.

All animals were relatively light, showed little fat storage in their tails,
and were sexually active from February to July. During this annual period
of sexual activity and minimum body weight, all cheirogaleids were active
during the D-phase. Most animals began to increase in weight in July and
attained their maximum annual body weight 3–6 wk later. The increased
body weight appeared largely due to fat stored in the tail. During this
period, the testes of male cheirogaleids became very small and withdrew

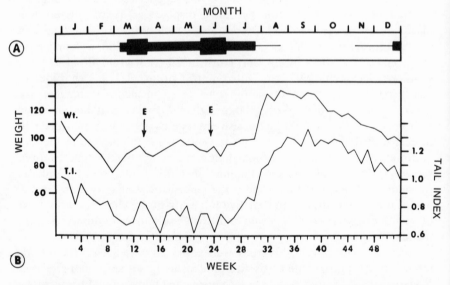

FIG. 1. The annual pattern of body weight change, tail fat storage, and sexual activity
in cheirogaleids maintained at $T_A = 18$–24°C, 12-h L-phase/ 12-h D-phase. A male
and female *M. murinus* which were housed together are shown: (A) Male, approximate size of testes, (no bar, testes not palpable). (B) Female, body weight in grams.
(E, estrus). The week of the observation is shown at the bottom of the graph,
the corresponding month at the top.

from the pendulous scrota (*M. murinus*) or genital region (*C. medius*), becoming impalpable after arriving at the superficial inguinal ring. All animals remained heavy from August to February, and many animals were observed at rest for prolonged periods (i.e., greater than 24 h).

Body Temperatures, March–April

All animals were active during the D-phase and at rest during the L-phase; the T_B of these cheirogaleids was consistently higher during the D-phase than during the L-phase (Table I). The T_B of *M. murinus* ranged from 37.0 to 39.2°C during the D-phase and from 29.0 to 38.4°C during the L-phase. The T_B of *C. medius* ranged from 37.0 to 38.0°C during D-phase and from 34.5 to 37.0°C during the L-phase. No animal was observed to have a T_B less than 6.0°C above the T_A during either the L- or D-phase.

Body Temperatures, September–December, *Microcebus murinus*

During the months of September through December, many animals remained at rest for prolonged periods, while other animals were observed at rest during the L-phase and active or awake during the D-phase. All animals which were captured while resting had remarkably low body temperatures.

From 2 to 8 h after the onset of the L-phase, the T_B of all captured animals ranged from 0.0 to 7.0°C above T_A. For 39% of the adult *M. murinus* observed during this period, T_B was no more than 1.0°C above T_A. The lowest T_A observed, 19.5°C, did not differ measurably from the T_A at that time. At T_B less than or equal to 30°C, all animals were lethargic, i.e., generally unresponsive to stimuli and incapable of rapid or coordinated movement when awakened. When animals were lethargic, their breathing was barely perceptible.

Many animals were awake or active during the D-phase; the T_B of these animals increased as their activity increased. Animals that were awake but not in motion had lower T_Bs than animals that were in motion (Table I). The T_B of the young males was consistently higher than the T_B of the adult animals. None of the young males was observed at rest during the D-phase, but 42% of the adult animals were at rest and lethargic when captured. One-third of these *M. murinus* observed at rest during the D-phase had T_Bs less than 3.0°C above T_A.

An approximation of the daily pattern of T_B change can be obtained by plotting all the capture temperature observations from September through December on a 24-h axis (Fig. 2). All *M. murinus* show low T_Bs during their L-phase period of inactivity. Animals which remained at rest during the D-

TABLE I. Mean Body Temperatures (±SDp) of Captive Cheirogaleids[a]

	March–April		September–December			
	L-phase rest	D-phase active	L-phase[b] rest	D-phase rest	D-phase awake	D-phase active
M. murinus, young males (3)	—	—	26.3 ± 3.5 (11)	(0)	35.4 ± 0.6 (6)	37.5 ± 1.0 (20)
M. Murinus, adults (9)	—	—	22.5 ± 3.6 (28)	25.6 ± 3.9 (24)	34.6 ± 1.6 (15)	36.1 ± 1.2 (17)
M. murinus, all (12)	35.9 ± 2.2 (32)	38.2 ± 0.5 (19)	—	—	—	—
C. medius, adults (3)	35.9 ± 1.1 (9)	37.5 ± 0.5 (6)	25.0 ± 3.7 (18)	24.5 ± 3.0 (10)	34.5 ± 1.6 (14)	36.1 ± 0.5 (5)

[a]All temperatures are in degrees Centigrade; numbers in parentheses refer to the sample size.
[b]Observations from 2 to 8 h after the onset of the L-phase. T_A = 18-24°C.

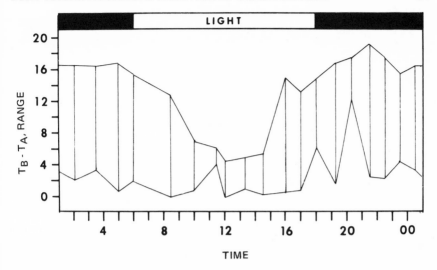

FIG. 2. *Microcebus murinus*, $T_B - T_A$ when T_A = 18–24°C. Each vertical line represents the $TB - TA$ range for nine individuals, time to the nearest 15 min. Data were gathered from September through December.

phase exhibited T_bs indistinguishable from the low T_bs observed during the L-phase. Animals which became active during the D-phase displayed relatively high T_Bs.

Observations of the T_B of animals housed 13–15 h in the sound-proof chamber demonstrated that *M. murinus* which are active during the D-phase exhibit a reduction of T_B during their L-phase period of inactivity (Fig. 3). After the onset of the L-phase, their T_Bs decreased at the rate of 0.05–0.07°C/min for several hours while they remained curled and at rest. Adult animals, who permitted their T_B to approach the T_A more closely than did the young males, attained their minimum T_Bs from 5 to 7 h after the onset of the L-phase. The young males attained their minimum T_Bs from 3 to 5 h after the onset of the L-phase. The minimum T_Bs observed represented T_B reductions of from 11.0 to 16.0°C and ranged from 0.8 to 6.9°C above T_A. All animals exhibited a spontaneous increase in T_B before the onset of the D-phase. Adult animals increased their T_B at the rate of 0.09–0.10°C/min until attaining a T_B of 35–36°C 2 h before the onset of the D-phase. Breathing is rapid, deep, and irregular during spontaneous arousal. *M. murinus* do not shiver during spontaneous arousal but will shiver violently if disturbed prior to spontaneous arousal.

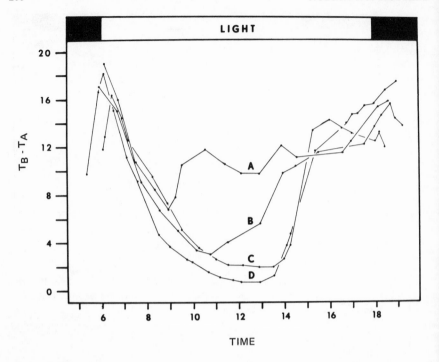

FIG. 3. *Microcebus murinus*, observations of individuals housed in the sound-proof chamber. Data were gathered in September and October. (A) Young male, body weight 64.1 g; T.I. = 0.6; T_A = 21°C. (B) Young male, body weight 104.6 g; T.I. = 0.9; T_A = 21°C. (C) Adult female, body weight 112.5 g; T.I. = 0.9; T_A = 21°C. (D) Adult female, body weight 136.7 g; T.I. = 1.4; TA = 20°C. Time given in hours, local time.

Body Temperatures, September–December, *Cheirogaleus medius*

The daily pattern of T_B change of *Cheirogaleus medius* was very similar to that described for *Microcebus murinus*. All animals displayed low T_Bs and were lethargic during their L-phase period of inactivity (Fig. 4). From 2 to 8 h after the onset of the L-phase, the T_B of captured animals ranged from 0.0 to 7.0°C above T_A. For 21% of the animals observed during this period, T_B was no more than 1.0°C above T_A. All *C. medius* were lethargic when their T_B was 29°C or less. The lowest T_B observed was 19.5°C, which was not measurably different from T_A at that time.

Many animals were awake or active when captured during the D-phase. The T_B of active or awake *C. medius* was very similar to the T_B of active or awake adult *M. murinus* (Table I). *Cheirogaleus medius* were observed at rest and lethargic in 30% of all D-phase observations. Almost half of these resting animals had T_Bs less than 3.0°C above the T_A.

Two *C. medius* were individually housed in the sound-proof chamber and their T_Bs were recorded for periods of 13 and 15 h. Both animals remained at rest during the L-phase. After the onset of the L-phase, their T_B decreased at the rate of 0.03–0.04°C/min until minimum T_B was attained 3–8 h later. The minimum T_Bs represented T_B reductions of 14.2 and 17.7°C and were 2.4 and 1.5°C above T_A. The T_B of one animal began to increase at the rate of 0.08°C/min 4-1/2 h before the onset of the D-phase (Fig. 4A). After attaining a T_B of 31°C, this animal remained at rest with only a slight increase in T_B until becoming active at the onset of the D-phase, 3 h later. The T_B of the second *C. medius* did not begin to increase until after the onset of the D-phase (Fig. 4B). This animal showed an increase of T_B at the

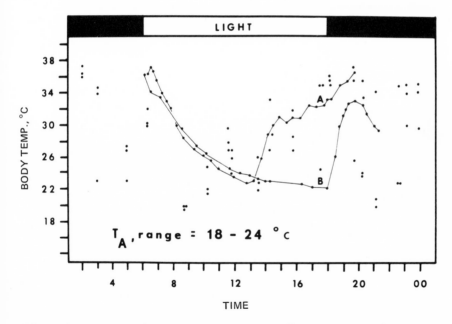

FIG. 4. *Cheirogaleus medius*, capture temperature readings are shown as dots; curves represent observations of individuals housed in the sound-proof chamber. Data were gathered from September through December. (A) Adult female, body weight 303.5 g; T.I. = 5.3; T_A = 20°C. (B) Adult female, body weight 302.0 g; T.I. = 3.7; T_A = 21°C. Time given in hours, local time.

rate of 0.12°C/min until it had attained a T_B of 33°C, 1 h 45 min after the onset of the D-phase. Then, without waking, the T_B of this animal began to decrease. Unfortunately, this experiment was terminated before it could be determined whether or not this animal was entering a prolonged period (greater than 24 h) of reduced T_B.

Body Temperatures at High Ambient Temperatures

During January, I observed the T_B of seven *M. murinus* and three *C. medius* which had been maintained at T_A of 30–31°C for several days. All of these animals were close to their maximum annual body weights and most had been observed to be lethargic during the L-phase the week before they were subjected to high T_A. Capture temperature observations were conducted during the L-phase; the T_B of *M. murinus* was recorded 19 times, *C. medius*, nine times. The mean T_B (±SD) was 35.8±1.1°C for *M. murinus* and 34.5±0.8°C for *C. medius*. None of the animals was lethargic when captured. The lowest difference between T_B and T_A was 2.2°C (*M. murinus*, single observation). The high ambient temperatures apparently inhibited the L-phase reduction of T_B. These results were not unexpected and were similar to the correlation between T_A and $(T_B - T_A)$ found by Morrison (1962) in a tropical microchiropteran, *Molossus* sp.

SUMMARY

Discussion

Under constant conditions of ambient temperature, humidity, diet, and day length, the daily patterns of body temperature change of captive cheirogaleids (*Microcebus murinus* and *Cheirogaleus medius*) vary at different times of the year. During March and April, the body temperatures (T_Bs) of these cheirogaleids were approximately 2° higher during their dark phase (D-phase) period of activity than during their light phase (L-phase) period of inactivity. From February through July, captive cheirogaleids were always active when observed during the D-phase. All animals were sexually active during these months, and their body weights fell to a minimum for the year. From September through December, the T_Bs of these cheirogaleids closely approached the ambient temperature (T_A) during their L-phase period of inactivity, and all animals were lethargic. During the D-phase, many animals remained lethargic, although some animals became active and displayed relatively high T_Bs. Sexual quiescence and attainment of maximum annual body weights also occurred during these months. Captive *M. murinus* and *C. medius* probably maintain relatively high T_Bs throughout their annual period of sexual activity, while they exhibit diurnal low T_Bs and, occasionally, prolonged periods of lethargy throughout their annual period of sexual quiescence. A parallel cycle of cytological changes in the follicular epithelium of the thyroid and the adrenal cortex has been observed by Perret (1972) in captive *M. murinus*.

The annual pattern of sexual activity and T_B change may be partially endogenous since both cycles occurred in captive animals deprived of obvious time clues. The daily pattern of T_B change may also be partially endogenous since spontaneous arousal from low T_B was observed during the L-phase in those animals housed in the sound-proof chamber. It has been suggested that high ambient temperatures (Bourliere *et al.*, 1965) as well as changing day length (Petter-Rousseaux, 1970; Martin, 1972*b*) are important stimuli for seasonal reproduction in captive *M. murinus*; however, annual patterns of sexual activity and correlated changes in body weight will occur under conditions of "low" ambient temperature (21°C) and constant day length (this chapter).

Annual patterns of sexual activity and weight fluctuation have been previously reported for both free-ranging and captive *M. murinus* and *C. medius* (Petter-Rousseaux, 1964; Martin, 1972*c*). Unfortunately, the annual pattern of sexual activity and weight change has not been adequately correlated with daily or annual patterns of T_B change. A pattern of seasonal lethargy and reduced T_B has been reported by Petter (1962) for free-ranging *C. medius*. As yet, no one has reliably reported seasonal lethargy in free-ranging *M. murinus*. Martin, who has performed an exemplary but brief field study of *M. murinus* (1972*a*), doubts that these animals show a seasonal pattern of inactivity and reduced T_B because *M. murinus* has been reported active in southern Madagascar during the cool-dry season when many other Madagascar mammals are lethargic (cf. Eisenberg and Gould, 1970). However, during certain months of the year, I have found that some captive *M. murinus* are active during the D-phase while others are lethargic and have depressed T_Bs. This may also prove to be true in the wild.

The adaptive significance of a daily or seasonal reduction in T_B in *M. murinus* or *C. medius* remains obscure. While a reduction of T_B and a concomitant reduction in metabolic heat production would serve to conserve both energy stores and water, the environmental constraints on these species are poorly known and the necessity of conserving these resources remains to be demonstrated. These Cheirogaleids may exhibit a reduction of T_B in response to daily or seasonal changes in microclimatic T_A and humidity or in response to daily or seasonal fluctuations in resource abundance. Field data on ecology and physiology are needed to test these hypotheses.

Cheirogaleids and Hibernation, Estivation, or Whatever

A thorough review of mammalian thermoregulation would be required to place the daily and seasonal patterns of T_B change of captive cheirogaleids into proper perspective. Such a lengthy review will not be attempted here, but some cautionary remarks about the use of physiological terms are warranted.

"Homeothermy" has been defined as the maintenance of a high, relatively constant T_B; mammals are said to be "heterothermic" when they exhibit a variable T_B (cf. Henshaw, 1970). The distinction is by no means clear. Most mammals exhibit a daily variation in T_B which corresponds to their daily pattern of activity and rest (Aschoff, 1970). For example, man's T_B varies by approximately 1°C per day; that of *Macaca mulatta* by approximately 2°C per day (Myers, 1971); that of *Cebuella* by 4°C per day (Morrison, 1962); that of *Loris tardigradus* by approximately 4–10°C per day depending on the T_A (Eisentraut, 1961); and, as shown above, the T_Bs of captive cheirogaleids vary by 2–20°C depending on the time of year. Terms used to define patterns of T_B change are equally difficult to define operationally. For example, the terms "hibernation" and "estivation" have been used to describe seasonal torpor (Martin, 1972a) or torpor that occurs at "low" or "high" T_A without regard to season (Folk, 1966). Similarly, Bartholomew and Hudson (1960) require a $T_B - T_A$ of 1° or less as a condition for hibernation, while Henshaw (1970) states that the $T_B - T_A$ of a hibernating mammal should be "a few degrees." Such arbitrary definitions delimit classes of phenomena which are probably graded expressions of a single physiological mechanism (cf. Bartholomew and Cade, 1957). The degree to which the T_B of captive *M. murinus* or *C. medius* approaches the T_A appears to be related to the T_A, time of day, time of year, and perhaps the age of the animal; defining a state of torpor in these animals as some arbitrary minimal value of $T_B - T_A$ would add nothing to our understanding of their physiology.

Clearly, many of the fundamental terms and concepts of mammalian thermoregulatory physiology are in need of review. Until such a review is completed, caution should be exercised when comparing the T_B changes of captive cheirogaleids with those of other small mammals and when discussing the possible evolution of mammalian body temperatures. Many mammalian taxa, both geologically ancient and geologically recent, contain extant species which exhibit highly variable body temperatures. The difficulties of interpreting physiological evolution are similar to the difficulties of interpreting behavioral evolution; neither physiology nor behavior leaves a fossil record. Perhaps only if constant correlations can be demonstrated between patterns of thermoregulation on the one hand and morphology (e.g., body size) and/or habitat (e.g., microclimate) on the other hand can we hope to propose reasonable models of the thermoregulation of extinct, ancestral primates.

ACKNOWLEDGMENTS

I gratefully acknowledge the assistance, encouragement, and advice of Ms. Lee W. McGeorge as well as the advice of Professors J. Buettner-Janusch, M. Cartmill, P. H. Klopfer, and J. A. Bergeron. This project was supported in part by funds from the Graduate School of Duke University and the Duke University Primate Facility.

REFERENCES

Aschoff, J., 1970, Circadian rhythm of activity and of body temperature, in: *Physiological and Behavioral Temperature Regulation.*(J. D. Hardy, P. A. Gagge, and A. J. Stolwijk, eds.), pp. 905–919, Thomas, Springfield, Ill.

Bartholomew, G. A., and Cade, T. J., 1957, Temperature regulation, hibernation and aestivation in the little pocket mouse, *Perognathus longimembris, J. Mammal.* **38**:68–72.

Bartholomew, G. A., and Hudson, J. W., 1960, Aestivation in the Mohave ground squirrel, *Citellus mohavensis, Bull Mus. Comp. Zool. (Harvard)* **124**:193–205.

Bourliere, F., A., Petter-Rousseaux, and Petter, J.-J., 1965, Regular breeding in captivity of the Lesser Mouse Lemur (*Microcebus murinus*), *Int. Zoo Yearbook* **3**:24-25.

Cartmill, M., 1972, Arboreal adaptations and the origin of the order Primates, in: *The Functional and Evolutionary Biology of Primates.* (R. Tuttle, ed.), pp. 97–122, Aldine, Chicago.

Charles-Dominique, P., and Martin, R. D., 1970, Evolution of lorises and lemurs, *Nature* **227**:257–260.

Eisenberg, J. F., and Gould, E., 1970, The tenrecs: A study in mammalian behavior and evolution, *Smithson. Contrib. Zool.* **27**:1–138.

Eisentraut, M., 1961, Beobachtungen über den Warmehaushalt bei Halbaffen, *Biol. Zbl.* **80**:319–325.

Folk, G. E., 1966, *Introduction to Environmental Physiology*, Lea and Febiger, Philadelphia.

Griffiths, J. F., and Ranaivoson, R., 1972, Madagascar, in: *Climates of Africa.* (J. F. Griffiths, ed.), pp. 461–499, Elsevier, New York.

Henshaw, R. E., 1970, Thermoregulation in bats, in: *About Bats, A Chiropteran Biology Symposium.* (B. H. Slaughter and D. W. Walton, eds.), pp. 188–232, Southern Methodist Press, Dallas.

Martin, R. D., 1972*a*, A preliminary field-study of the Lesser Mouse Lemur (*Microcebus murinus* J. F. Miller 1777), *Z. Tierpsychol. Beiheft* **9**:43–89.

Martin, R. D., 1972*b*, Adaptive radiation and behaviour of the Malagasy lemurs, *Phil Trans. Royal Soc. Lond. Ser. B* **264**:295–352.

Martin, R. D., 1972*c*, A laboratory breeding colony of the lesser mouse lemur, in *Breeding Primates* (W.I.B. Beveridge, ed.), pp. 161–171, S. Karger, Basal.

Morrison, P., 1962, Thermoregulation in mammals from the tropics and from high altitudes, in: *Comparative Physiology of Temperature Regulation, Part 3.* (J. P. Hannon and E. Viereck, eds.), pp. 389–419, Arctic Aeromedical Laboratory, Fort Wainwright, Alaska.

Myers, R. D., 1971, Primates, in: *Comparative Physiology of Thermoregulation*, Vol. 2: *Mammals* (G. C. Whittow, ed.), pp. 283–326, Academic Press, New York.

Perret, M., 1972, Recherches sur les variations des glandes endocrines, et en particulier de l'hypophyse, au cours du cycle annuel, chez un Lémurien malgache *Microcebus murinus* (Miller, 1777), *Mammalia* 36(3):482–516.

Petter, J.-J., 1962, Recherches sur l'écologie et l'éthologie des lémuriens malgaches. *Mém. Mus. Natl. Hist. Nat. Paris Sér. A* 27:1–146.

Petter-Rousseaux, A., 1964, Reproductive physiology in the Lemuroidea, in: *Evolutionary and Genetic Biology of Primates*, Vol. 2 (J. Buettner-Janusch ed.), pp. 91–132, Academic Press, New York.

Petter-Rousseaux, A., 1970, Observations sur l'influence de la photopériode sur l'activité sexuelle chez *Microcebus murinus* (Miller, 1777) en captivité. *Ann. Biol. Anim. Bioch. Biophys.* 10(2):203–208.

Russell, R. J., in preparation, Time interval behavioral observations of prosimians.

von Weidholz, A., 1932, Bemerkungen zum Sommerschlaf der Zwerglemuren, *Zool. Gart. Leipzig (N.F.)* 5:282–285.

PART IV
BEHAVIOR AND ECOLOGY

Observations on Behavior and Ecology of *Phaner furcifer*

11

J.-J. PETTER, A. SCHILLING, AND G. PARIENTE

Phaner furcifer is found throughout the west of Madagascar, to the north of the Onilahy and Fiherenana rivers. In the east of the island, however, we have located it only in the region of Hiaraka in the Masoala peninsula. In the north, we have recorded its presence at an altitude of over 1000 m, near the forest station of Roussettes in the Montagne d'Ambre. The animals from the north and east are rather larger than those found in the west and should probably be regarded as constituting a distinct subspecies. The climatic and environmental ranges encompassed by the various populations of *P. furcifer* are thus so great that it is difficult to understand why the species is not found in areas such as Ankarafantsika in the western forest region, or why, in the east, its distribution extends no further south than Cape Masoala. On the other hand, its absence from the south of the island is presumably related to the great aridity of that region.

HABITAT

The animals were studied primarily in an area of the western forest some 60 km north of Morondava, near Analabe and close to the shores of the

J.-J. PETTER, A. SCHILLING, and G. PARIENTE   Laboratoire d'Ecologie Générale, Muséum National d'Histoire Naturelle, 91 Brunoy (Essonne), France.

Mozambique Channel. The study area is situated in some 8000 hectares of undisturbed primary forest bordering a sisal concession; it is near Lake Andranovala, which represents a local widening of the Kirindy River (Petter *et al.*, 1971, Fig. 4).

Climate

In spite of the proximity of the Mozambique Channel, this region is dry and hot, with rainfall concentrated in the period from December to March; between September and December, there is relatively severe desiccation. During the dry part of the year, there is no water supply to the numerous tributaries of the Kirindy, which rapidly dry up after March. During October 1970, this dryness enabled us to study in a zone which becomes partially flooded during the wet season, and which is strongly dissected by the courses of small streams flowing into the Kirindy. At this time, these stream beds were about to dry up but had not completely done so. The presence of standing water resulted in an enormous concentration of mosquitoes in the area.

The temperature of the forest reached 38°C toward midday, but fell to 19°C by 5 a.m. This drop of 19° caused significant nocturnal condensation. For the most part the sky was clear during our study, but from the beginning of November, large clouds accompanied by lightning began to form after 5 p.m. The first rain fell on November 8. Our study in this area was carried out during the nights and the coolest parts of the days between 24 October and 10 November.

Vegetation

The forest in which we observed *Phaner* is quite low, the large trees being no more than 10–15 m tall. Most of the trees lose their leaves from September to December, and observation of the animals is easiest during this period. The forest is characterized by the presence of numerous baobabs, around 12 per hectare, and by the dissemination of large trees throughout a primarily bushy vegetal cover.

These bushes, together with lianas, form a dense and almost impenetrable lower stratum. The baobabs (*Adansonia* sp.) are undoubtedly an important factor in the survival of the lemurs during the last part of the dry season, since the water they conserve allows an early reappearance of their leaves. Among the bushes, *Terminalia* and *Rhopalocarpus* spp. are much frequented by *Phaner*. They occur in densities of about 100 and 400 plus per hectare.

The forest surrounding Analabe is, in general, less degraded by man than that occurring farther to the south and north, and should be strictly protected. As in all of western Madagascar, there seems to have been progressive aridification over the last 50 yr. Certain large trees such as *Adansonia* and *Hernandia* sprout fewer and fewer new leaves, and the streams are dry for longer periods than formerly.

Fauna

In the protected study area, the diurnal lemurs *Propithecus verreauxi verreauxi* and *Lemur macaco rufus* (=*Lemur fulvus rufus*) occur in abundance; among the nocturnal forms, besides *Phaner* there are *Lepilemur mustelinus ruficaudatus, Microcebus coquereli*, the red and gray forms of *M. murinus*, and *Cheirogaleus medius*. The last two are rare; *P. furcifer* is by far the most abundant.

Other mammals include at least three rodent species; at least two insectivorans, *Tenrec ecaudatus* and *Setifer setosus*; two carnivorans, *Crytoprocta ferox* and *Mungotictis striata*, both very abundant; and numerous suids.

DAYTIME NESTS

Certain areas of the forest contain numerous spherical nests, 25–30 cm across. These are usually between 5 and 10 m above ground level and situated in a fork sheltered by tangled vines. These nests are made of some 30–50 twigs, intricately interwoven, each about 25 cm long and 1–4 mm in diameter at the cut ends. A round hole about 4–6 cm wide provides easy entry for the animal. Of 30 nests observed in the study area, 26 were empty, although some of these were fresh and in excellent condition. Of the remainder, two were occupied by *Phaner*, and the other two by *M. coquereli*. We have never observed these nests under construction, but their total absence in forests rich in *Phaner* but lacking *Microcebus* leads us to conclude that they are made by the latter, although they may be inhabited by both. In any event, no *Phaner* was observed to return regularly to a nest; one nest was inhabited by two individuals for several days, but during the 3 days following, it sheltered only one.

Given the small number of nests compared with the number of individuals of *Phaner*, it is evident that most *P. furcifer* do not rest in nests even where they are available. Two individuals were elsewhere observed in holes in dead trees; several others disappeared at daybreak into the broken branches of baobabs.

ACTIVITY RHYTHM

Phaner is extraordinarily vocal throughout its period of activity, and in general vocalizations are most frequent when the animals are most active (Petter *et al.*, 1971). Calling tends to diminish gradually during the night, but it never stops, and it is rare that 10 min passes during which the observer hears nothing. Any disturbance touches off a vocal reaction; thus certain prolonged series of calls warn of the passage of a boar or of the presence of a rapacious bird. During the period of our study, the earliest call regularly occurred at 6:20 p.m. (on only one occasion at 6:10 p.m.), but general calling did not begin until around 6:30 p.m.

The time when visible activity starts is similarly regular, and it has been possible to relate this to the intensity of ambient light (Pariente, 1971, quoted in Petter *et al.*, 1971). Activity begins when ambient luminosity falls below 8 lux. During the following 10 h, during which time light conditions within the forest are relatively stable, at about 300 μlux.

The time when activity ceases is less precise. We have never seen *Phaner* active beyond 5:20 a.m., when the light has reached a level of several lux. However, activity generally stops well before this, at around 4:20 a.m., when ambient luminosity is about 0.01 lux and after which time the light level increases most rapidly (see Fig. 9 in Petter *et al.*, 1971). On October 28, the sun first began to appear above the horizon at 5:25 a.m. and set at 6:11 p.m.; one can thus see that ambient light levels, rather than sunrise or sunset strictly defined, are the actual stimuli in the activity rhythm of *Phaner*.

All individuals leave their sleeping places with about the same rapidity; they abruptly emerge, run along a branch, jump into a tree, and from there, by several jumps combined with rapid running, cover about 30 m in a few seconds before emitting their first call. The vocalization appears to be elicited by the presence of another individual in the area.

Between 6:20 and 6:45 p.m., it is easy to see numerous *Phaner* silhouetted against the still-light sky. In the 15 min between 6:30 and 6:45 p.m., the animals seem to appear from everywhere and are very noisy.

As soon as they leave their sleeping places, the animals go directly to certain trees to feed and generally remain there for a few minutes before continuing. For about an hour, the *Phaner* move about a good deal; after that, movements diminish considerably, although we have observed *Phaner* actively feeding and moving at all hours of night.

LOCOMOTION

Phaner runs quadrupedally along branches and jumps from the end of one branch to that of another. In general, *Phaner* moves around at a height of between 3 and 4 m, the level at which most horizontal branches occur in this type of forest. However, we have observed individuals descending to the ground or moving in the larger trees at heights of above 10 m. Progression is rapid and consists of bouts of running interspersed with jumps. *Phaner* runs as well vertically as horizontally on lianas or small trunks, along which it descends head first like the other cheirogaleines.

Its great agility combined with a powerful spring permits *Phaner* to jump horizontally across distances of 4–5 m, but since the animal will often allow itself to fall at the same time, the total distance covered may be as much as 10 m. *Phaner* can move on the finest twigs, and does so notably when licking flowers. We have seen it adopt the most varied positions for feeding, sometimes even suspending itself from its hind feet.

Phaner possesses keeled, pointed nails as do some prosimians, which allow it to climb up large vertical tree trunks. This type of locomotion strongly distinguishes these forms from those primates which climb solely by gripping the support, and recalls that of the callithricid ceboids. It is interesting that it correlates with dietary regime in that these animals eat gums or parasitic exudations which are often found on trunks of large diameter.

AUDITORY COMMUNICATION

The call of *P. furcifer* is greatly variable in both intensity and tonality, but is highly distinctive. The basic vocalization is a sort of "Kia!" It may be low and single, rapidly repeated and sharper, or repeated regularly every few seconds for a very long time (we once counted a sequence of 1600 repeats).

The "Kia" seems to be a warning signal. It is preceded by soft grunts which may, with increasing volume, cover all the intermediate sounds between the low grunt and the typical "Kia." This cry is emitted each time an individual hears another in its vicinity. The second animal responds in kind, and there usually follows a prolonged duet. One of them generally calls more strongly than the other, with longer series of sharper, more closely spaced cries. After this, he will often physically pursue his antagonist. Sometimes a minor battle will ensue, accompanied by a series of short, sharp, rapidly repeated vocalizations, thus: "Csi-csi-csi-csi-csi." The intensity of these will decrease, and one of the opponents may let himself fall, to flee among the bushes. The winner then continues to emit several series of strong "Kia" calls, while the other, some distance away, may answer more feebly, vocalize irregularly over a long period, or not respond at all.

OLFACTORY COMMUNICATION

Male *Phaner* possess a large glandular zone situated below the neck (Rumpler and Andriamiandra, 1971). This bare round patch which may exceed 1 cm in diameter is easily visible by the light of a torch and often assists in distinguishing the sexes.

During the course of our observations, we have frequently seen an animal, head and neck stretched out, flattened against a branch and engaged in licking the bark. It is not impossible that in this attitude the individual also marks the zone being licked, which it is in an excellent position to do. Only once did we see a *Phaner* definitely rubbing its neck against a trunk; this was observed after the individual concerned had loudly and energetically chased another. While his victim fled among the low bushes, the pursuer rubbed his neck gland for several seconds against a slightly sloping trunk. Situated at a height of some 3 m and facing toward the ground, the animal had its four feet spread and its head stretched out, and its entire ventral surface seemed to rub the bark.

HOME RANGE

On several occasions, we have been able to observe an animal leaving its nest and traveling to its principal feeding places. Each time, the same itinerary was followed with only slight variations (Fig. 1).

It appears that the movements of *Phaner* are visually oriented, especially since their travel is so rapid. On the next to last day of our observations, we cut down a favorite feeding tree to obtain samples for analysis. As soon as night had fallen, we witnessed the arrival of two animals at this spot at an interval of 5 min. Running and jumping rapidly through the branches, they appeared completely disoriented at failing to find the familiar tree. They stayed immobile for at least a minute, looking at the place where the tree had stood; the first of them returned twice to search again before finally moving off towards other feeding places.

We were able to identify within the study area a large number of trees to which *Phaner* repeatedly returned to seek food. The living range in which we were best able to follow and observe the animals consisted of about a hectare of forest in which about a dozen individuals could be found at any given moment. The same zone may thus be exploited by several *Phaner*. Although on some occasions we saw two or three individuals feeding in the same tree, *Phaner* is most often solitary in its travels, and generally the "Kia" call will be emitted when another animal approaches within 10 m or so.

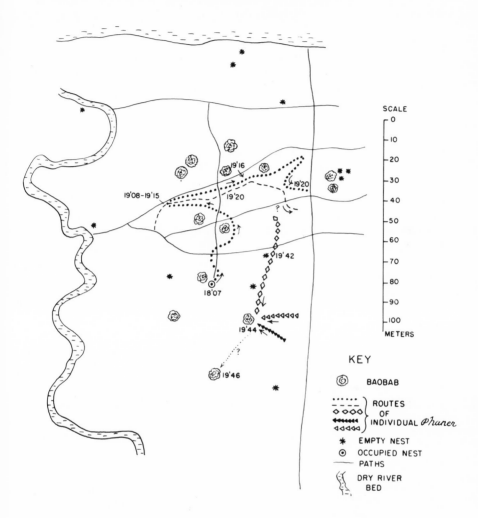

FIG. 1. Routes followed by several *Phaner* between 1800 and 2000 h on the night of 6 November, 1970.

Our observations are not detailed enough to permit us to say whether or not *Phaner* possesses a territory in the strict sense, or whether the vocal chases and combats we saw represent struggles between neighbors. Such battles are frequently observed; moreover, their traces are often visible, and several captured animals had badly torn ears.

POPULATION DENSITY

Because of the numerous possible sources of error in each approach, population counts were conducted both by direct visual observation and by locating the sources of vocalizations, triangulating with three observers. Vocalizations were elicited by various noises including whistles and calls replayed on a tape recorder; when the number and locations of the animals had been approximated in this way, the results could be compared with the direct observations. In an area of about 2 hectares we counted 17 *Phaner* to 2 *Lepilemur* and 1 *Microcebus coquereli*; this is equivalent to densities per km² of about 850 *Phaner*, 100 *Lepilemur*, and 50 *M. coquereli*. More detail on densities and distributions of these three species is given in Petter *et al.*, 1971 (see especially their Fig. 13 and Tables 1–3). When a greater variety of forest types than represented in our 2-hectare area is considered, densities are arrived at of about 680 *Phaner*, 210 *M. coquereli*, and 250 *Lepilemur* per km². For a number of reasons, however, these figures may be about 20–25% low, and it should be borne in mind that population density undoubtedly varies with the number of exploitable trees in a particular tract of forest.

A similar census carried out about 200 km farther south, on a tributary of the Mangoky (the forest of Antserananomby, see Sussman, this volume), yielded an average density of *Lepilemur* (260/km²) about the same as that recorded at Analabe, but the density of *Phaner* observed (550/km²) was slightly lower, although again, this count was probably lower than the actual total. It is clear that, while *Lepilemur* prefers primary forest, *P. furcifer* is more often found in areas of secondary forest, although it appears that this species does not occupy zones in which a continuous forest canopy is lacking.

DIET

During October and November, *Phaner* sometimes feeds on nectar, but subsists primarily on vegetable resins and on the secretions of insects. The different places at which the animals stopped during the course of a night's movement were essentially branches or trunks where such nourishment was available. The range of each animal contained numerous bushes of the species *Rhopalocarpus lucidus*, almost all of the fine branches of which were parasitized by final-stage larvae of homopteran insects, protected by coverings of snail's-shell form which were adherent to the twigs. From the openings of these exuded a syrupy liquid, on which *Phaner* spent much time feeding.

Terminalia sp., another abundant low tree, was also greatly frequented. This tree often possesses swellings on the trunk at the level of the lowest branches, which probably represent a reaction of the host to the attacks of coleopteran larvae. We have found such larvae, in their tunnels, eating away the dead wood on the outside of the swellings. From these exude thick translucent secretions which *Phaner* licks with passion. Often the animal will also dig with its incisor teeth at the abnormal bark to expose further reserves of the gum.

We have also observed *Phaner* attacking the healthy bark of young trees of this genus. They scrape it with their incisors and even bite it with their canines, holding their heads sideways. The sap which oozes out is then licked off. *Phaner* will also spend considerable time near the buds in the terminal branches of baobabs.

The trees we have mentioned are, by far, those most frequently visited by *Phaner*, but we have seen animals occupied for long periods in licking the flowers of *Crateva greveana*, one of the few trees flowering in our area at the beginning of November. Clinging to the finest terminal branches, they carefully and methodically lick the clusters of flowers. Presumably they visit other flowering trees in season.

Elsewhere (Petter *et al.*, 1971) we have considered the correlations between this remarkable diet and certain anatomical specializations displayed by *Phaner*. These latter include extensive adaptation of the incisors, premolars and molars; considerable development of the digestive tract; sharp, keeled nails; and the long, pointed tongue. The reader is referred to that report for an extended discussion.

REPRODUCTION

We were unable to obtain any direct information on reproduction in *Phaner* during our short study, but according to local informants, *Phaner* gives birth at around the time of the first rains, i.e., about November 15. Further, *P. furcifer* is said to bear a single young, which at first is sheltered in a hole in a tree, then is carried on the mother's stomach, and finally rides on her back.

DISCUSSION

Little direct interaction was observed during this study between the various nocturnal lemur species living in the study area. On one occasion, we saw a *Phaner* chase a *M. coquereli*, but usually individuals of the different species may be close together without showing any animosity. Neither have we noticed any evidence of competition between *Phaner* and *Lepilemur*. The latter, less numerous in our study area than in many other regions, must be limited there by the scarcity during the dry season of resources such as buds and young leaves. In contrast to its highly vocal congeners elsewhere in Madagascar, *Lepilemur* in this western part of the island is very discreet and calls no more than once or twice during the course of an entire night. In fact, as far as the general pattern of vocalizations throughout the forest is concerned, *Phaner* seems to play the role here which *Lepilemur* plays elsewhere on the island.

If any direct competition exists between *Phaner* and *Microcebus coquereli* or any of the other sympatric nocturnal lemurs, the only serious aspect probably lies in the exploitation of homopteran insects during the dry season. Both *P. furcifer* and *M. coquereli* may feed on the various gums and exudations, but only *Phaner* seems to be strongly adapted to this specialized regime; *M. coquereli* is forced during the dry season to adopt a mixed rather than a purely insectivorous diet, which may explain the large size of its home range (and therefore its low density). Conversely, it is certainly dietary specialization which permits *Phaner* to survive in such high density during a period of severe desiccation lasting from September to December.

ACKNOWLEDGMENTS

This study was carried out with the cooperation of the Ministry of Foreign Affairs and the Service des Eaux et Forêts of the Malagasy Government. Valuable assistance was rendered by the Zoological Laboratory of O.R.S.T.O.M., Tananarive. We also gratefully acknowledge the extensive and varied aid rendered to us by Monsieur de Heaulme, on whose property most of our study was carried out.

REFERENCES

Petter, J.-J., Schilling, A., and Pariente, G., 1971, Observations éco-éthologiques sur deux lémuriens malgaches nocturnes: *Phaner furcifer* et *Microcebus coquereli*, *Terre Vie* **25**(3):287–327.

Rumpler, Y., and Andriamiandra, A., 1971, Etude histologique des glandes de marquage de la face antérieure du cou des Lémuriens Malgaches, *C. R. Soc. Biol.* **165**:436–442.

Lemur catta: Ecology and Behavior

12

NORMAN BUDNITZ AND KATHRYN DAINIS

From April 1972 to May 1973, we studied *Lemur catta* in a gallery forest reserve in Berenty, Madagascar. We directed our observations toward two aspects of lemur behavior. One study dealt with feeding behavior and population dynamics. For this study we gathered data on feeding activities, made several censuses of the reserve, and analyzed the vegetation. Our information indicates that, although the size of the population did not change during the course of our study, the density of animals was different in various parts of the reserve. These differences seem to be correlated with differences in the distribution of resources.

Our second study dealt with the social behavior of *Lemur catta*. This study included a series of intensive observations on the role of the males in one lemur troop. Although our observations were concentrated on the males (for purposes of practicality), we were able to show that the adult females formed the core of the troop. Some of the males changed troops during the course of the year, thus effecting gene flow throughout the population.

GEOGRAPHICAL DISTRIBUTION

The geographical distribution of *Lemur catta* has been examined most recently and thoroughly by Sussman (1972). He found that *L. catta* was re-

NORMAN BUDNITZ and KATHRYN DAINIS   Department of Zoology, Duke University, Durham, North Carolina 27706

stricted to southern and southwestern Madagascar. This species lives in three basic habitats: continuous canopy forests, brush and scrub forests, and mixed forests. This last habitat occurs where a continuous canopy forest merges into brush and scrub habitat.

We did our field study of *Lemur catta* in a private reserve in Berenty in southern Madagascar. This reserve is a gallery forest which, under natural conditions, would merge into desert-like forest of the Didiereaceae. Gallery forests are most commonly found along rivers running through savannahs or semiarid regions (Richards, 1957). They derive their moisture from the water in the ground rather than from direct precipitation. The reserve is located along the Mandrare River in the semiarid southern sector of Madagascar. Generally, the primary vegetation for this sector is adapted to very dry conditions. The average rainfall in Berenty is about 500 mm per year, most of this precipitation coming in January, February, and March. In some years there is no rain at all. The vegetation is dominated by members of the endemic family Didiereaceae (in particular, various species of the genus *Alluaudia*). Associated plant species are often characterized by thorns, succulent leaves, odoriferous compounds, or latex, as in the families Asclepiadaceae and Euphorbiaceae.

The margins of the Mandrare River, however, can support a completely different type of vegetation because of the permanent supply of water in the soil. The Berenty reserve is comprised of four main types of vegetation. Close to the river is the continuous canopy forest dominated by *Tamarindus indica*, reaching 20–25 m in height, and other large trees such as *Celtis gomphophylla*. The character of this extremely rich forest changes somewhat as the distance from the river increases (and as the amount of water in the soil decreases). A large section of the reserve is characterized by this more open forest. Tamarinds are still found in large numbers, but the size of the trees is smaller, the average height dropping to about 16 m. The continuous canopy breaks down and there are large open spaces where thorny vines such as *Capparis sepiaria* grow into almost impenetrable tangles.

A third section of the reserve is what Sussman (1972) might call brush and scrub forest. This region is drier than the open forest and the trees are even more reduced in height. There are still tamarinds to be found, but most of the other plants in this area are quite different from either the closed canopy or open forests. Bushes and small trees (*Azima tetracantha, Maerua filiformis, Salvadora angustifolia*, and *Guisivianthe papionaea*) dominate, and there is no canopy at all. An indicator species for this region is a succulent vine, *Xerosicyos perrieri*, which is quite important to *Lemur catta* (see below).

The fourth section of the reserve is a small piece of the subdesert forest characterized by the Didiereaceae, Asclepiadaceae, and Euphorbiaceae mentioned above. This type of forest was probably the natural boundary of the gallery forest, but it has now been cleared and replaced by a plantation of sisal (*Agave rigida*). *Lemur catta* can be found in the first three regions of the reserve, the closed canopy and open forests and the brush and scrub, but we never observed them in the desert-like region.

USE OF THE HABITAT

Sussman used Richards' (1957) categories of forest stratification for analyzing the lemurs' use of the different forest levels. Table I presents Sussman's data (1972) and Jolly's data (1966) with data we collected for two troops of *Lemur catta* in Berenty. Our observations were made at 15-min intervals from dawn to dusk, approximately 0500–1830 h. During the first few minutes of each interval, we observed as many individuals as possible and recorded their position in the forest and their activity (moving, grooming, inactive, and so forth). We then summed all the observations and figured the percentages for each forest level. Short Tail's troop (Troop ST) lived in a section of the reserve which was dominated by the closed canopy and open forests. Dudley's troop (Troop DUD) lived in an area of brush and scrub and open forest. The troops studied by Jolly lived in the same piece of forest used by Troop ST. The two troops studied by Sussman (one in Berenty and one in Western Madagascar) lived in habitats which combine characteristics of all three types of vegetation. The data for Troops ST and DUD show some marked differences. Troop ST invariably spent most of their time in level 4. They seemed to spend more time in the lower levels in December when temperatures were higher. Troop DUD showed a similar tendency to spend most of their time in the upper levels during the cooler months of October and November and they moved down in December and January. This move to the lower levels was much more dramatic than the move by Troop ST. Troop DUD spent as much as 65% of their time in levels 1 and 2. This difference between the two troops is quite important because it reflects the differences in the habitats in which they lived. Troop ST could find shade in any level of the forest, but Troop DUD had to stay close to the ground under dense shrubs in order to escape the intense midday heat whenever they were in the brush and scrub part of their range.

Jolly's data are very similar to those of Troop ST as would be expected. Sussman's data, which look very similar to the data we collected for Troop DUD in December and January, are a little more difficult to reconcile because both of his troops had areas of closed canopy forest within their home ranges. However, both of these troops did make extensive use of areas of either brush and scrub or degraded vegetation during a period when temperatures were quite high. This could account for the high percentage of time these troops spent in the lower levels of the forest.

TABLE I. Percentage of Time Spent in Each Forest Level by Various Troops of *Lemur catta*, Including Dates and Temperatures

Level	Sussman AC-1 1970 Jul–Nov	BC-1 1970 Nov	Jolly 1963 Mar–Apr	1963 Jun–Sep	1964 Mar–Apr	Troop ST 1972 5 Oct	6 Oct	23 Oct	1 Nov	20 Nov	29 Nov	18 Dec	28 Dec	Troop DUD 1972–1973 16 Oct	19 Oct	7 Nov	15 Nov	5 Dec	14 Dec	2 Jan	13 Jan
1. Ground	30	36	15	20	15	14	12	9	18	4	4	11	16	16	9	11	17	34	17	36	19
2. Shrub (1–3m)	10	17	10	10	10	1	4	16	4	7	4	16	11	16	20	6	11	29	15	29	26
3. Small tree (3–7 m)	22	23	15	10	15	9	20	14	4	9	7	18	23	5	6	6	11	13	18	12	19
4. Canopy (5–15 m)	26	24	60	60	60	73	62	61	74	80	86	54	50	61	64	77	60	23	49	23	36
5. Emergent (15+ m)	12								Not significant at Berenty												
Minimum temperature (°C)	11.8–15.5	19.3				18	18	18	19	19	19	24	20	20	21	21	20	22	22	21	23
Maximum temperature (°C)	32.1–35.5	34.7				32	31	38	31	31	27	32	34	31	31	29	31	34	35	36	30

Sussman (1972) pointed out that *Lemur catta* spent 65–71% of its traveling time (movement of the troop from one location to another) on the ground. This is an important observation because it shows a marked difference from other lemurs. For example, Sussman found that *Lemur fulvus rufus*, which is sympatric with *L. catta* in southwestern Madagascar, spent less than 1% of its traveling time on the ground, but more than 80% in the closed canopy.

GROUP SIZE, COMPOSITION, AND POPULATION DYNAMICS

Table II shows the results of our censuses of the reserve. The "first census" includes our best first count of each troop. In five cases, this first count was made in May 1972; six other troops were first counted in August or September 1972; and one troop was seen for the first time in early December 1972. Therefore, this first census is spread over a period of about 6 months. Our last counts, which make up the "last census" of Table II, were made between late February and mid-May 1973.

Counting troops of *Lemur catta* is not an easy task. We always waited until the troop began to travel. If we were lucky, the animals would move in single file at some point and we could count and identify each animal as to sex and age class (adult, juvenile, infant). We also noted any animals we could recognize. At first we relied on obvious marks such as short tails, cropped ears, and so forth; but after several months, we could recognize about 60 animals by their facial features. We feel certain that we have counted every troop which has a home range within the boundaries of the reserve.

The average troop size of *Lemur catta* at Berenty ranged from about 12.8 animals in mid-1972 to 17.3 animals in late 1972 after the infants of the year were born. We found troops with as few as five and as many as 22 members. The group of five was indeed a troop, consisting of one male, three females, and one juvenile. We observed this troop on many occasions and it never lost its integrity. From September 1972 to February 1973, when the males were changing troops (see below), we saw solitary males and all-male groups of two, three, and four animals. However, we do not count these groups as troops, because they were never very long-lasting. These males were in the process of joining some other troop.

The area of the reserve at Berenty which we censused thoroughly was 94.2 hectares. Our first census of the adult and juvenile population was 153 animals in 12 troops. Between the end of August and the end of November 1972, the infants of that year were born. In all, we noted the birth of 47 infants, which would bring the population up to 200 animals. However, by May 1973 our census showed 152 *L. catta* in the same 94.2 hectares. Thus the pop-

TABLE II. Census Data for *Lemur catta* in Berenty

		Troop ST	Troop OE	Troop WT	Troop CE	Troop HO	Troop JL	Troop CEN	Troop DUD	Troop BR	Troop MI	Troop 5S	Troop BAT	Total
First census May–Dec. 1972	Adults	12	13	10	6	17	14	8	14	11	10	4	4	123
	Juveniles	2	4	3	3	3	3	3	4	0	2	1	2	30
	Infants	0	0	0	0	0	0	0	0	0	0	0	0	0
	Total	14	17	13	9	20	17	11	18	11	12	5	6	153
Last census Feb.–May, 1973	Adults	10	10	9	7	11	12	10	11	10	11	5	4	110
	Juveniles	2	0	2	2	1	2	0	2	0	0	1	2	14
	Infants	5	2	1	2	1	4	2	3	3	2	1	2	28
	Total	17	12	12	11	13	18	12	16	13	13	7	8	152

ulation of this reserve seems to be stable, at least according to these results. In our first census we counted 123 adults and 30 juveniles (no infants). In our last census we found 110 adults (a loss of 13) and 42 juveniles and infants (an increase of 12). These juveniles were actually subadults at this time, but for clarity we will continue to refer to them as juveniles. These data show that the mortality of infants and juveniles is quite high, but that once an animal reaches maturity its chances of surviving increase.

The actual causes of death of these animals are as yet unknown. We can only offer some anecdotal suggestions. No one has ever observed a single instance of predation on *Lemur catta*. During our observations the lemurs either paid no attention to snakes or watched them passively. Certain aerial predators—*Gymnogenys radiata, Milvus migrans*, and *Corvus alba* (two hawks and a crow)—evoked screams from the lemur troop when they flew close to the treetops, but we never saw any overt attempts at predation by these birds. Man, the worst enemy of the lemurs in most of Madagascar, was not a predator in Berenty. In November–December 1972, we noticed several lemurs bothered by bouts of sneezing or coughing which persisted for several days at a time. It is possible that this problem could prove fatal for some animals, but we did not know any of the afflicted individuals well enough to record their subsequent fates.

Perhaps the most likely cause of death for infants is falling out of trees. We observed many infants playing, wrestling, and generally learning how to maneuver in the arboreal environment. On numerous occasions, we saw infants fall from as high as 20 m. In every case, the baby hit the leaf litter of the forest floor, acted stunned, shook itself, then ran to the nearest tree and climbed to safety. We never saw a young lemur seriously hurt in a fall. Of course, we have no way of knowing what the long term effects of falling might be, but we never observed any immediate injuries.

The troop of five animals mentioned above increased to seven by the end of our study. This increase was due to the addition of one male and one infant during the male-exchange season (see below). The troop of 22 had only 12 members by the end of our study. They lost several males due to male-exchange and six out of eight infants and juveniles disappeared. This variability in troop size throughout the forest makes it difficult to compare our figures with the data collected by Jolly (1966, 1972), Klopfer and Jolly (1970), Sussman (1972), and Sussman and Richard (1974). These authors were able to census only two or three troops or did their censusing during the period of male-exchange when troop size was not stable. These factors led Jolly (1972) to hypothesize that the population of *Lemur catta* at Berenty was either increasing or facing resource shortages. She noted an increase in intratroop spacing behavior, peripheralization of subordinate males, and "time-plan" sharing of various resources by two or more troops. Our 14-month study at Berenty seems to indicate that the population was not really increasing. These behavioral changes most likely reflect seasonal changes in resource availability and changes in troop size due to births and male-exchange.

Our figures of population size differ markedly from those of Jolly (1966) and Sussman (1972). These authors extrapolated their estimates from studies of small areas of forest where the density of lemurs was relatively high. Jolly, for example, estimated that the Berenty reserve contained about twenty troops, or 350 animals. This would be a density of $350/km^2$. Sussman estimated densities in two forests to be $215/km^2$ and $250/km^2$. Our census is very much lower than these estimates. We found a density of only $167/km^2$.

This lower overall density is important, because it reflects a striking difference between the use of environment by lemurs living in very different parts of the forest. We studied the two troops mentioned above, Troops ST and DUD, with the hope of elucidating some of these differences. These two troops were of approximately equal size (about fifteen adults and juveniles) during the months of this part of our study. Troop ST, which lived in the closed canopy and open forest habitats of the reserve, had a home range of 8.1 hectares. Troop DUD, on the other hand, which lived in the brush and scrub and open forest habitats, had a home range of 23.1 hectares, an almost threefold increase. We also made estimates of home ranges for some of the other troops in the reserve. Although our data for these other troops are not as extensive as for Troops ST and DUD, we definitely did confirm the tendency for smaller home ranges in richer environments. One of these troops lived in the same area as one of Jolly's troops and our estimate of home range (6.0 hectares) agrees very nicely with hers (5.7 hectares). Sussman's estimates of 8.8 and 6.0 hectares for his two troops indicate that these lemurs were probably living in relatively rich environments.

The richness of environment discussed above will be dealt with in detail elsewhere (Budnitz, in preparation). In this chapter we will simply point out some important differences between these two habitats. For instance, in the closed canopy part of Troop ST's home range, over 50% of the ground was shaded by the crowns of tamarind trees, one of the lemurs' basic food resources. The highest density of tamarind cover in Troop DUD's home range was approximately 20%. The size of these trees also varied, Troop ST's tamarinds being quite a bit larger than those of Troop DUD. Troop ST had access to the river whenever they needed to drink, an important factor in the hot season when they went to the river twice a day. Troop DUD had no open water for drinking. They could lick dew in the morning and they could get at water that was caught in tree holes after a rain, but during long, dry spells they had to spend a great deal of time in the brush and scrub part of their range eating succulent plants. One of these plants, the cucurbit *Xerosicyos perrieri*, has round, succulent leaves which the lemurs ate regularly. Thus Troop DUD had to cover more ground in order to visit their less densely distributed food supply and to find sufficient water.

Charles-Dominique and Hladik (1971) estimated populations of *Lepilemur mustelinus leucopus*, a small, nocturnal lemur, in various parts of the reserve in Berenty. They found a density of 450 animals/km² in the gallery forest and 350/km² in the subdesert forest. In breaking down their figures for the gallery forest, they found a density of 810/km² along the river bank and only 270/km² in a drier part (which may correspond to our "open forest" habitat). Thus *L. m. leucopus* shows a similar tendency toward lower densities in drier regions. Of course, this animal is quite different from *Lemur catta*, so perhaps these data are not strictly comparable. Whereas Troop ST and DUD both rely heavily on tamarinds, as do the *L. m. leucopus* in the gallery forest, the lepilemurs in the subdesert eat mainly *Alluaudia* leaves. Also, lepilemurs are nocturnal and caecotrophic. This latter adaptation makes it possible for these animals to restrict their diet almost exclusively to leaves. Comparative data on *Propithecus verreauxi verreauxi*, the other diurnal lemur found in Berenty, would be most helpful, but no information is available at present.

SOCIAL BEHAVIOR

Most of our observations of social behavior in *Lemur catta* were made on Troop ST at Berenty between May 1972 and May 1973. This troop consisted of 14 individuals at the beginning of the study: 6 adult females, 4 adult males, 1 subadult female, 1 subadult male, 1 juvenile female, and 1 juvenile male. Five infants were born in August 1972. We considered this troop to be average in both size and composition.

Lemur catta troops exhibit a high incidence of intratroop aggression in the form of chasing, cuffing, and scent marking. These intense behavioral acts are almost always accompanied by a spat call from the defensive animal. Jolly (1966) reported agonistic interactions ranging from 6/h to a high of 22/h during the weeks preceding mating. She described an animal as being dominant if it "won consistently in spats and other aggressive encounters," and we have used this definition in determining status of individuals in Troop ST.

A *Lemur catta* troop is organized around a core group consisting of the adult females, their infants, younger juveniles, and, in some instances, the dominant male or males. Females are dominant over males and juveniles in all cases. During troop progressions, this core group moves first. The subordinate males (Jolly's "Drones Club") bring up the rear and often lag quite far behind. When the troop arrives at a feeding site, the core group selects the preferred places and the subordinate males feed on the periphery or wait in an adjoining tree until some of the others have moved out of the feeding tree. The same pattern is observed at drinking holes in trees. Sometimes a hole is licked dry before the entire troop has finished. Juveniles may move with the troop in either a front or rear position and are allowed to feed and drink near the females. The fact that males, in general, have access only to food and water which females and infants do not choose to utilize may be important in the more arid habitats where food and water may be limiting at times.

In most situations, a *Lemur catta* troop has no clear leader. They often progress in a loosely knit, wandering manner rather than in single file; the front position is usually occupied by more than one animal during the course of movement. While the lead position is often held by a female or dominant male, any animal may lead. In fact, the most subordinate male in Troop ST was sometimes seen at the front.

Although this lack of consistent leadership is true in general, there are situations in which the most dominant female assumes the role of leader. The only incidences where this was observed occurred when the troop deviated from its day-to-day routine—in movement toward an invading neighbor troop, or on excursions outside the territory. It might be that the dominant female controls daily progressions as well, but by less direct means that are not obvious to human observers.

Jolly found a fairly clear linear dominance order among female *Lemur catta* on the basis of 22 observed spats between individuals. We found a linear dominance hierarchy among the six females of Troop ST (see Table III). This was based on 252 agonistic encounters observed over a period of 1 yr. These encounters represent an average of 42 observed encounters per female with a range of 30–49. Thus each adult female seemed to be participating as a winner or a loser in approximately the same number of agonistic interactions. The female hierarchy was strictly adhered to and was stable throughout the year. All females were always dominant over all males. At no time was a female ever submissive toward an adult animal it had once dominated. There was no evidence of any adult female undergoing a change in status or leaving a troop to join another (in any of the eight troops that were well known). Based on our 1 yr of observations, it appears that a female spends her entire life with the troop into which she is born.

The general comportment of all females is similar, and it is difficult to identify the status of a female by her actions. All females directed agonistic behavior toward those animals of lesser status. The dominant females exhibited the same sorts of aggressive behavior as the more submissive females; only a female's selection of victims could be used to distinguish her. The status of a female was not easily distinguished by her friendly interactions, either. Our data indicate that a female may contact or groom any animal in the troop; there may be preferences in grooming partners, but there is not sufficient evidence to document this. Female spats are usually instantaneous squabbles over a specific action or object such as right of way or a tamarind

TABLE III. Agonistic Interactions in Short Tail's Troop, July 1972 to May 1973

| | Winners | | | | | | | | | | | | |
| | Females | | | | | | | Males | | | | |
Losers	MAD	5F	SL	FF	SH	W	BF	JL	LE	CT	CK	5M	JS
Females													
MAD	—												
5F	3	—		6									
SL	6	3	—										
FF	14		10	—			1						
SH	2	11	21	7	—								
W	15	7	8	12	1	—							
BF	4	5		2		5	—						
JL	2	7	3	3		3	1	—					
Males													
LE	5	15	7	10	5	10	1		—				
CT	6	4	4	4	9	4	2		30	—	6	16	
CK	3	1	2	1	6	1	1		18	20	—	16	
5M	11	3	4	5	9	5	4		26	13	2	—	
JS	3	2						3			1		—

pod. Within an hour of a spat, it is not uncommon to find the winner and the loser curled up together during a siesta.

Jolly found a clear dominance order among the males of her Troop 1 in 1964. The subordinate males were easily identified by their lowered head and tail carriage, their habit of lagging behind the troop, and their avoidance of other animals. The dominant males carried themselves erect and moved through the troop in a manner reminiscent of a swaggering neighborhood bully.

Males surpassed females in agonistic encounters in both amount and variety. In addition to chasing, cuffing, and so forth, males use a variety of different scent markings in their ritualized "stink fights" (Jolly, 1966). Males also tend to engage in more agonistic encounters than females, with a peak during the breeding season. Like the females, however, these encounters seldom involve physical contact and no animal has been observed being injured in a fight except during the breeding season.

In our observations of Troop ST, we confirmed Jolly's 1963–1964 observations on male comportment and position in the troop, although we did not find consistency throughout the year due to the pattern of male movement from troop to troop and the maturation of subadult males. In May 1972, Troop ST included four adult males (Dark Face, Light Eyes, Short Tail, and Crooked Tail) and one subadult male (Fifth Male) who was 1½ yr old. Limited data suggest that the four adults adhered to a linear dominance hierarchy (see Table IV). Dark Face, the most dominant, engaged in behavior patterns described by Jolly as those associated with dominant animals. He was allowed to approach and groom the females, although he was subordinate to all adult females. He often moved with the core group and could displace any other male with impunity. On one occasion we observed Dark Face break up a stink fight between Short Tail and the subadult Fifth, by strolling over and cuffing Short Tail away.

Territory was primarily defended by the females. In general, a territorial fight consisted of two opposing groups of females running at each other, making distinctive vocalizations, but rarely engaging in direct physical contact. These attacking groups were often led by a dominant female. Sometimes the juveniles would accompany the females, but they usually remained on the periphery as observers rather than as participants. The males were always last to arrive at the scene of the fight and often remained safely in the background. Occasionally a male of one troop would initiate a stink fight with a male of the opposing troop, but it was not obvious whether these males were defending territory or advertising their individual superiority.

Like the other males, Dark Face did not participate in the actual fighting but remained at the scene sniffing the battle grounds thoroughly for up to an hour after his troop had moved away. On returning to his troop, which may have moved some distance from him, Dark Face would characteristically move from female to female, sniffing and waving his tail, an aggressive visual and olfactory signal. Often he was cuffed away by a female but remained undeterred from his "rounds."

At the other extreme, Crooked Tail, the most subordinate animal, interacted very little with the rest of the troop. He usually moved at the end of the troop and was quick to move out of the way if another animal approached. He was often chased by others and spent most of his time alone or with Short

TABLE IV. Agonistic Interactions Between Males in Short Tail's Troop, 5 July to 18 October 1972

	Winners					
	DF	LE	ST	CT	5M	JS
Losers						
DF	—					
LE	1	—				
ST	3	4	—		2	
CT	1	5	1	—		
5M	5	1	3	1	—	
JS						—

Tail, the next lowest male in the hierarchy. The actual positions of Light Eyes and Short Tail were not clear, except that they were between Dark Face and Crooked Tail. The subadult male did not appear to have a consistent position in the troop. Some days he would comport himself like a dominant male, swaggering and occasionally winning in spats with other males. At other times he was submissive and would spat and retreat if another animal as much as looked at him. Most of the time he vacillated between these two extremes. At 2 yr of age, Fifth went through a period of several months when he and the subordinate males, Short Tail and Crooked Tail, engaged in long stink fights. These fights would continue off and on for up to a week, with no obvious winner. These took place until October, when a major shift in troop composition occurred.

In September and October 1972, we observed a period of general unrest in the forest at Berenty. This was the period just following the birth of most of the year's infants. Howling, a form of troop advertisement, generally given by a single male at dawn or dusk (Jolly, 1966), increased in both amount and intensity (that is, more troop members howling at one time and howling occurring throughout the day). Females often joined in these bouts of howling with barks and loud, wailing mews, the total effect being quite different from the vocalizations heard at any other time of year. Jolly (1972) described similar vocalizations in Berenty in October 1970. We observed single males and small groups of two, three, and four males throughout the forest, unassociated with any troop. Scent marking and male aggression increased during this time and intertroop encounters were common. Jolly also mentions these phenomena in her 1970 observations.

The first evidence of a change in male relationships in Troop ST occurred toward the end of September, when Dark Face and Short Tail began spending time with each other. While they had interacted little or not at all during previous observation periods, they now were often found mutual grooming or resting side by side. The more subordinate Short Tail was now able to initiate these activities with Dark Face, the dominant male, although he did so hesitantly and with defensive spat calls. By October 1, these two males began to spend several hours a day watching a neighboring troop, Troop WT. The following is an excerpt from our field notes describing this behavior:

> Dark Face and Short Tail approach Troop WT, leaving their own troop out of sight. They stop when Troop WT can be heard but not seen by us. They sniff the area thoroughly and do a lot of scent marking, interspersed with several periods of sitting and staring toward Troop WT. Troop ST mews in the distance and Dark Face answers. A male of Troop WT approaches to 5 m and marks at Dark Face and Short Tail. Dark Face and Short Tail retreat, but remain watching Troop WT.

After an hour of this behavior, Dark Face and Short Tail returned to their own troop. Dark Face remained the dominant male. He groomed the dominant female, broke up a fight between Fifth Male and Short Tail, and displaced several of the males.

Dark Face and Short Tail continued to behave in this manner, spending several hours a day watching the neighboring Troop WT. Then, after October 22, they were never again observed with Troop ST.

We followed Dark Face and Short Tail with their new troop. Troop WT had decreased from a total of five males in July 1972 to three males in September. We do not know what happened to these males. Shortly after Dark Face and Short Tail made their switch, another male disappeared from Troop WT, leaving only two males to oppose them as intruders.

For the first 2 wk after leaving their old troop, Dark Face and Short Tail spent most of their time together on the periphery of Troop WT. If they tried to approach Troop WT closer than about 20 m, they were vigorously repulsed by the two resident males. These two males were on constant guard against Dark Face and Short Tail and tried to drive them away with chasing and scent marking. The females of Troop WT seemed to ignore Dark Face and Short Tail most of the time, although they would join in the chasing if they were disturbed at a favored food source.

At the end of the first week with Troop WT, Dark Face and Short Tail confronted their old troop during a territorial dispute. Dark Face and Short Tail were driven from the area by Light Eyes, Crooked Tail, and Fifth Male. It was clear that Dark Face had lost his former status as dominant male and that neither Dark Face nor Short Tail would be accepted back into their old troop.

In Troop ST, Light Eyes became the dominant male. On November 21, 1 month after Dark Face and Short Tail had left, two foreign males were observed trying to enter Troop ST in the same manner as Dark Face and Short Tail had entered Troop WT. One of the foreign males disappeared in February 1973, but the other, Clark, remained a member of Troop ST until we left in June.

The males of Troop WT gradually relaxed their guard, and Dark Face and Short Tail were allowed closer to the troop center. By December 1972, Dark Face seemed to have regained some of his former status. He consistently won in spats with the less dominant of the two Troop WT males, and on February 12, we first observed him grooming a Troop WT female. Short Tail remained the most subordinate male. However, it was very clear that both animals had been accepted as regular troop members. This temporal pattern (ranging from September 1972 to February 1973) was observed in other troops and has particular significance because it begins soon after the infants are born and it ends just before the breeding season begins.

Assorted data indicate that the above occurrences are not isolated events. Troop changing by male *Lemur catta* was a widespread and probably annual phenomenon in the Berenty reserve. In each of the eight troops at Berenty that were well censused, we saw males change troops. Sussman, in November 1970, and Jolly, in September–October 1970, both made special note of the peripheralization of males in *Lemur catta* troops at Berenty. It is possible that they were observing foreign males trying to enter a new troop rather than resident troop members being driven away as they hypothesized. The evidence is not conclusive though, because long-term data for individual males are lacking from other sites and different years.

Evans and Goy (1968) observed captive *Lemur catta* and found that their animals were polyestrus if they were not impregnated the first time around. The average length of 17 cycles was 39.3 days, and nonpregnant females had two consecutive cycles (which included three estrus periods) each year. Thus the breeding season for an individual female could last up to 3 months. The dates of onset of estrus ranged from October to February (which would correspond to March to July in Madagascar, which is in the Southern Hemisphere). The average gestation period for eight full-term pregnancies was 134.5 days.

We missed most of the breeding season in Berenty in 1972, but we did record the appearance of infants from mid-August through early November. In Troop ST, the five infants were all born in August within 1 wk of each other. Most of the infants in the reserve were born during the month of September. The numbers of births in October and November were very much smaller, indicating that most of the females probably bred successfully during their first estrus.

The onset of the breeding season in 1973 was mid-April. The reserve was in an uproar, with lemurs howling and chasing and mating. Jolly (1966) described this behavior in detail. She noted that dominance status of the males did not affect their access to estrous females. We have confirmed her observations with the following clarification. The young adult males, that is, the males who are just reaching sexual maturity ($2\frac{1}{2}$ yr), only get to mate with females when the older males are occupied by some other activity. For example, Fifth Male mated once when the other males had fallen out of a tree while fighting over the estrous female. Fifth also mated on another occasion when two females were in estrus at the same time and the other males were occupied with only one of them.

The linear dominance hierarchy which was suggested by the data in Table IV broke down during the breeding season. Thus the data for the males in Table III, which includes the breeding season, does not show this hierarchy. The older adult males all seemed to have equal access to the females in estrus, including the males who had moved into the troops during the male-exchange period. Clark, the newcomer to Troop ST, was observed copulating on three different occasions. Crooked Tail was also observed mating three times, while Light Eyes, the dominant male after Dark Face left, was observed copulating only twice. On no occasion was a male seen mating with a female belonging to a troop other than his own. The male equality in copulation is therefore critical as a method of accomplishing gene flow throughout the population.

SUMMARY

The influence of vegetation, and in particular food resources, on the behavior of *Lemur catta* was studied from April 1972 through May 1973 in Berenty, Madagascar. Troops were observed in two habitats, one a resource-abundant habitat in a gallery forest and the other a scrubby transition zone between the forest and the desert vegetation where resources were more scattered. The availability of resources was shown to be an important factor in determining the density of *L. catta* in a given area. Population size remained constant through the year, although 47 infants were born during the study. A clear relationship was demonstrated between the sizes of lemur home ranges and the distribution of resources. Troops living in the poor environment required home ranges nearly three times as large as troops utilizing the rich environment.

The social organization of *Lemur catta* troops is centered around a core group of females and their infants and young juveniles. Females appear to remain in the troop of their birth and are dominant over all males without exception. Defense of territory seems to be primarily a female responsibility. Males, in general, were peripheral to the core group and seldom entered into territorial disputes. They did not necessarily remain with one troop their entire lives. Male-exchange was observed in every troop censused in Berenty. This male-exchange provides an effective method for achieving gene flow in a species whose members probably breed only within their own territories.

ACKNOWLEDGMENTS

We gratefully acknowledge the de Heaulme family for helping to make our work in Berenty such a meaningful and pleasant experience. We also thank the government of the Malagasy Republic and all the officials who helped us. To Dr. Peter H. Klopfer and Dr. John Buettner-Janusch, we give special thanks for their moral support and intellectual guidance. This work was supported in part by National Institute of Mental Health Fellowship No. 5 FO1 MH46274-02 and by Duke University Biomedical Sciences Support Grant No. 5S05-RR-07070-08.

REFERENCES

Budnitz, N. A., in preparation, Ph.D. Thesis, Duke University, Durham, N.C.

Charles-Dominique, P., and Hladik, C. M., 1971, Le *Lepilemur* du sud de Madagascar: Ecologie, alimentation et vie sociale, *Terre Vie* **25**:3-66.

Evans, C. S., and Goy, R. W., 1968, Social behavior and reproductive cycles in captive ring-tailed lemurs (*Lemur catta*), *J. Zool. (Lond.)* **156**:181-197.

Jolly, A., 1966, *Lemur Behavior*, University of Chicago Press, Chicago.

Jolly, A., 1972, Troop continuity and troop spacing in *Propithecus verreauxi* and *Lemur catta* at Berenty (Madagascar), *Folia Primatol.* **17**:335-362.

Klopfer, P. H., and Jolly, A., 1970, The stability of territorial boundaries in a lemur troop, *Folia Primatol.* **12**:199-208.

Richards, P. W., 1957, *The Tropical Rain Forest*, 4th ed., Cambridge University Press, Cambridge.

Sussman, R. W., 1972, An Ecological Study of Two Madagascan Primates: *Lemur fulvus rufus* Audebert and *Lemur catta* Linnaeus, Ph.D. dissertation, Duke University, Durham, N.C.

Sussman, R. W., and Richard, A., 1974, The role of aggression among diurnal prosimians, in: *Primate Aggression, Territoriality, and Xenophobia* (R. L. Holloway, ed), pp. 49-76 Academic Press, New York.

A Preliminary Study of the   13
Behavior and Ecology of
Lemur fulvus rufus
Audebert 1800

ROBERT W. SUSSMAN

This is a preliminary report on the ecology and behavior of *Lemur fulvus rufus* (Fig. 1A, B). The brown lemur, *Lemur fulvus*, is almost exclusively an arboreal quadruped, and in general behavior and morphology resembles arboreal monkeys more than any other prosimian. Although subspecies of *L. fulvus* are among the most widely distributed and abundant of the Madagascar lemurs, there have been no published reports on long-term field studies of this species. Information for this study was collected during an 18-month field study (September 1969–January 1971) in the southwest of Madagascar. The focus of the study was to compare sympatric populations of *L. f. rufus* and *Lemur catta*, and a number of papers on that aspect of the research have been completed (Sussman, 1972, 1973, 1974).

    The taxonomy of *Lemur fulvus* is a matter of controversy (Schwarz, 1936; Buettner-Janusch, 1963; Petter, 1965; Buettner-Janusch *et al.*, 1966; Nute and Buettner-Janusch, 1969; Petter, 1969; Albignac *et al.*, 1971; Rum-

ROBERT W. SUSSMAN   Department of Anthropology, Washington University, St. Louis, Missouri 63130

FIG. 1A. Male *Lemur fulvus rufus.*

FIG. 1B. Female *Lemur fulvus rufus.*

pler, Chapter 3). A number of authors consider *L. fulvus* and *Lemur macaco* to be populations of a single species that exhibits continuous morphological variation over a wide geographical range. However, cytological research and backcrossing experiments suggest the maintenance of *L. macaco* as a separate species (Rumpler, Chapter 3). Furthermore, very little is known about the geographical distribution and possible interactions between populations of these two groups in their natural habitat. Populations of *L. fulvus* are found in all forested areas around the coast of Madagascar except in the extreme south where semiarid conditions and desert-like vegetation occur (Fig. 2). The species is divided into seven subspecies: *Lemur fulvus fulvus*, *L. f. rufus*, *L. f. flavifrons*, *L. f. albifrons*, *L. f. sanfordi*, *L. f. collaris*, and *L. f. mayottensis* (found only on the Comoro Islands). *L. f. rufus* is the most widely distributed of the subspecies, found in the forests throughout the west coast of the island (Fig. 3).

STUDY AREAS

Groups of *Lemur fulvus rufus* were studied intensively in three forests near the Mangoky River (approximately 21°46' south latitude, 44°7' east longitude): Antserananomby, Tongobato, and a forest on the southern bank of the Mangoky (Fig. 4). These forests are located within 15 km of each other and were probably once part of a continuous forest range. The flora and structure of the forests are very similar. All three are primary deciduous forests made up of a *Tamarindus indica* (kily) consocation. The kily trees make up a closed canopy about 7–15 m high. There is a scattered, discontinuous emergent layer consisting of *Acacia rovumae*, *Ficus soroceoides*, *Terminalia mantaly*, and other species of tall trees. These scattered tall trees or stands of trees rise above the canopy layer and are about 15–20 m in height. A subordinate tree layer, 3–7 m high, consists of certain species of smaller trees (e.g., *Tisomia* sp., *Flacourtia ramontchi*, *Poupartia caffra*) and the saplings of the taller trees. The emergent and subordinate layers are thus made up of rarely occurring species which are scattered throughout the forest but occur more frequently on the forest edge. Throughout the forest, trees are generally spaced about 5 m apart, making it easy to move about below the canopy layer.

The mean annual rainfall in the southwest of Madagascar is around 700 mm. However, the rainfall is seasonal, generally falling from December to March, with little or no rain between April and November. In the rainy season, access to the forests is usually difficult or impossible due to the condition of the roads during this period. Maximum and minimum temperatures recorded in the study area were 40°C and 10°C.

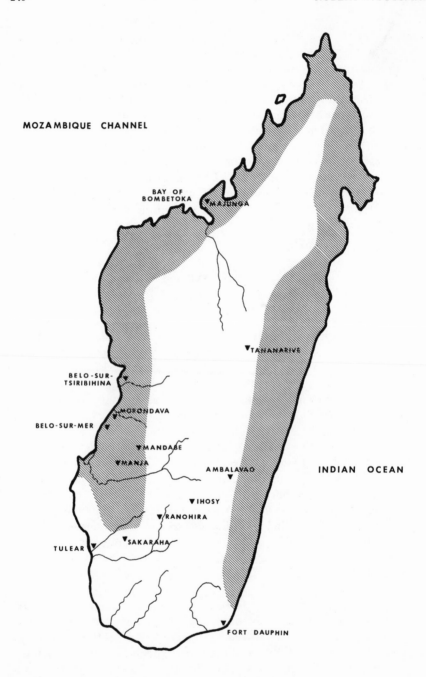

FIG. 2. Distribution of *Lemur fulvus*.

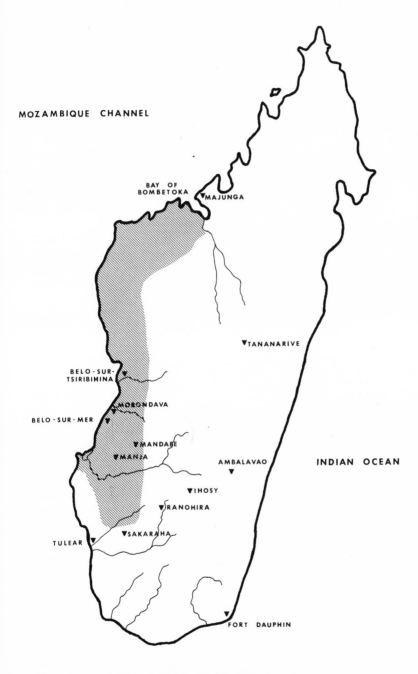

FIG. 3. Distribution of *Lemur fulvus rufus*.

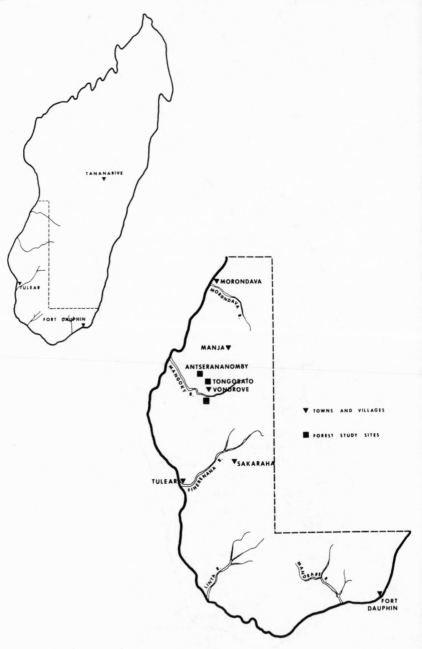

FIG. 4. Study sites.

The three study forests differ mainly in the configuration of the surrounding terrain. The forest at Tongobato is surrounded by cultivated fields and degraded vegetation. The forest at Antserananomby is bordered on the east by the Bengily River, a tributary of the Mangoky River. The Bengily River contains no water during the dry season. To the north and northwest of this forest, the continuous, closed canopy is replaced by primary brush and scrub vegetation. The study forest on the southern bank of the Mangoky River is a narrow strip of gallery forest about 50–60 m wide which runs along the bottom of a steep hill that parallels the river.

Lemur catta and *Lemur fulvus rufus* coexist in two of the forests (Antserananomby and the forest on the bank of the Mangoky River). *L. catta* is not found in the forest at Tongobato. *Propithecus verreauxi verreauxi* inhabits all three forests. Four nocturnal primates are found in the forests of this region: *Lepilemur mustelinus, Phaner furcifer, Microcebus murinus,* and *Cheirogaleus medius.* Many of the same genera of mammals, birds, and reptiles are found in all three forests. Although no incidents of predation were seen, a viverrid, *Cryptoprocta ferox,* and two hawks, *Gymnogenys radiata* and *Buteo brachypterus,* are common to the area.

METHODS OF OBSERVATION

Intensive study of *Lemur fulvus rufus* was carried out between October 1969 and November 1970. However, very few observations were made during the rainy season. The animals were observed for a total of about 400 h. Most observations were made at Antserananomby (about 240 h) and Tongobato (about 120 h). A survey of the distribution of this subspecies was also carried out in the southwest of Madagascar, and another survey of the distribution of *L. f. rufus* and *L. f. fulvus* was done with Ian Tattersall in the northwest of the island in July and August 1973.

During the intensive study, it was difficult to follow *Lemur fulvus rufus* throughout the day. *L. f. rufus* is almost totally arboreal, and it was often not possible to observe the animals continuously. The ease with which the animals became habituated to the observer was dependent on the amount of hunting done in the forest, the number of hours a particular group had been observed, and how recently it had been observed. At Tongobato, the animals were hunted and never became as well accustomed to me as did those at the other forests. At Antserananomby and the forest bordering the Mangoky River, the animals quickly became used to my presence and were approached easily. Once animals were accustomed to the observer, the tolerance distance was well under 10 m, and in many instances they came as close as 1 m. In these forests, it was often possible to obtain as many as 10–12 h of more or less continuous observation, although not always on the same group.

Observations on the animals were recorded in a notebook in journal form and on prepared data sheets. Individual Activity Records (Crook and Aldrich-Blake, 1968) were kept on animals every 5 min. Quantitative data were collected on six categories of activity (feeding, resting, grooming, movement, travel, and other) and on the particular stratum of the forest in which the activity took place. Statistical comparisons could then be made between the quantitative data collected on *Lemur fulvus rufus* and *Lemur catta* and on the populations of *L. f. rufus* in different forests. The methods of data collection and statistical analysis are described in more detail in Sussman (1972) and Sussman *et al.* (in preparation).

GROUP SIZE AND COMPOSITION

Groups were counted throughout the study period at each of the forests, and census data on each group were checked and rechecked. At all three forests, census data were collected by more than one observer.[1] In most cases, two observers worked together. Although *Lemur fulvus rufus* is an arboreal primate, a number of factors alleviated the usual difficulties which are encountered in following and counting arboreal primates: groups are quite cohesive, individuals usually remain in close proximity to each other, and the entire group travels together, seldom dividing into subgroups. Furthermore, the high density of the population and the small size of the home and day ranges (see below) facilitate the collection of census data, as does the fact that there are distinct differences between the sexes.

There are conspicuous differences in coloration between the sexes. The male is gray and has a pronounced white face mask with a bright red-orange tuft of hair on the top of the head, whereas the female is red-orange and has white patches over the eyes and a black bar across the forehead and between the eyes (Fig. 1.). These distinctions are obvious even in juvenile animals, although it is not always possible to determine the sex of infants.

Each individual was placed into one of three age categories: infant, juvenile, or adult. Births occur once a year, usually within a 2-wk period. The young all mature at approximately the same rate and the achievement of different stages of maturity is synchronous. Animals that were still being carried by their mothers were classified as infants. This included animals up to about 4 months of age. Animals that were not being carried but had not reached full size were considered juveniles. In the field, however, it is often difficult to distinguish older juveniles from adults. Adult size is attained at about 2 yr of age.

---

[1] I was assisted in collecting census data by Linda Sussman, Alain Schilling, Folo Emmanuel, and Bernard Tsiefatao.

Table I includes the census data for 18 groups of *Lemur fulvus rufus* in the three forests. Groups ranged in size from 4 to 17 animals. The average size of all groups censused was 9.4 animals, with the average group containing 3.4 adult males, 4.2 adult females, and 1.7 juveniles and infants. Thus the adult sex ratio was 1 male to 1.22 females. The ratio of male to female juveniles also approached a 1:1 ratio. The ratio of infants to adult females at Tongobato was 1:2.4. At Antserananomby, the ratio of juveniles (mainly 1-yr-olds) to adult females was 1:2.8. There were no infants at the time that census data were collected at Antserananomby.

TABLE 1. Composition of Groups of *Lemur fulvus rufus*

| Name of group | Adult | | Juvenile | | Infant | Total |
	Male	Female	Male	Female		
Antserananomby[a]						
AF-1	4	6	1	1	0	12
AF-2	4	5	1	0	0	10
AF-3	4	3	0	1	0	8
AF-4	2	5	2	1	0	10
AF-5	4	6	0	2	0	12
AF-6	3	2	0	0	0	5
AF-7	3	4	1	1	0	9
AF-8	2	2	0	0	0	4
AF-9	2	2	1	0	0	5
AF-10	4	5	2	0	0	11
AF-11	5	7	1	2	0	15
AF-12	4	4	1	0	0	9
Totals	41	51	10	8	0	110[a]
Means	3.42	4.25	0.83	0.66	0	9.17
Tongobato						
TF-1	5	8	0	0	4	17
TF-2	2	3	1	0	1	7
TF-3	4	5	0	1	2	12
TF-4	3	4	0	0	1	8
TF-5	3	2	0	1	1	7
Totals	17	22	1	2	9	51
Means	3.40	4.40	0.20	0.40	1.80	10.20
Mangoky						
MF-1[b](June 1970)	4	3	1	0	0	8
Totals	4	3	1	0	0	8
Means	4	3	1	0	0	8
Overall Totals	62	76	12	10	9	169
Overall Means	3.44	4.22	0.66	0.55	0.50	9.38

[a]There was also an extra male group consisting of two individuals. Therefore, the total number of animals censused at Antserananomby was 112.

[b]This group was also censused in November 1969 and contained the same individuals as shown here. The juvenile, however, was still an infant at the time.

Groups in which I could recognize individuals remained stable through-out the study. The group studied on the banks of the Mangoky was censused first in November 1969 and again in June 1970, and contained the same indi-viduals during both counts. It is likely, however, that groups divide when they reach about 15–17 animals. At Tongobato, a group of 17 animals, group TF-1, was seen early in the study. This group consisted of 13 adults and 4 infants. The group was not seen later in the study, and it is likely that it split after the infants began to move independently of their mothers. At Antserananomby there was one group of 15 animals. This group was actually in the process of dividing. The whole group usually slept together, but often formed into sub-groups that fed or rested in separate areas during the day. The subgroups, however, had not stabilized, and different individuals associated with each other on different days.

DAILY CYCLE OF ACTIVITY

Figure 5 represents the mean percentage of animals engaged in each of the six recorded activities during periods of several hours throughout the day (from 0600 h to 1825 h) at Antserananomby and Tongobato. Animals were active very early in the morning and late in the afternoon. There were two peaks of feeding, one between 0600 h and 0930 h and one between 1700 h and 1825 h. *Lemur fulvus rufus* rested for 60% or more of the time between 1000 h and 1600 h (6 h) at Antserananomby and between 0930 h and 1500 h (5.5 h) at Tongobato. The activity to rest ratios in the two forests were 0.79 (44/56) at Antserananomby and 1.00 (50/50) at Tongobato.

Generally, the activity patterns of the animals at both of these forests were similar with two exceptions: there was a higher percentage of "other" activities during the first period of the day at Antserananomby, and a higher percentage of movement and travel throughout the day at Tongobato. At Antserananomby, the temperature at night was lower than at Tongobato; the mean minimum temperatures at the two forests during the study were 13.7°C and 19.3°C, respectively. The animals at Antserananomby sunned in the morning—they sat in unshaded branches of the trees with their arms out-stretched and their ventral surfaces facing the sun. At Tongobato, the high proportion of movement and travel probably resulted from the difficulty in habituating the animals to the observer because the animals were often hunted in this forest.

The following is an example of the activities on a typical day at Antse-rananomby. The animals awake at around 0600 h, sometimes a bit earlier de-pending on the temperature. They sleep later on very chilly mornings, and in clumps of two to five animals on cold nights. After waking, the group travels

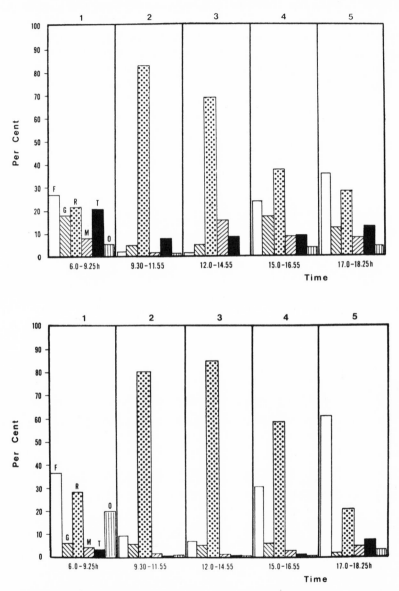

FIG. 5. Mean percentage of animals engaged in each of the six recorded activities at Tongobato (above) and Antseranomby (below). F, Feeding; G, grooming; R, resting; M, moving; T, travel; O, other.

from the kily in which it slept, at the edge of the forest, into the mid-forest to sun in barer and taller trees (between 0630 and 0730 h). The animals feed both before and after traveling, and again after sunning. The group then travels back to a large, vine-covered kily tree about 20 m from the night sleeping site and begins its afternoon rest (usually between 0930 and 1030 h). During the afternoon, the animals usually rest alone or in pairs. The group awakes between approximately 1530 and 1630 h, feeds in the kily, and then moves slowly to feed in a tall *Acacia* and then in another kily on the border of the dry river bed. Just before dark it returns to the *Acacia* and feeds in this tree until well after dark, finally settling high in the branches of the tree for the night. In a day similar to the one just described, the group moved a total of 145 m, the longest single movement being 60 m.

UTILIZATION OF HABITAT

Group Spacing, Home Ranges, and Population Density

The home ranges of groups were very small and overlapped extensively (Figs. 6 and 7). Home ranges were determined by plotting the location of known groups on a prepared map whenever the group was sighted. At Tongobato and on the banks of the Mangoky, the average size of home ranges was 1.0 hectare; at Antseranañomby, it was 0.75 hectare. In the latter forest, 12 groups (112 animals) occupied an area of about 9 hectare, and groups were never seen outside of the portion of the forest which contained a continuous canopy. The population density at Antseranañomby was around 1200 animals/p km², and at Tongobato it was around 900/p km². The biomass of *Lemur fulvus rufus* was roughly 25 kg/hectare at Antseranañomby and 20 kg/hectare at Tongobato.

The high population density, the small home ranges of groups, and the fact that the boundaries of ranges overlap extensively keep groups of *Lemur fulvus rufus* in close proximity to each other. There were many encounters between groups. Agonistic encounters often developed over a particular feeding site or when one group surprised another as it entered a tree in which the first was feeding or resting. However, as far as I could tell, there were no disputes in which particular territorial boundaries were defended. The normal pattern of group travel and spacing throughout the year is related to seasonal changes in the flora. This will be discussed in the section on intergroup encounters.

Groups were very cohesive; members usually remained within the same tree and, for the most part, within visual contact. When groups traveled, the animals usually exchanged a particular vocalization—a low, gargling, gutteral grunt which might be called a "progression grunt" (cf. Struhsaker, 1967).

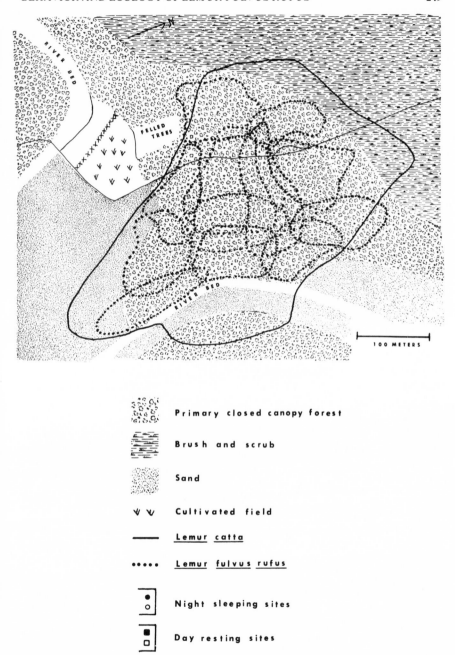

Primary closed canopy forest

Brush and scrub

Sand

∀ ∀ Cultivated field

——— Lemur catta

••••• Lemur fulvus rufus

Night sleeping sites

Day resting sites

FIG. 6. Antseranomby: home ranges of *Lemur catta* and *Lemur fulvus rufus*. One group of *L. catta* and 12 of *L. f. rufus* are represented.

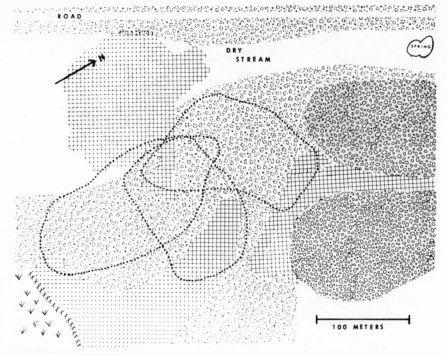

FIG. 7. Tongobato: home ranges of three groups of *Lemur fulvus rufus*.

Groups moved little within the period of a day. It was difficult to follow a group throughout the whole day, but I was able to follow three groups for 2 consecutive days each. The average distance groups traveled on these days was between 125 and 150 m. However, in one instance at Tongobato, I followed an unhabituated group for 1200 m as it attempted to get away from me. In similar circumstances, a group at Antserananomby circled the entire forest, ending up where it began. In both cases, the groups passed through the home ranges of a number of other groups.

Utilization of Forest Strata

Lemur fulvus rufus spent most of its time in the continuous canopy of the forest. The mean percentage of animals observed at each of five strata of the forest was calculated.[2] The animals spent over 70% of their time in the

---

[2] The five strata were as follows: level 1, the ground; level 2, small bushes; level 3, the subordinate tree layer; level 4, the continuous canopy layer; level 5, the emergent layer.

continuous canopy layer (level 4) and over 95% in the three highest forest strata (levels 3, 4, and 5). Animals were seen on the ground in less than 2% of the observations. The use of the vertical strata was similar in all three forests.

The animals were most specific in the choice of strata used for group travel. Over 80% of this activity took place in the continuous canopy. When groups moved horizontally through the trees, the animals ran along the fine end branches of the larger trees. These branches were generally horizontal and formed a contiguous series of pathways throughout the closed canopy. The animals would jump from one tree to another only if there was a gap in the fine series of branches. Approximately 80% of resting and grooming activities also took place in the closed canopy, but these activities usually occurred in the larger branches and crotches of larger branches of the tall trees. Over 50% of feeding took place in the continuous canopy and over 90% in levels 3, 4, and 5.

Diet

Lemur fulvus rufus appears specialized in its choice of diet, especially during the dry season. At Tongobato, where the animals were studied immediately before and after the peak of the rainy season, *Tamarindus indica* accounted for about 50% of their diet. At Antserananomby, during the dry season, it accounted for over 85%. Animals fed on the leaves, pods, stems, flowers, sap, and bark of this tree. At Tongobato, the leaves of the kily alone made up 42% of the diet, and at Antserananomby, kily leaves accounted for over 75% of the diet.

The amount of fruit, flowers, and young leaves eaten depended on the season and the distribution of trees other then the dominant kily tree within the forest. Groups supplemented the predominantly kily diet with parts of those trees which happened to be within their small home ranges. At both Tongobato and Antserananomby, over 80% of the diet was made up of only three species of trees (*Flacourtia ramontchi, Tamarindus indica,* and *Terminalia mantaly* at Tongobato, and *Acacia* sp., *Ficus soroceoides*, and *Tamarindus indica* at Antserananomby). All the species of plants on which the animals were observed to feed are listed in Table II.

During the entire study, the animals were seen to eat a total of only 15 different species of plants. A few species of trees made up a very large portion of the diet, and kily leaves were the main staple. However, *Lemur fulvus rufus* is found in forests which are not dominated by kily trees, and thus is not specifically dependent on the products of this species.

The animals fed most frequently in the finest, terminal branches on the top or edges of the tall trees. They would move out to the fine net of tiny branches and then stand or hang on with three extremities while reaching a

TABLE II. Plants Eaten by *Lemur fulvus rufus* at All Three Forests[a]

Plant species	Part eaten	Forests
Acacia rovumae Olia	Leaves, sap	A, T
Acacia sp.	Leaves	A, T
Alchornea sp.	Leaves, buds	T
Ficus cocculifolia	Fruit	M
Ficus soroceoides Bak	Fruit, leaves	A, M
Flacourtia ramontchi	Fruit, leaves	T
Grevia sp.	Fruit, leaves, stems, buds	M
Lawsonia alba Lamk	Leaves	T
Papilionaceae	Leaves	A
Quisivianthe papinae	Leaves, flowers	A
Rinorea greveana H. Bn.	Fruit	A, M
Tamarindus indica	Fruit, leaves, stems, bark, sap, flowers	A, T, M
Terminalia mantaly	Leaves, buds, sap	A, T
Tisomia sp.	Leaves	A
Vitex beravensis	Leaves, fruit	A, T

[a]A, Antserananomby; T, Tongobato; M, forest on the southern bank of the Mangoky River.

leaf, fruit, or branch and pulling it toward them. The animal would then pull the edible part to its mouth and pick it off with a jerk of the head. When feeding on fruits, the animals tilted their heads back while chewing. Large fruits were held in the hands, squirrel fashion, and chewed.

Water was obtained by licking the dew off leaves early in the morning, and from that contained in the fruits and leaves the animals ate. An analysis of the nutritional composition of kily leaves shows that they contain approximately 70% water (Charles-Dominique and Hladik, 1971). The animals also obtained water by lapping it directly from hollows in trees in which rainwater had collected.

INTERACTIONS BETWEEN ANIMALS

Intragroup Encounters

Agonistic encounters between members of the same group were infrequent, approximately 0.10 per hour, and usually occurred when an animal approached, tried to sit in contact with, or groom another. A few occurred while animals were attempting to gain access to waterholes in trees. During agonistic encounters, one animal would simply lunge at or attempt to grab or slap the other, and many ended with a brief chase. Most encounters were quite mild, seldom involving more than momentary physical contact. Sometimes a grunt accompanied a threat, and in the most intense encounters, high squeaks were given. I could not discern any dominance hierarchy in groups of *Lemur fulvus rufus*.

Intergroup Encounters

Encounters between groups of *Lemur fulvus rufus* have been described in detail in Sussman and Richard (1974), and only a brief summary will be given here. The home ranges of groups overlap extensively and the borders are not defended; they are very small. In most cases, groups probably maintained spatial separation by means of frequent vocalizations (grunts grading into high-pitched rasps) which were given spontaneously or when groups were moving. These vocalizations were exchanged by neighboring groups. When encounters did occur, they most often developed when groups were moving or when one group surprised another as it entered a tree in which the first group was feeding or resting.

The frequency of group encounters depended on the season and was highest when new leaves and fruit were forming on the kily trees. At these times, the groups would utilize those portions of their home ranges which were closest to the center of the forest where the density of the kily trees was highest. At other times of the year, the groups utilized the portions of their home ranges near the edge of the forest and were thus usually farther apart. However, they did come together to feed in particular species of trees which occur rarely throughout the forest and in which there was ripe fruit.

This pattern of group spacing allowed groups to exploit the evenly distributed products of the kily trees. It also allowed a number of groups to take advantage of the seasonal products of rarely occurring species of trees. Group encounters do not seem to be important in maintaining group boundaries, but do appear to be important for the maintenance of group integrity.

Reactions to Other Nonhuman Primates

There were no aggressive encounters between groups of *Lemur fulvus rufus* and those of *Lemur catta* or *Propithecus verreauxi*. These species frequently fed or rested in the same trees, and individuals of different species often mixed freely. In most cases, however, they ignored each other. On many mornings, *L. catta* would sun in the highest branches of a large kily tree while a group of *L. f. rufus* was feeding in the branches below. On a few occasions, a *L. f. rufus* entered into a quick agonistic interaction with a *L. catta*. However, no contact between animals was seen, and no species consistently displaced the other. The behavioral repertoire during these encounters was the same as during those between conspecifics. In a few instances, a large *P. verreauxi* would jump into a tree and shake the branches containing some *L. f. rufus.* This would usually cause the smaller animals to look up or, depending on the violence of the shaking, to move to another branch or tree. No grooming was seen between different species of lemurs.

All three diurnal species would react to the warning signals of one another and draw attention to the presence of human or nonhuman predators. Each has a specific call to ground and to aerial predators.

I never saw an interaction between *Lemur fulvus rufus* and any of the nocturnal species of lemur.

Reactions to Predators

The most dangerous predator on populations of *Lemur fulvus rufus* is man. If the animals were approached by a man to whom they were not accustomed, they would face the intruder, grunt and wag their tails, and then usually leave, running through the dense, closed canopy. If the animals remained aroused, their grunts graded into high-pitched rasps which usually elicited similar vocalizations in nearby groups.

Besides man, there are probably three animals which prey on *Lemur fulvus rufus* and the other diurnal species of lemur in the southwest of Madagascar: a large hawk (*Gymnogenys radiata*) and two carnivores (*Cryptoprocta ferox* and *Fossa fossa*). *L. f. rufus, Lemur catta,* and *Propithecus verreauxi* would always move into the densest foliage of the closed canopy and give particular loud vocalizations when *G. radiata* flew overhead. All three species would give warning calls in unison. These calls were never given to another species of bird nor to the large fruit bat (*Pteropus rufus*) that frequently flew over the trees in the early evening.

The reaction of *Lemur fulvus rufus* to ground carnivores is similar to that given to humans. The animals wag their tails and grunt at the potential predator while watching it very closely.

Reactions to Other Birds and Mammals

Generally, the lemurs ignored nonpredatory birds and mammals, although they would frequently react to their warning signals. In many cases, birds and lemurs fed in the same trees. The lemurs frequently fed in trees in which fruit bats rested during the day. I saw only one agonistic interaction between *Lemur fulvus rufus* and a bird. In this case, a bird (species unidentified) dived at a *L. f. rufus* that was being chased by one of its conspecifics. The bird probably saw an excellent opportunity to obtain nest material.

DISCUSSION

Lemur fulvus rufus is a folivorous, arboreal primate and it shows a number of adaptations which parallel those of folivorous Old and New World primates. Hladik and Hladik (1972) and Eisenberg *et al.* (1972) suggest that folivores have smaller home ranges, move less throughout the day, and have higher population densities than do frugivores of an equivalent size class. These criteria are true for *L. f. rufus* in relation to other diurnal lemurs with more frugivorous diets. In fact, the population density of *L. f. rufus* reaches as high as 1200 animals/km². To my knowledge, this density is higher than that for any other species of primate. Madagascar seems to be a microcosm in which the diurnal prosimians have a number of adaptations parallel to those of the Old and New World monkeys. However, group sizes and day and home ranges are generally smaller than those of similarly adapted monkeys, and population densities and biomass are generally higher. For example, in Ceylon, the total biomass of folivores (mainly primates) in some areas of Polonnaruwa is reported to be approximately 27 kg/hectare (Hladik and Hladik, 1972), and that of herbivores during the period of maximal concentration at Wilpattu is reported to be 29.4 kg/hectare (Eisenberg and Lockhart, 1972). The total biomass for *L. f. rufus, Lemur catta,* and *Lepilemur* at Antserananomby is about 32 kg/hectare, and this does not include the biomass of the highly folivorous *Propithecus*, whose density in this forest is very high.

Eisenberg *et al.* (1972) have also suggested that both frugivores and folivores tend towards either unimale or age-graded-male group structures, with folivores especially tending toward unimale systems. They propose that

> not only terrestrial primate species adapted to arid climates but also most primate species adapted to forests are characterized by either uni-male troops or age-graded-male troops. The true multi-male system is a less frequently evolved specialization, and the term "multi-male" should be restricted to those species having large troops that inlcude several functionally reproductive adult males, as well as nonreproductive males of different ages. (p. 867)

> The multi-male system is apparently a specialized form of social grouping that represents a particular adaptation to terrestrial foraging by intermediate-sized primates. It is readily derived from an age-graded-male system. The uni-male system or the age-graded-male system is favored in arboreal species, both frugivores and folivores. (p. 873)

Thus these authors assume that the normal group structure of primates is a unimale system and that, under certain conditions, some species develop age-graded-male systems. Only in very special circumstances do species develop a multimale group structure. The social organization of *Lemur fulvus rufus* is characterized by very cohesive groups averaging 9.4 individuals. There is no noticeable dominance hierarchy, and the adult male-to-female ratio is approximately 1:1. Groups contain on the average over 3 adult males, and out of the 18 groups censused, 14 had more than 2 adult males. As far as could be determined, the adult males in these groups were functionally reproductive. There were no uni-male groups.

I believe it premature to make arbitrary categories or *a priori* decisions about the evolution of primate social structure. If we find groups which do not fit our categories, do we ignore them or should we try to account for them in complex ways? Eisenberg *et al.* (1972) state that those arboreal primates with groups having more than one adult male are very likely age-graded-male groups in the process of splitting into smaller unimale groups. Although this might be true in some cases, it is not necessarily true in others. It is obviously not so in groups of *Lemur fulvus rufus.* It is likely that group structure in various species of primates did not go through one evolutionary sequence which is represented in different stages in living primates. Multimale groups probably occurred in a number of ways in response to a number of different environmental conditions. All species with multimale systems, therefore, should not be lumped together for the purposes of analysis at the expense of understanding the different social organizations they actually represent; however, neither should they be ignored or explained away because they do not fit our preconceived schema.

SUMMARY

The ecology and behavior of groups of *Lemur fulvus rufus* in three forests are described. The 18 groups censused ranged in size from 4 to 17 animals, with a mean group size of 9.4 The number of adult males was approximately equal to that of adult females. Groups were very cohesive and there were no noticeable dominance hierarchies. Home ranges were between 0.75 and 1 hectare, and the population densities at two of the forests were approximately $900/km^2$ and $1200/km^2$. Boundaries between groups were not defended, and ranges overlapped extensively. The frequency of encounters between groups and the pattern of group movement were related to the seasonal productivity of plant species. The animals were highly folivorous, feeding mainly on kily leaves. They also ate kily fruit and flowers when available. Animals supplemented their predominantly kily diet with the seasonal products (mainly fruits and flowers) of other, more rarely occurring species of trees. *L. f. rufus* spent over 70% of the time in the canopy layer of the forest.

Details of feeding, drinking, diurnal activity cycle, and interactions between *Lemur fulvus rufus* and other species of mammals and birds are described. A 5-min sampling technique was employed to quantify observations on activity cycles, utilization of the forest strata, and the amount of time spent feeding on various species and parts of plants. Details of the statistical methods used to analyze these data are explained elsewhere.

ACKNOWLEDGMENTS

I would like to thank the various officials of the Malagasy government who have assisted me during my research in the field. I am also indebted to many people of the villages of Manja and Vondrove, whose kind hospitality is deeply appreciated. The study was supported in part by Research Fellowship MH46268-01 of the National Institute of Mental Health, by a Duke University Graduate Fellowship, and by National Science Foundation Research Grant No. BG-41109.

REFERENCES

Albignac, R., Rumpler, Y, and Petter, J.-J., 1971, L'hybridation des lémuriens de Madagascar, *Mammalia* **35**: 358–368.

Buettner-Janusch, J., 1963. An introduction to the primates, in: *Evolutionary and Genetic Biology of Primates*, Vol. I (J. Buettner-Janusch, ed.), pp. 1–64, Academic Press, New York.

Buettner-Janusch, J., Swomley, B. A., and Chu, E. H. Y., 1966, Les nombres chromosomiques de certains lémuriens de Madagascar et les problèmes de différenciation des espèces, *Cah. ORSTOM, Sér. Biol.* **2**:3–12.

Charles-Dominique, P., and Hladik, C. M., 1971, Le *Lepilemur* du sud de Madagascar: Ecologie, alimentation et vie sociale, *Terre Vie* **25**:3–66.

Crook, J. H., and Aldrich-Blake, P., 1968, Ecological and behavioral contrasts between sympatric ground dwelling primates in Ethiopia, *Folia Primatol.* **8**:192–227.

Eisenberg, J. F., and Lockhart, M. C., 1972, An ecological reconnaissance of Wilpattu National Park, Ceylon, *Smithson. Contrib. Zool.* **101**:1–118.

Eisenberg, J. F., Muckenhirn, N. A., and Rudran, R., 1972, The relation between ecology and social structure in primates, *Science* **176**:863–874.

Hladik, C. M., and Hladik, A., 1972, Disponibilités alimentaires et domaines vitaux des primates à Ceylan, *Terre Vie* **26**:149–215.

Nute, P. E., and Buettner-Janusch, J., 1969, Genetics of polymorphic transferrins in the genus *Lemur, Folia Primatol.* **10**:181–194.

Petter. J.-J., 1965, The lemurs of Madagascar, in: *Primate Behavior.* (I. DeVore, ed.), pp. 292–319. Holt, Rinehart and Winston, New York.

Petter, J.-J., 1969, Speciation in Madagascan lemurs, *Biol. J. Linn. Soc.* **1**:77–84.

Schwarz, E., 1936, A propos du "*Lemur macaco*" Linnaeus, *Mammalia* **1**:25–26.

Struhsaker, T., 1967, Auditory communication among vervet monkeys (*Cercopithecus aethiops*), in: *Social Communication Among Primates* (S. A. Altmann, ed.), pp. 281–324, University of Chicago Press, Chicago.

Sussman, R. W., 1972, An ecological study of two Madagascan primates: *Lemur fulvus rufus* Audebert and *Lemur catta* Linnaeus, Ph.D. thesis, Duke University.

Sussman, R. W., 1973, Socialization, social structure, and ecology of two sympatric species of *Lemur*, Paper delivered at the Annual Meeting of the American Association of Physical Anthropologists, April, 1973.

Sussman, R. W., 1974, Ecological distinctions in sympatric species of *Lemur,* in: *Prosimian Biology.* (R. D. Martin, G. A. Doyle, and A. C. Walker eds.), pp. 75–108, Duckworth, London.

Sussman, R. W., and Richard, A., 1974, The role of aggression among diurnal prosimians, in: *Primate Aggression, Territoriality, and Xenophobia* (R. L. Holloway, Jr., ed.), pp. 49–76, Academic Press, New York.

Sussman, R. W., O'Fallon, W. M., Buettner-Janusch, J., and Sussman, L. K., in preparation, A method for statistically analyzing data on activity cycles.

Field Observations of Social Behavior of *Lemur fulvus fulvus* E. Geoffroy 1812

14

JONATHAN E. HARRINGTON

Lemur fulvus is a widely ranging prosimian found in nearly all of the remaining forested areas of Madagascar except the dry forest of the extreme South. As stated in Sussman (Chapter 13), the taxonomy of the *L. fulvus–macaco* group is in an uncertain condition and may need revision as more information on the morphology, distribution, behavior, cytogenetics, and biochemistry of these animals becomes available.

The observations reported here were obtained from two groups of *L. fulvus fulvus*, in which nearly all individuals could be identified, at Ankarafantsika, northwestern Madagascar, between February and July 1969. Special attention was given to olfactory communication, which has been described elsewhere (Harrington, 1971, 1974).

L. fulvus and *L. macaco* have also been studied in the wild by Petter (1962), Jolly (1966), and Sussman (1972, 1974, and Chapter 13).

JONATHAN E. HARRINGTON   Laboratory of Animal Behavior, The Rockefeller University, New York, New York 10021. Present address: Department of Biology, University of Missouri, St. Louis, Missouri 63121.

STUDY AREA

The study area was located in the Forest Reserve of Ankarafantsika near the village of Ampijoroa in northwestern Madagascar (Fig. 1). The native vegetation in this area is a deciduous forest with a continuous canopy about 15 m in height. The flora and fauna of the region have been described by Petter (1962), Perrier de la Bathie (1936), and François (1937). Nearly all the rainfall occurs from November to March. During the dry season most trees of the canopy lose their leaves.

About half of the study area was occupied by native forest with a canopy height of 8–15 m, with a few kily (*Tamarindus indica*) trees emerging to about 30 m. The foliage was rather sparse, and there were many vines which provided pathways for vertical movement. The other half of the study area was occupied by introduced species of trees, mainly teak (*Tectona grandis*),

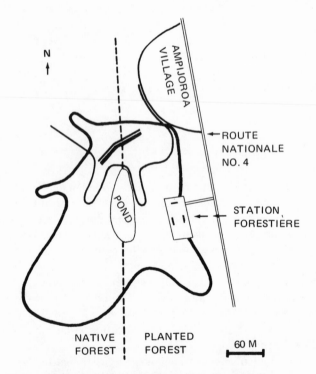

FIG. 1. Map of study area. Heavy line, Group 1 home range; light line, group 2 home range; double line, territorial boundary when both groups are in the area of overlap at the same time.

Eucalyptus, Albizzia, kapok (*Ceiba pentandra*), *Hura crepitans,* and mango (*Mangifera indica*), which had been planted around the forestry station and the village of Ampijoroa. A heavy undergrowth of native trees and shrubs had been allowed to grow up in much of the planted forest.

A government forester was in residence at the forestry station, and the hunting of lemurs was prohibited. Therefore, the animals in the study area were already habituated to the frequent passage of the Malagasy villagers beneath the trees when I began my observations.

METHODS

All observations were made by approaching the animals quietly on foot in full view. Because the animals were not habituated to humans who tried to observe them for extended periods of time, many alarm responses to my presence were seen at the beginning of the study. These responses were rarely seen after the fourth week of observations.

Most observations were made from a distance of 10–30 m. When thoroughly habituated to me, the animals would approach as close as 2 m. I used 7 × 30 binoculars and took notes by hand. Some observations were made at all hours between sunrise at about 600 h and sunset at about 1800 h. However, relatively fewer observations were made from sunrise to about 700 h, when I was usually searching for the animals, and during the midday rest period of the animals from about 1000 to 1500 h.

L. fulvus were first observed on February 24, 1969. Group 1 was first identified on March 5. All of the animals in this group could be reliably identified by April 14 except for the two juvenile males, PA and MO. These two animals could not usually be distinguished from each other throughout the study, and in most observations they are identified only as JM (juvenile male).

A neighboring group, Group 2, was identified on April 16. All individuals in this group could be identified by May 29. The field study ended on July 3.

Group 1 was observed for 213.5 h during which conditions were good enough to permit observation of social behavior. Of this total, 158.5 h was done after April 14, when all individuals (except the two JM) could be identified. Group 2 was observed for 32.8 h, of which 24.6 h was done after May 29, when all individuals could be identified.

Quantitative records were obtained from both groups on anogenital sniffing, scent marking, grooming, contact, agonistic interactions, play, and sexual interactions. Nearly all occurrences of these behaviors which were seen were recorded; a time-sampling technique was not used. Beginning March 25, the movements of the groups were plotted on a map of the study area. Some observations were also made of feeding, drinking, locomotion, activity cycles, height above the ground, and other behavior.

Most of the data from Group 2 are not presented here because of the relatively short period of observation of this group. However, it is my impression that the behavior of Group 2 did not differ from that of Group 1 in any major ways.

Data on interactions between identified individuals in Group 1 are taken from the 158.5 h of observation of this group after April 14, when all individuals except the two JM could be identified. For certain other kinds of data, e.g., changes in sexual activity over the mating season, the entire 213.5 h of observation on this group from March 5 to July 3 has been used.

ECOLOGY AND NONSOCIAL BEHAVIOR

Only a brief summary will be given here, since ecology and nonsocial behavior were not a primary focus of the study.

L. fulvus is generally considered to be diurnal. The animals in this study were active from sunrise to sunset, with a rest period from about 1000 to 1500 h. However, they were nearly always active when I found them in the morning and when I left them at night, even when it was quite dark. On two nights when the moon was full they were seen moving and feeding at 2100 h, about 3 h after sunset. Furthermore, loud vocalizations and sounds of group progression were occasionally heard at later hours of the night. *L. fulvus* may have a considerable amount of nocturnal activity whose nature and extent are still unknown.

L. fulvus moved primarily on horizontal branches, but also moved quite readily on vertical or oblique branches and vines. They also progressed by vertical clinging and leaping between tree trunks less than 30 cm in diameter. They were almost completely arboreal, spending less than 2% of their time on the ground.

Leaves, buds, flowers, and fruits were eaten. Usually an animal pulled a branch to its mouth with one hand and ate the food item directly from the branch. No evidence of insect-eating was seen.

Water was obtained in three ways: (1) by licking dew or rainwater from leaves; (2) by lapping water from hollows in trees; (3) by dipping one hand into a tree hollow, removing the hand, and licking water from the backs of the flexed fingers. The animals were never seen to drink ground water, although the home ranges of both groups bordered on a small pond.

GROUP COMPOSITION

The composition of the two groups is given in Table I. Both groups had 12 members; Group 1 had seven males and five females, and Group 2 had five males and seven females. Three age classes were distinguished: juveniles, subadults, and adults. (In the analysis of most of the data, juveniles and subadults have been classified together as "young.") Juveniles could be distinguished by their small size. The juveniles in this study would have been born around October of the previous year. Subadults were nearly as large as adults, but the males had smaller scrota, and both sexes had smaller perianal glandular areas than adults. These animals would have been born around October 2 yr previous to the period of the study. All other animals were classified as adults. This age classification is in agreement with the characteristics of captive-born *L. fulvus* at Duke University, North Carolina (personal observation).

Males could be distinguished from females by their large furred scrota and their somewhat longer cheek fur. Unlike most other subspecies of *L. fulvus*, *L. f. fulvus* has relatively little sexual dimorphism in its pelage features.

Individuals could be distinguished from each other by such features as coat color, which ranged from reddish-brown to gray; iris color, which ranged from pale yellow to deep orange; the pattern of nicks on the ears; and various facial markings.

TABLE I. Composition of Groups

Males			Females	
		Group 1		
Adult	RU		Adult	YE
	YEM			PYE
	LE			BE
	RA			
Subadult	YM		Subadult	RLA
Juvenile	PA		Juvenile	LF
	MO			
		Group 2		
Adult	LFM		Adult	LS
	EG			FO
	OE			YEF
	BU			MA
			Subadult	PF
Juvenile	JR		Juvenile	BFF
				CL

HOME RANGE AND MOVEMENTS

Figure 1 shows the observed home ranges of the two groups from March 25 to July 3. A smooth line has been drawn which encloses all of the sightings of Group 1. This group was never observed heading into or out of this area, and therefore the enclosed area of about 7 hectares probably represents the entire home range of this group over the 14-wk period. The entire home range of Group 2 could not be determined as it was frequently seen entering or leaving the heavily forested area to the northwest where it was impossible to follow. The observed ranges of both groups were quite compact except for a few single excursions outside their usual ranges, represented by the "fingers" on the map.

No groups other than Groups 1 or 2 were positively identified in the area shown on the map. To the south and west, the range of Group 1 was bounded by areas of grass with scattered trees in which *L. fulvus* were never seen. Thus it is likely that Group 1 had no contact with groups of *L. fulvus* other than Group 2.

CONTACT AND GROOMING

The most frequent social interaction in the groups of *L. fulvus* consisted of sitting peacefully in contact, often with mutual grooming. During the midday rest period, a group typically broke up into clumps of from two to six animals, the animals within each clump sitting in very close contact with their tails wrapped around each other and grooming each other.

Like other prosimians, *L. fulvus* grooms with its tongue and its teeth. The lower incisors are modified to form a tooth comb used for grooming the fur. The hands are used to hold the fur near the part being groomed, but never for fine manipulations of the fur and skin such as occur in anthropoid primates. Grooming bouts usually lasted several minutes, with both members of the pair grooming each other intensively. An animal often presented its ventral surface, forearm, or head to another to be groomed.

Table II shows the numbers of contacts observed between identified individuals during rest periods in Group 1 after April 14. Since individual *L. fulvus* usually sit in quite close contact with each other during rest periods, only actual body contacts were counted rather than other measures of proximity such as sitting within an arm's reach of another animal. Momentary contacts between moving animals are not counted. Contacts occurring during sexual behavior (including anogenital sniffing and scent marking) are also not counted in these data. Contacts occurred within subgroups of from two to six

TABLE II. Contacts Between Identified Individuals in Group 1 (April 14–July 3)

		Adult males				Adult females			Young			
		RU	YEM	LE	RA	YE	PYE	BE	YM	RLA	LF	JM^a
A. By individuals												
Adult males	RU											
	YEM	3										
	LE	0	1									
	RA	1	3	0								
Adult females	YE	15	4	2	5							
	PYE	11	2	1	3	1						
	BE	4	2	0	0	10	2					
Young	YM	8	5	7	5	4	3	5				
	RLA	5	6	5	7	15	16	11	9			
	LF	8	5	2	2	39	4	6	5	10		
	JM^a	12	4	10	5	8	20	17	14	19	11	2

B. By age-sex class (expected figures in parentheses)

	Adult males	Adult females	Young
Adult males	8 (35)		
Adult females	49 (70)	13 (18)	
Young	96 (116)	148 (87)	70 (50)

C. Among adults only

	Adult males	Adult females
Adult males	8 (20)	
Adult females	49 (40)	13 (10)

^{a}Includes both juvenile males PA and MO.

animals. If two animals were seen to break contact and resume contact after 10 min or more, a new contact was recorded.

Grooming was seen during 219 (57%) of the 384 contacts observed. Because a short bout of grooming could easily be missed during a prolonged contact period, this figure is probably an underestimate of the number of contacts that also involved grooming.

Contacts between adult females and young, and among young, were more frequent than expected, and other contacts were less frequent. If the figures for contacts among adult animals only are examined separately, contacts among adult males are much less frequent than expected and contacts among adult females and between adult males and adult females are somewhat more frequent.

The frequency of contacts between individuals of different age–sex classes did not change over the period of observation, which included the mating season.

Both of the subadult animals YM (male) and RLA (female) had many contacts with adults of the same sex, in contrast with the relatively fewer number of contacts with animals of the same sex among fully adult animals. Despite their nearly adult size, the subadults seemed to be treated by the adults more as juveniles than as adult animals.

SEXUAL BEHAVIOR

Around the fourth week in April, the males of Group 1 began to show sexual responses to the females. Males began to follow the females around, sniff the females' anogenital regions or places where females had recently anogenitally marked or sat, scent-mark after sniffing female anogenital scents, and try to sit close behind the females with their arms clasped tightly around the females' waists ("clasping"). The males' movements were quick and "frantic," unlike the normal relaxed movements of animals seeking contact with each other. The males sometimes gave short, soft grunts while seeking contact with females, and sometimes had erections.

Olfactory signaling or marking was prominent in sexual behavior. Males very often, and females occasionally, scent-marked with the anogenital area. Adults of both sexes have a conspicuous patch of naked glandular perianal skin, and males also have glands on the scrotum which produce a copious secretion easily visible in the field (Montagna, 1962). Males also head-marked with the forehead and hand-marked with the palms of the hands. Males often sniffed the anogenital region of the female or a spot where a female had anogenitally marked or sat, and females occasionally sniffed anogenital scents of males. Sniffing of female anogenital scent by a male was often followed by anogenital marking, head marking, or hand marking by the male. Anogenital marking by a male was often directed to the lower back of a female. Head marking by a male was often directed to a branch where a female had recently anogenitally marked or sat. Thus it seems that males deposited their anogenital scents on the backs of females, and picked up the anogenital scents of females on their own foreheads.

Figure 2 shows the changes in frequency over the whole study period of four kinds of "olfactory" responses to females by males: (1) sniffing female anogenital scent (i.e., the female's anogenital region or a spot where she has

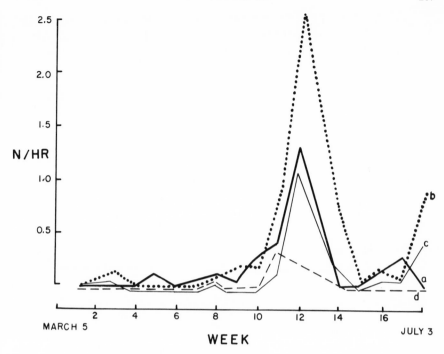

FIG. 2. Male sniffing and scent-marking responses to females (Group 1), March 5–July 3. a, Male sniffs anogenital scent, without marking; b, male sniffs female anogenital scent, then marks; c, male head-marks in female anogenital secretion; d, male anogenitally marks onto female.

just anogenitally marked or sat) without subsequent scent marking by the male; (2) sniffing female anogenital scent with subsequent scent marking by the male; (3) anogenital marking on a female's back; (4) head marking where a female has recently anogenitally marked or sat. Figure 3 shows the changes in frequency of clasping of females by males, including three copulations. All of these activities show a sharp peak around the third week in May, week 12 of the study. These data suggest that *L. fulvus*, like all other prosimians for which information is available, has a sharply defined mating season (Petter-Rousseaux, 1964).

Anogenital marking consisted of conspicuous rubbing movements of the anogenital region. Although females anogenitally marked much less than did males during sexual behavior, the anogenital scent of females was obviously stimulating to the males. Apparently the females could deposit anogenital scent on branches by just sitting down, without making conspicuous scent marking movements.

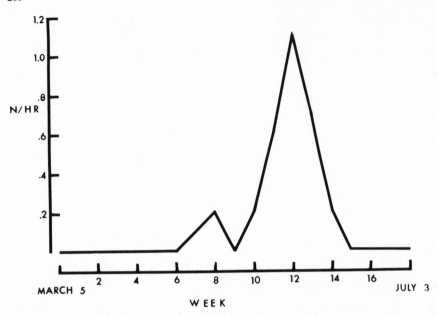

FIG. 3. Clasping of females by males in Group 1, March 5–July 3. Included are three copulations in week 12.

The females' part in sexual behavior was less active than the males'. In the two groups over the whole course of the study, males scent-marked during sexual behavior a total of 104 times (59 anogenital markings, 35 head markings, and 10 hand markings). Females scent-marked during sexual behavior 18 times; all of these were anogenital markings. Females did not anogenitally mark onto the backs of males. Males sniffed female anogenital scents 149 times; females sniffed males 16 times. Females did not present their genital regions to males. Typically, a male would come up to a female abruptly and begin sniffing her anogenital region and trying to clasp her. The female would either remain sitting quietly or get up and go away, sometimes stopping to sit down or to anogenitally mark as she went. The male would follow her, stopping to sniff intensively the spots where she had sat or anogenitally marked. On three occasions, a female cuffed a male who was trying to clasp her, and once a male cuffed a female, but no serious aggressive behavior was seen between pairs of animals engaged in sexual activity.

Three copulations were seen, all between the same two animals on the same day. RA (male) and YE (female) copulated at 1454, 1519, and 1700 h on May 20, during the week in which other indicators of sexual activity reached a peak. Each copulation was preceded by less than 5 min of grooming, anogenital sniffing, scent marking, and following, during which the pair remained at least 3 m from other animals. The animals did not form a consort pair, but intermingled with the group between copulations. The copulations took place with both animals sitting upright on a branch, the male behind the female with his arms clasped around her waist. Pelvic thrusts were short and fast (2–3/s). Intromission lasted less than 10 s for each copulation.

I saw little agression associated with sexual behavior. A pair often engaged in sexual behavior in the midst of the group without interference from other animals. Sometimes one male replaced another, usually without any agonistic interaction. On one occasion only I saw a male (RU) repeatedly chase away another unidentified male who was approaching him as he tried to clasp BE. This intermale chasing resembled the behavior observed during the mating season of *L. catta* as described by Jolly (1966). Jolly observed the behavior many times, while I saw it only once. However, three animals in the two groups acquired new cuts in their ears during the study: LS (female), May 19; FO (female), June 2; YEM (male), June 12. These wounds could have been the result of fighting in a sexual or other context that I did not see.

Figure 4 shows for each of the adult males and females of Group 1 the percentage of the total number of four kinds of male–female interactions: clasping of females by males, contact between females and males, sniffing of female anogenital scent by males, and reciprocal scenting between females and males. Clasping and contact are the interactions defined above. Sniffing of female anogenital scent by males includes all instances of a male sniffing the anogenital region of a female, or a spot where she had recently anogenitally marked or sat, with or without subsequent scent marking by the male. Mutual scenting includes all instances where a male anogenitally marked on a female, or head marked on a spot where a female had recently anogenitally marked or sat.

There are no significant differences among the three females in the numbers of sniffing, mutual scenting, or clasping interactions with males. The differences among the three females in the numbers of contacts with males (which excludes contacts involving sexual activity) are significant. In contrast, there are significant differences among the four adult males in the numbers of all four kinds of interactions with females. RU has the highest scores in all four interactions. YEM has the lowest scores in all but contact. RA and LE generally occupy the intermediate positions. However, the three observed copulations involved RA (male) and YE (female).

AGONISTIC BEHAVIOR

Most agonistic interactions were quite mild. Usually an animal just thrust its head toward another animal ("head threat"), sometimes grunting or baring its teeth at the same time. An animal might also aim a cuff of its hand

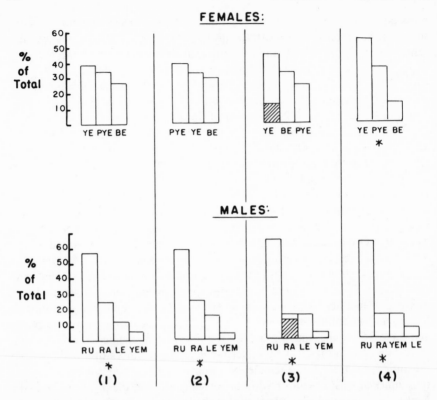

FIG. 4. Clasping, sniffing and scent marking, and contact interactions between identified males and females of Group 1, April 14–July 3. (1) Male sniffs female anogenital scent, with or without subsequent scent marking ($n = 82$). (2) Mutual scenting (male anogenitally marks onto female, or head-marks in female anogenital secretions, $n = 28$). (3) Male clasps female (shaded area indicates copulations) (n = 25). (4) Male–female contact (n = 49). *$P < 0.01$, χ^2 test.

in the direction of another animal, which might or might not make contact. I saw only two examples of high-intensity combat in which two animals were locked together and biting each other; on both of these occasions the animals fell about 5 m to the ground while locked together. Scent marking during intragroup agonistic encounters was not seen.

Most agonistic interactions were also one-sided. Typically an animal mildly threatened another with head or hand, and the threatened animal just went away without returning the threat.

A total of 55 agonistic interactions was seen in Group 1 over the course of the study. Table III shows the 26 interactions after April 14 in which both individuals could be identified, and in which the agonistic gesture was not re-

TABLE III. One-Sided Agonistic Interactions Between Identified Individuals in Group 1 (April 14–July 3)[a]

	Attacked	
Attackers	Adult	Young
Adult	3	16
	(8)	(7)
Young	0	7
	(7)	(4)

[a]Of the 16 agonistic gestures directed against young by adults, 9 were given by adult females and 7 were given by adult males (expected figures in parentheses).

ciprocated. The great majority of these were among young, or were directed to young by adults of either sex.

Table IV shows the situations in which the 55 agonistic interactions occurred in Group 1. Nine of these interactions were associated with sexual behavior; all of these were mild head or hand threats, except the one instance of vigorous chasing of one male by another male described above. The nine agonistic interactions over access to a waterhole in a tree all occurred on the same day, and included the two instances when animals fell to the ground while locked together and biting each other. The most common sort of agonistic interaction (20) was one in which an animal, usually a juvenile or subadult, approached a contact and grooming group, and one of the adults in the group threatened the young animal away.

Because agonistic interactions were rather infrequent and were mostly directed against young animals by adults, I was not able to discern any dominance hierarchies based on agonistic interactions in either of the groups. It was my impression that more subtle indicators of dominance such as are found among many anthropoids, e.g., supplanting behavior and facial signaling, were also infrequent. It is of course possible that more prolonged and

TABLE IV. Situations in which Agonistic Interactions Occurred[a]

Situation	Number
Attacked animal seeks contact or grooming with attacker	20
Conflict over waterhole (all on May 14)	9
Attacked animal interferes with couple in sexual behavior	5
Female attacks male during sexual behavior	3
Male attacks female during sexual behavior	1
Didn't see preliminaries	17
	55

[a]Includes all 55 agonistic interactions seen in Group 1 from March 5 to July 3.

careful observations of such behavior might reveal the existence of structured dominance relationships within groups of *L. fulvus*.

The frequency of agonistic interactions did not change over the mating season. However, it is possible that high-intensity aggression occurred which was not seen, perhaps at night, since many adults of both sexes did have nicks in the ears and new nicks appeared in three animals during the study. The prevailing impression throughout the study was that intragroup agonistic behavior plays a small part in the behavior of *L. fulvus*.

PLAY

Animals played by grappling at each other with hands and feet, lunging with open mouth, play-biting, jumping, and chasing. Table V shows the number of instances of play between identified individuals observed after April 14 in Group 1. Animals played in groups of up to five. Play was seen only among young animals or between adult females and young.

RELATIONS BETWEEN NEIGHBORING GROUPS

There was considerable overlap between the home ranges of Groups 1 and 2 (Fig. 1). However, the two groups appeared to observe a boundary (double line in Fig. 1) when both groups were in the overlap region at the

TABLE V. Play Interactions Between Identified Individuals in Group 1 (April 14–July 3)[a]

	Adult males	Adult females	Young
Adult males	0		
	(7)		
Adult females	0	0	
	(13)	(3)	
Young	0	16	54
	(21)	(16)	(10)

[a]Expected figures in parentheses.

same time. Exclusion seemed to be accomplished in three ways: (1) simple avoidance, (2) territorial defense behavior, and (3) loud vocalizations.

1. In the most common situation, the two groups simply avoided entering each other's exclusive areas, while resting in close proximity to each other. On 10 different days, the two groups spent up to 5 ½ h during the midday rest period resting near the village along the double line in the figure, with as little as 5 m between the nearest members of the groups. During this period, animals occasionally gave long, slow grunts which were answered by members of the other group. Once LFM, an adult male from Group 2, went into the trees among the members of Group 1, disappeared for about 5 min, and then returned to Group 2. This was the only instance of mixing between the groups observed during the study. Generally, the animals rested in separate groups in close proximity to each other, there were no intergroup interactions except the long grunts, and neither group tried to penetrate beyond the grove of trees where they were resting.

2. Territorial defense behavior was seen on 3 days when the groups encountered each other in the overlap region to the north of the pond while moving around and feeding late in the afternoon. In each of these encounters, the two groups faced each other with about 5 m between the nearest members of the two groups. Adults of both sexes gave loud shrieking (= "rasping" of Sussman, Chapter 13) and grunting vocalizations, swung their tails rapidly from side to side, scent-marked, and ran quickly back and forth in the trees. After about 10 min of this behavior, both groups moved away, each in the direction of its own home range. No contacts were seen between the members of different groups. On these three occasions, *L. fulvus* appeared to show territorial defense behavior, defending its territory by a combination of visual, vocal, and olfactory signals.

3. It is possible that loud, shrieking vocalizations also play a part in maintaining spacing between groups over long distances, as in some forest-living anthropoid primates (Marler, 1968). These vocalizations were often heard when the group was alarmed by a man or during territorial encounters between groups. Similar vocalizations were often given by a group during undisturbed progression late in the afternoon or at night. These vocalizations were almost never given by undisturbed groups before about 1600 h. The shrieks were often answered by similar vocalizations from one or perhaps more unseen *L. fulvus* groups. Shrieks were particularly frequent around nightfall at 1800 h, at which time the animals were still actively moving and feeding. Although there are no data on movements of groups relative to each other during shrieking, it seems plausible that these vocalizations may function in maintaining spacing between groups, particularly at night.

RESPONSES TO POTENTIAL PREDATORS

Man

When I approached a group of *L. fulvus*, their first response was a low grunting. After habituation had occurred in the first 4 wk of the study, the group would usually settle down to its usual activities after a few minutes of grunting. In nonhabituated groups, the animals would go through a graded series of vocalizations from the low grunts to louder grunts to shrieks. They also scent-marked, ran around quickly in the trees, and swung their tails. All of these responses are similar to those given by groups during territorial encounters.

After a few minutes of these responses, the group would sometimes go away, but more often it would approach and mob me. The animals would surround me in a semicirele, the nearest animal as little as 2 m away, and vocalize, scent-mark, etc. Mobbing lasted from a few minutes to an hour. The three species of lemur which were active in the daytime at Ankarafantsika, *L. fulvus*, *L. mongoz*, and *Propithecus verreauxi*, responded to each other's mobbing vocalizations by approaching and joining in the mobbing. I was often mobbed by two or even three species at the same time.

Adult female *L. fulvus* showed considerably more alarm responses to me than did adult males, although both sexes showed some. Of the 133 scent markings given by members of Groups 1 and 2 when the group was alarmed by me, 101 were given by adult females. The females also initiated most mobbing approaches to me, and approached me more closely during mobbing than did other animals.

Birds

The largest bird of prey at Ankarafantsika was *Gymnogenys radiata*, a large harrier hawk. This hawk was seen circling around or sitting near trees containing *L. fulvus* on three occasions. The lemurs gave loud barks followed by a series of chattering vocalizations; the latter vocalizations were never heard at any other time in the wild. The lemurs also moved quickly up or down in the canopy, improving their cover, when this hawk was nearby. They also swung their tails, urinated, and defecated. However, they did not scent-mark or give any of the graded series of grunting and shrieking vocalizations that they gave when alarmed by a man.

If a smaller *Buteo* hawk or any of various other large birds down to about the size of a crow flew over, the lemurs usually looked at the bird intently and sometimes gave soft grunts, but none of these birds evoked the responses that were given to *G. radiata*.

RELATIONS WITH OTHER SPECIES OF LEMURIFORMES

Lemur mongoz

L. mongoz and *L. fulvus* were seen resting, feeding, and progressing together on 8 different days for periods up to 4 h. Typically, the groups of *L. mongoz*, which contained from two to four individuals, would follow the *L. fulvus* groups during group progression, the first *L. mongoz* a few meters behind the last *L. fulvus*. When the *L. fulvus* stopped to feed, the *L. mongoz* stopped and did the same, often in the same branches. The two species also responded to each other's mobbing signals.

L. mongoz and *L. fulvus* are quite similar in size, general appearance, visual and vocal signals, arrangement of scent glands, and scent-marking behavior. Experiments with captive lemurs have also shown that male *L. fulvus* respond quite strongly to the scent of male *L. mongoz* (Harrington, 1971). However, members of the two species did not attempt to contact each other or sniff each others' scent marks, and no behavioral interactions other than those described were seen in the wild.

Propithecus verreauxi

P. verreauxi and *L. fulvus* did not associate together during group progression. However, they were often seen feeding within a few meters of each other in the same trees. They also associated with each other when mobbing me.

Usually members of these two species ignored each other except when mobbing. However, on seven occasions, a *P. verreauxi* displaced a *L. fulvus* from its position on a branch. *P. verreauxi* progressed through the trees by long bipedal leaps, apparently ignoring the *L. fulvus* in their path, and if a *P. verreauxi* leaped toward a *L. fulvus*, the *L. fulvus* usually got out of the way. The only other interactions observed between these species were two occasions when a *L. fulvus* pulled the tail of a *P. verreauxi*, one occasion when a *L. fulvus* head-threatened a *P. verreauxi*, and one occasion when a *P. verreauxi* head-threatened a *L. fulvus*.

DISCUSSION

Behavior within the Group

An analysis of the distribution of behaviors among members of the different age–sex classes in Group 1 showed the following: Contact (often including grooming) was most frequent among young or between adult females and young, but also occurred among members of all other age–sex classes. Play, another "friendly" interaction, occurred exclusively among young or between adult females and young. Agonistic gestures were most frequent among young, or directed toward young by adults; however, some agonistic behavior did occur among adults. Shrieks during group progression, territorial defense responses, and alarm responses to man were given by adults of both sexes; females were more active than males in responses to man. Sexual activity—clasping, following, anogenital sniffing, and scent marking—occurred almost entirely between adult males and females, and most of this behavior consisted of responses to the females by males, the females being mostly passive.

In all of these activities, the figures for subadults (2-yr-olds of nearly adult size) were similar to the figures for juveniles, and all these animals have therefore been classified together as "young."

Data on several kinds of heterosexual interactions were analyzed to see if there were significant differences among *individual* adults in the frequency with which they were performed. The interactions were clasping, anogenital sniffing by the male with or without subsequent scent marking, mutual scenting, and contact. There were significant differences among the four adult males in all interactions, and no significant differences among the three adult females except for contact. Only one male was seen copulating, and he was considerably exceeded by another male in other indices of sexual activity. These limited data suggest that there may be a certain amount of competition among the adult males of the group for access to females, but any firm conclusion based on these data would be premature.

Aggressive Behavior in Social Lemurs

Perhaps the most striking difference between the social behavior of *L. fulvus* and that of *P. verreauxi* and *L. catta* is the relative peacefulness of relations within the *L. fulvus* groups. The animals spent most of their time during rest periods in close contact and grooming groups. Agonistic interactions were mild and infrequent. Most agonistic interactions occurred between adults and young, and not among adults. No dominance hierarchies based on agonistic interactions could be discerned. Scent marking was not seen during agonistic encounters, and agonistic visual and acoustical signals were not highly developed. There was no change in the frequency of agonistic behavior during the mating season. These observations are similar to those reported for *L. f. rufus* (Sussman, 1972; Sussman and Richard, 1974).

Intragroup agonistic behavior plays a much more important part in the life of *P. verreauxi* and *L. catta*. (Jolly, 1966; Sussman, 1972, 1974; Richard, 1974; Sussman and Richard, 1974). Both of these species show a marked increase in agonistic behavior during the mating season, which results in a restructuring of the dominance hierarchy as previously low-ranking males attempt to gain access to females. In *P. verreauxi*, there are also incursions by strange males into social groups, resulting in fierce fighting. Agonistic behavior in these species involves actual fighting, in which animals inflict severe wounds on each other, and also elaborate olfactory, visual, and acoustical signaling. Outside of the breeding season, both of these species still engage in sufficient agonistic behavior to reveal the existence of a dominance hierarchy.

Timing of Mating Season

The observations suggested that Group 1 had at least one peak of sexual activity in 1969, around the third week in May. Insufficient data were obtained to show if the neighboring group, Group 2, also had a similar peak. However, the frequency of sexual activity in Group 2 was also high about this time, although no copulations were seen. The data thus suggest that *L. fulvus*, like other prosimians (Petter-Rousseaux, 1964) has a sharply defined mating season, perhaps with synchrony between adjacent groups.

The mechanism for estrous synchrony within and possibly between groups in *L. fulvus* is not known, but olfactory communication is certainly a likely possibility, as has also been suggested for *L. catta* and *P. verreauxi* (Jolly, 1966). The responses of males to female anogenital scents increased greatly during the mating season. Males responded by sniffing the females' anogenital scents and by scent marking; thus male olfactory signals were made available to stimulate the female. Scents were deposited not only on branches, but also on the fur of the animals themselves. Males anogenitally marked onto the backs of the females, and head marked onto spots near where females were sitting. Thus each sex presumably had its back or head fur anointed with anogenital scents of the opposite sex, where they could possibly exert continuous primer effects on reproductive physiology as well as behavior.

Intergroup Spacing Behavior

The two groups appeared to maintain spatial separation from each other by three means: (1) simple avoidance (although in close proximity), (2) territorial defense, and (3) long-distance shrieking vocalizations. Although there was considerable overlap between the home ranges of the groups, the groups appeared to observe a territorial boundary when both were in the area of overlap at the same time. Sussman (1972 and Chapter 13), observed a quite different situation among 15 groups of *L. fulvus rufus* at Antserananomby and Tongobato in the western deciduous forest. Here the home ranges were much smaller in size, about 0.75–1.0 hectare, and adjacent home ranges overlapped much more extensively. Although Sussman observed behavior similar to what I have called territorial defense, he also frequently observed groups moving around in close proximity to each other in the areas of overlap, making no attempt to exclude each other.

These observed differences in intergroup spacing behavior may be associated with differences in population density in the two areas, which in turn may be correlated with different ecological conditions. Sussman's observations were made in relatively undisturbed forest, while about half of the 7–hectare home range of my Group 1 consisted of introduced trees. Furthermore, the range of Group 1 was bordered on the south by unsuitable lemur habitat where no groups were seen. It thus appears that *L. fulvus* under conditions of low population density such as existed in my study area may defend a considerable part of its home range as a territory; while under more crowded conditions such as in the area studied by Sussman, it will tolerate much more overlap with neighboring groups.

ACKNOWLEDGMENTS

This field study was made possible by the generous cooperation of the Direction des Eaux et Forêts and the Ministre de l'Agriculture of the Government of the Malagasy Republic. I especially wish to thank MM. Andriamampianina, Natai, Ramalanjoana, Ramanantsoavina, and Razanajatovo.

I also gratefully acknowledge the help of the following people during the field work and the preparation of this paper: Dr. R. Albignac, Mr. N. Budnitz, Dr. J. Buettner-Janusch, Mrs. V. Buettner-Janusch, Dr. A. Jolly, Dr. P. Klopfer, Dr. P. Marler, Dr. J.-J. Petter, M. J.-A. Randrianarivelo, M. G. Randrianasolo, and Dr. R. Sussman.

The research was supported by NIH HD02319, RR00388, a Sigma Xi Grant-in-Aid of Research, a NASA Predoctoral Traineeship, and a NSF Predoctoral Fellowship.

REFERENCES

François, E., 1937, Plantes de Madagascar, *Mém. Acad. Malgache* **24**:1–75.

Harrington, J. E., 1971, Olfactory communication in *Lemur fulvus*, Ph.D. Thesis, Duke University.

Harrington, J. E., 1974, Olfactory communication in *Lemur fulvus*, in: *Prosimian Biology* (R. D. Martin, G. A. Doyle, and A. C. Walker, eds.), pp. 331–335, Duckworth, London.

Jolly, A., 1966, *Lemur Behavior*, 187pp. University of Chicago Press, Chicago.

Marler, P., 1968, Aggregation and dispersal: two functions in primate communication, in: *Primates: Studies in Adaptation and Variability* (P. C. Jay, ed.), pp. 420–438, Holt, Rinehart, and Winston, New York.

Montagna, W., 1962, The skin of lemurs, *Ann. N.Y. Acad. Sci.* **102**:190–209.

Perrier de la Bathie, H., 1936, *Biogéographie des Plantes de Madagascar*, Société d'Editions Géographiques, Maritimes et Coloniales, Paris.

Petter, J.-J., 1962, Recherches sur l'écologie et l'éthologie des Lémuriens malgaches, *Mém. Mus. Natl. Hist. Nat. Paris. Sér. A.* **27**(1):1–146.

Petter-Rousseaux, A., 1964, Reproductive physiology and behavior of the Lemuroidea, in: *Evolutionary and Genetic Biology of Primates*, Vol. 2 (J. Buettner-Janusch, ed.), pp 91–132 Academic Press, New York.

Richard, A., 1974, Patterns of mating in *Propithecus verreauxi verreauxi*, in: *Prosimian Biology* (R. D. Martin, G. A. Doyle, and A. C. Walker, eds.), Duckworth, London.

Sussman, R. W., 1972, An ecological study of two Madagascan primates: *Lemur fulvus rufus* Audebert and *Lemur catta* Linnaeus, Ph.D. Thesis, Duke University.

Sussman, R. W., 1974, Ecological distinctions in sympatric species of *Lemur*, in: *Prosimian Biology* (R. D. Martin, G. A. Doyle, and A. C. Walker, eds.), pp. 75–108 Duckworth, London.

Sussman, R. W., and Richard, A., 1974, The role of aggression among diurnal prosimians, in: *Primate Aggression, Territoriality, and Xenophobia* (R. Holloway, ed.), pp. 49–76 Academic Press, New York.

Preliminary Notes on the Behavior and Ecology of *Hapalemur griseus*

15

J.-J. PETTER AND A. PEYRIERAS

In this chapter we present some preliminary observations on wild-living *Hapalemur griseus* which should be of interest despite their brevity, since this species has previously been the subject only of anecdotal observation. Elsewhere (Petter and Peyrieras, 1970), we have attempted to synthesize this material with more copious data derived from captive studies of the animal; the reader is referred to that study for more comprehensive information.

GEOGRAPHICAL DISTRIBUTION

Hapalemur griseus ("Bokombolo" in Eastern local vernacular and "Bekola" in Western) is widely distributed in Madagascar and occurs over a considerable range of altitudes. Increasingly it is being reported from zones where it was previously unknown; e.g., the peninsula of Ampasindava, the

J.-J. PETTER Laboratoire d'Ecologie Générale, Muséum National d'Histoire Naturelle, 91 Brunoy (Essonne), France. A. PEYRIERAS Laboratoire de Zoologie, O.R.S.T.O.M., Tananarive, Madagascar

northwest coast, and Tsaratanana (J. M. Betsch, personal communication, 1967). We have found it at altitudes of up to 1000 m in the forests of Cape Masoala.

A large form (local name, "Bandro") is, on the other hand, very localized; its present distribution is limited to a small area of semiaquatic vegetation surrounding Lake Alaotra. This lake, like other expanses of fresh water in Madagascar, has been drying up for several centuries and is now much reduced in area. Its depth today is no greater than 2.5 m, whereas in 1900 it was over 8.0 m. On the south and southwest, the lake is bordered by some 75,000 hectares of marshes, but the vegetation they support is regularly burned (particularly in September and October), and the future of the fauna inhabiting it is very somber.

HABITAT

We have observed *H. griseus* primarily in the neighborhood of Maroantsetra in the northeast of Madagascar. The town is surrounded by degraded forest, much of which is partially flooded for several months of the year (December–September). In general, the region is covered with bushy vegetation; the most common plant being *Aframomum danielli*, whose long stems, bearing large leaves, may attain 3 or 4 m in height. In the wettest areas this vegetation often forms dense clumps. However, in the midst of this low bush one finds, sometimes as closely spaced as 50–100 m, considerable patches of bamboo, often 20–30 m in length and 4–5 m across. The large individual bamboos comprising these may be 20 cm in diameter and up to 15 m high. It is in these clumps of bamboo that *Hapalemur* lives, and all except the most isolated of these are exploited. The animals are numerous in forests of this type throughout the northeast coast of the island and in the Ampasindava peninsula on the northwest coast. *Hapalemur* is also occasionally found in the dense forest at Périnet, in the primary forest to the west of Maroantsetra, and in the Masoala peninsula. In such conditions, however, as in all areas where they are hunted, the animals are less numerous and it is rare that an observer can maintain contact with them for more than a few moments at a time. In any event, it seems fairly clear that the distribution of this subspecies of *Hapalemur* is closely tied to that of bamboo.

Our observations of the larger form near Lake Alaotra have been too short to allow us to give more than a preliminary indication of the environment of this population. Lake Alaotra is now totally isolated from the forest; the densely populated surrounding area has been deforested throughout a radius of several kilometers from the lake shores. Dense need beds line the eastern and western shores of the lake, however, and it is in these that the large subspecies of *H. griseus* lives. Hunted on the eastern side of the lake, the animals are most easily observed on the west, away from human interference; we were able to accumulate 20 h of observation in this area. Our contacts with the animals were usually quite brief; the lemurs were rarely in exposed positions, and generally attempted to hide or to flee at our approach, jumping through the vegetation at water level or moving into the bases of papyrus clumps. As in the case of the other subspecies, individuals usually fled downwards towards the water level where the vegetation was more dense.

LOCOMOTION AND POSTURE

Hapalemur griseus may be considered as intermediate between the lemurids and the indriids in its locomotion and postures. Like all lemurids, it can run quickly on the ground. During such activity the anterior portion of the body is somewhat depressed, since the forelimbs are relatively short. Due to these proportions, the animal generally assumes a sitting position with its trunk held vertically when feeding on the ground. The long hindlimbs and feet enable it to jump considerable distances from one vertical support to another. The Alaotra subspecies, although heavier, has a mode of locomotion comparable to that of the smaller form: it never comes to the ground, but is able to move horizontally over the dense vegetation which in places covers the surface of the water; most often, however, it jumps in a vertical posture from one reed stem to another. These vertical supports are usually only 3 or 4 m high, the distances jumped are not large, and the jumps themselves are generally clumsy; one jump provoked by our approach resulted in the individual falling into the water. According to local fishermen, this larger form swims very well, dog-fashion, with the head just protruding from the water; even females bearing infants on their backs can cross canals over 15 m wide.

ACTIVITY RHYTHM

At Maroantsetra, our principal place of observation, we have been able to see *H. griseus* at all hours of the day. Their activity commences by sunrise, when they are already feeding in the tops of the bamboo. As soon as the sun has risen, the animals descend to the shadow of the center of the bamboo clump, where they remain quite close to the ground for the remainder of the day. It may be that *Hapalemur* living in dense forest has a slightly different rhythm involving more activity during the day. At Maroantsetra, the animals do not recommence activity until about 4 p.m., when they climb the bamboo to feed again or move to a different clump. This activity continues until night-fall, and perhaps even lasts for a couple of hours beyond sunset. We have seen groups feeding in the bamboo tops between 8 and 9 p.m. However, it is between 5 and 7 p.m. that the animals are most active, running rapidly through the bamboo tops or jumping from stem to stem. The rhythm of Alaotra *Hapalemur* is similar: periods of activity in the morning and evening separated by rest in the cover of the reeds.

DIET AND FEEDING BEHAVIOR

The early morning and evening are devoted to feeding. *Hapalemur* may also eat during the nights, but we are not certain of this. All of our observations lead us to believe that Eastern and Western forms of *H. griseus* feed in the wild almost solely on the shoots and leaves of bamboo. To eat, an individual sits on one of the larger branches at the base of a tuft, generally toward the end of a stem, and nips off a young twig with its teeth. Then, taking it by its ends, the animal explores the twig, passing it through his mouth from side to side while his jaw moves rapidly. The twig becomes more or less lightly crushed, according to the frequency of its passage across the premolars. Next the individual slices the twig, and from time to time bringing to his mouth the morsels held in his hand, eats the tender portions. The remains are thrown away when the edible parts have been consumed. This subspecies also seeks the young shoots of the bamboo, which it exposes by discarding the surrounding leaves. The base of the young leaf is extracted from its sheath by the mouth, but is transferred to a hand before being eaten. The animal starts with the pale, tender base, which it eats bit by bit, biting sections off with its premolars. Subsequently it tackles the green part, thrusting it into its mouth with a rapid back-and-forth movement of the hand, which, holding the end of the shoot, carries it alternately to the left and right sides of the mouth. Thus the young leaf is consumed, diminishing each time by the width of the jaw.

Hapalemur is also vegetarian at Lake Alaotra. We have seen several individuals eating *Phragmites* leaves, and according to local fishermen they eat not only the leaves and shoots of this plant, but also the buds and pith of *Papyrus*.

SOCIAL STRUCTURE AND HOME RANGE

In the case of *H. griseus* in the East, one most frequently finds groups of three to five individuals within a single bamboo clump; rarely does one see a larger number. From a distance it is almost impossible to determine the sex of individual animals, but several series of captures suggest that one adult male, one or two adult females and one or two juveniles would represent a typical group. Although these animals would thus seem to live most often in small family groups, we have no idea as to the permanence of these associations; local informants claim that larger groupings are sometimes seen in January.

Although nearly all bamboo clumps are exploited by *Hapalemur*, not all are inhabited simultaneously. It appears that a group may live in two, three, or four clumps, the number largely depending on their size. Animals within a group always stay together, and although they may be found in any of their home clumps, they prefer to spend their days in the clump remote from human activity. In traveling between clumps, the animals move rapidly through the vegetation close to ground level.

Comparable information on the Alaotran form is lacking, although it is clear that, at least in July, the animals live in small, spatially separated groups. According to local fishermen, groups at this time of year number three or four individuals, while a few months later assemblages of a dozen animals are seen; during February, when the water is at its highest, group size reportedly reaches 30–40. At this season the animals are very vocal.

REPRODUCTION

In the Maroantsetra region, the birth season occurs during January and February. In one group at Lake Alaotra in July, we observed an individual of about half the size of an adult; comparison with other *H. griseus* would suggest an age of about 6 months. This agrees with the information supplied by locals, who say that births take place during January and February. The young is always single, and is carried from the moment of birth on the mother's back; we have never seen an infant carried ventrally.

PREDATION

The principal enemy of *Hapalemur griseus* is man. However, living as it does in a degraded milieu, this form will not suffer from destruction of the forest as long as bamboo is available. In fact, in areas burned and abandoned long ago, where the bamboo has entirely replaced the original forest, the density of animals appears to be greater than in the undisturbed habitat. This is the case, for example, in the regions of Sambava and Vohemar, near Tamatave, between Fenerive and Vavatena, and in the area of Mahanoro–

Marolambo, where these animals are exceptionally abundant. It may also be that, man aside, the Maroantsetra area contains fewer predators than elsewhere, since the larger carnivores are almost exclusively forest animals.

The shelter furnished by the bamboo provides good protection against rapacious birds. In the Maroantsetra region, the species *Tyto alba* and *Buteo brachypterus* are common, and probably present a danger to young *Hapalemur*. Although rarer, *Gymnogenys radiata* may also pose a hazard to adults. Among the reptiles, the widely distributed boid *Acranthophis madagascariensis* is certainly a potential predator on young and adult alike.

As a somewhat larger form, the Alaotra subspecies probably has fewer natural predators. Its semiaquatic habitat is less accessible to some predators, but the vegetation itself provides less cover from aerial attack. According to locals, the small carnivore *Galidia elegans* is common along the lake shore, as is the large boid *Sanzinia*.

However, it is man who poses the greatest danger to *Hapalemur*. During a visit to the Lake Alaotra region in October 1969, we witnessed huge fires set in the reed beds. The animals fleeing from the flames were, in huge numbers, either killed on the spot or captured for later consumption. In one village we passed through, seven gentle lemurs had been eaten on the night of our arrival.

REFERENCE

Petter, J.-J., and Peyrieras, A., 1970, Observations éco-éthologiques sur les lémuriens malgaches du genre *Hapalemur*, *Terre Vie* **24**(3):356–382.

Field Observations on *Indri indri:* A Preliminary Report

16

J. I. POLLOCK

The following preliminary report describes the habits of the largest extant lemur, *Indri indri* (Gmelin, 1788). As a preliminary account, it contains assertions which can only be more fully substantiated elsewhere (Pollock, in preparation).

INTRODUCTION

Unlike those of its fellow indriids, *Propithecus* and *Avahi*, the distribution of *Indri* is limited to a part of the humid eastern region of Madagascar. The population density of *Indri indri*, the only member of its genus, varies widely within this small region, confounding attempts to estimate the total species' population. Among all lemur genera, only *Daubentonia* has as restricted a distribution as *Indri*, which inhabits some of the most southern and elevated true rain forests in the world. *Indri* and sympatric lemurs probably experience the lowest temperatures encountered by any tropical primate with winter minima sometimes approaching 0°C.

J. I. POLLOCK   Department of Anthropology, University College, Gower Street, London WC1E 6BT, England

approx I foot

FIG. 1. Adult male *Indri* in "mid-leap" position with the lower half of the body turning toward the landing support.

FIG. 2. *Indri* in posture of no pelvic support, and feeding on the young mature leaves of *Ocotea* sp.

The most complete physical description of *Indri*, by Hill (1953), empha-sizes the great variability in pelage coloration. Infants are very dark or com-pletely black and develop whiter areas, especially on the lower dorsal region, sides of arms, eyebrows, throat, and forehead, between 4 and 6 months of age. Although the intensity of coloration in adults differs individually, the pattern of variegation is consistent. A deep brown covers the chest and abdomen and the heel of the foot possesses a yellow or orange hue.

The characteristic indriid skeletal structure is well documented by Hill (1953). The relatively low intermembral index reflects the specialized vertical clinging and leaping locomotor system defined by Napier and Walker (1967) and analyzed more completely by Walker (1967). Habituated animals in par-tially degraded forest during the present study confirmed Walker's observa-tion of the almost exclusive use of vertical trunks during leaping sequences. However, in undisturbed primary forests under conditions where a high pro-portion of trees possess breast-level diameters as great as 50–150 cm, unha-bituated *Indri* consistently make use of angled boughs of small diameter at the lower levels of high canopy while fleeing the observer. In this environ-ment, they also descend and frequently use strictly vertical supports at heights of and below the discontinuous intermediary canopy layers (between 15 and 20 m).

The most striking feature of the *Indri* skeleton, which is otherwise very similar to that of *Propithecus*, is the rudimentary tail, visible only as a short stump that is elevated during defecation. Grand and Lorenz (1968) concluded that the tail of *Tarsius* provides an essential force during vertical resting posi-tions without pelvic support, since animals with grease-coated tails slip down vertical posts. Despite adopting postures identical to those of *Tarsius*, how-ever, *Indri* remain in such attitudes (Fig. 2) for many minutes.

The visceral characteristics of all indriids, including hypertrophied sali-vary glands, voluminous abdominal viscera with a capacious stomach, elon-gated caecum, lengthy uniform intestine, and elaborately complex colon, cer-tainly reflect their folivorous–frugivorous diet. Compared with other indriids, *Indri* exhibits the greatest development of these alimentary specializations.

Habitat

The topographic structure of the eastern region of Madagascar consists of a series of steep crests running approximately north to south and budding off successions of secondary folds. These geological "waves" have ampli-tudes of up to several hundred feet and lengths from a few hundred to 2000 m. The prevailing easterly winds contribute to the great variety of atmos-pheric conditions found between the valley bottoms, eastern and western slopes, and mountain tops.

Mean monthly rainfall and temperatures at the forestry station of Ana-lamazoatra (altitude 928 m) near Perinet, a main study area, are shown in Table I. The mountainous structure found in most forests is less acute in this peripheral region, which lies 15 km inside the western limits of the rain forest. Between 1933 and 1964, the absolute minimum and maximum temperatures recorded in Perinet were 5.3°C and 33.6°C. No data are available for higher regions of the forest, but the study area near Fierenana (Fig. 3) was considerably wetter and colder throughout the year. In all study areas the seasonality of both temperature and rainfall was marked; afternoon tropical downpours during the warm austral summer (November to March) contrasted with long periods of light rain in other months.

The rain forest flora of Madagascar had been generally but incompletely described by Perrier de la Bathie (1921), Humbert (1936) and Humbert and Cours Darne (1965). A recent treatment of the subject by Koechlin (1972) summarizes contemporary knowledge and lists some of the more common plant genera. The large number of species, their physiognomical similarities, and the differing characters of individual trees found at different heights on the slopes make plant description and identification difficult. A small proportion of the floral species is described by Cabanis *et al.* (1970).

The few (unpublished) phenological studies available illustrate the irregularity of fruiting patterns not only between but also within species. Very few, if any, common plants fruit every year.

During the study it became clear that the trends associated with varying altitudes in different forests of the eastern domain were also present on a single mountain. Hence the wet, mossy, undergrowth-free valleys containing the straightest and tallest trees with continuous canopy become gradually replaced near the crests by small, densely packed trees, often with twisted shapes and covered with lichens and orchids. Exposed rock or huge boulders are common near the valley streams, and glades of bamboo are found in thick patches in the higher regions. Tree ferns are common at all altitudes.

The structure of the eastern rain forest is, therefore, locally heterogeneous, and the limited distribution of some plant species, the various gradients of temperature, rainfall, and exposure to wind, and the diverse physical organization of the forest make for a notably complex environment rich in botanical diversity.

Study Sites

Three study sites were chosen: the forest of Analamazoatra near Perinet (Fig. 4), latitude 18°56', longitude 48°24', altitude 928m; the forest of Vohidrazana near Maromiza, 10 km southeast of Perinet, 1200 m; the forest of Sahamanga near Fierenana, 60 km north of Perinet, 1300+ m (see Fig. 3).

TABLE I. Mean Monthly Temperatures and Rainfall at the Forest of Analamazoatra[a]

	J	F	M	A	M	J	J	A	S	O	N	D	Mean
Rainfall (mm)	306.0	320.0	261.9	92.4	61.4	76.9	77.8	67.1	51.1	43.9	112.2	237.6	142.4
Mean maximum temperature	27.0	27.1	26.1	25.4	23.2	21.3	20.3	20.6	22.2	24.1	26.3	27.1	24.2
Mean minimum temperature	16.9	16.6	16.7	15.4	12.9	11.3	10.5	10.3	11.1	12.4	14.7	16.3	13.8
Mean temperature	22.0	21.9	21.4	20.4	18.1	16.3	15.4	15.5	16.7	18.3	20.5	21.7	19.0

[a]Overall mean annual rainfall 1708.2 mm. All temperatures in °C. Rainfall data from 1928 to 1960. Temperature data from 1941 to 1960.

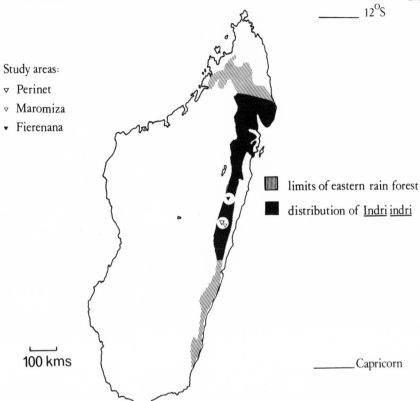

Study areas:

▽ Perinet

▽ Maromiza

▼ Fierenana

limits of eastern rain forest

distribution of Indri indri

100 kms

FIG. 3. Approximate limits of the distribution of *Indri indri* in Madagascar. After Hill (1953), Peyrieras (personal communication), and personal observation.

Habituation

At Perinet, a forest of low stature previously subjected to small amounts of selective logging, two groups (Group P and Group V) containing young animals of different ages were habituated by repeated location and following. The initial responses of alarm vocalizations and fleeing diminished after 1 wk and feeding was at that time occasionally observed before the animals were lost. After 3 wk, animals would approach and feed within a few meters, although individual variation in the habituation response was marked, adult females and young feeding confidently nearby and adult and old males watching nervously from afar. The habituation process continued throughout the study, and all animals would eventually feed within 1 ft of the observer.

Habituation of lemurs may result from accumulative as well as sustained

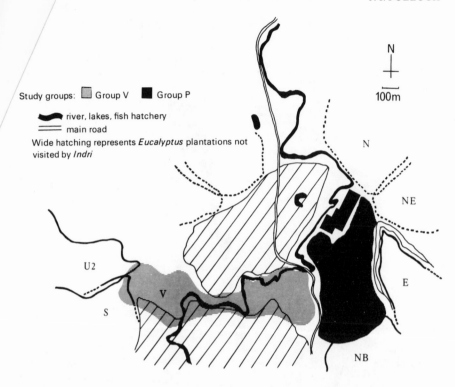

FIG. 4. Local distribution of *Indri* groups in the forest of Analamazoatra near Perinet. Known (continuous lines) and approximate (discontinuous) boundaries are shown for ten *Indri* groups.

contact with humans. Group 4 at Maromiza habituated almost perfectly over a period of 15 months as a result of the 3-day visits every 6 wk.

Study Schedule

The study was organized into eight 6-wk circuits from the beginning of September 1972 until the end of July 1973. Each circuit was composed of 20 days' intensive quantitative observations on the two habituated groups at Perinet (P and V, Fig. 4), 3 or 4 days at Fierenana where all lemurs were studied in an effort to gauge qualitatively the ecological parameters present in primary forests, and 3 successive days at Maromiza to collect vocalization data from a whole population of *Indri*. The remaining days in each circuit were allocated to travel, census work, the investigation of other forests, and observations on other lemur species.

GROUP STRUCTURE AND COMPOSITION

The censusing of forest-living primates presents a variety of problems, as several previous authors (e.g., Chalmers, 1968; Aldrich-Blake, 1970) have noted. Particularly in environments such as that of *Indri*, where observer movement is limited, atmospheric conditions are poor, and vegetation density is high, the identification and, if possible, the control of visibility bias, favoring obvious at the expense of discrete activities, are necessary. Such control was, unfortunately, impractical during the study reported here because of the long periods of time involved in locating the animals.

Although the absence of obvious morphological dimorphisms in *Indri* impedes the direct determination of sex in all but well-habituated animals, sexing is facilitated by the sexual distinction in behavior, both toward potential predators (e.g., the observer) and in relations with other group members. These behaviors will be fully discussed elsewhere (Pollock, in preparation). In all habituated groups, physical observations confirmed behavioral allocations to sex.

Three age categories of immature animals were recognized; since, as in other lemurs studied to date (Petter-Rousseaux 1964, 1968, 1969), reproduction in *Indri* is seasonal, the category of any individual can be recognized by its size or state of physical development. Animals in their first or second years can be readily distinguished from each other and from adults by the criterion of size; by the third year the immature animal is generally of intermediary size and has probably not yet achieved full sexual development. Table II shows the composition, actual or inferred, of the groups censused after the 1972 birth season. In the total of 18 groups there were 55 animals (mean group size 3.1; modal size 3.0). Table III presents the results of recensusing in June–July, 1973; unfortunately, adverse weather conditions during this period severely reduced the chances of counting many groups.

In the eight groups accurately recensused in 1973, the total number of animals increased from 27 to 28. Of all the six groups that were recensused and which had originally contained infants, all possessed the expected juvenile. This contrasts strongly with Richard's (1973) observation of 100% infant mortality ($n = 4$) in her southern study area and 84% infant mortality ($n = 6$) with the northern subspecies of *Propithecus verreauxi*.

Apart from one two-male group, 14 out of the 20 contained an adult female, the other five possessing individual(s) of undetermined sex. Seventeen of the 20 groups included one adult male, the remaining three with individual(s) of undetermined sex. Twelve of the 20 groups had infants in 1972 and one an animal of 1 yr of age and another older individual of small size. An infant was born to one group of four the next year, the only known birth of 1973. These data, combined with the apparent absence of sexual competition within a group and knowledge of mother–offspring relationships, strongly

TABLE II. *Indri* Group Composition in June 1972[a]

Location		Group	A♀	A♂	A*	J2	J1	I	UK	Total
Perinet		P	1c	1c	1(♂c)	1(♂c)	1(♀c)	—	—	5
		V	1c	1c	—	—	—	1(♂c)	—	3
		NB	1c	1b	—	—	—	1	1	4
		NE	—	—	—	—	—	—	2	2
		W	—	1b	—	—	—	—	>1	>2
		U	—	—	—	—	—	—	>3	>3
		U2	—	—	—	—	—	—	2(♂c)	2
		S	—	1b	—	—	—	—	1	2
	Total	8	3	5	1	1	1	2	>10	>23
Fierenana		1	1c	1b	—	—	—	1	—	3
		2	1c	1b	—	—	—	1	—	3
		3	1c	1b	—	—	—	1	—	3
		4	1c	1b	—	—	—	1	—	3
		5	1c	—	—	—	—	1^b	1	3
		6	1c	1c	—	—	—	1^b	—	3
	Total	6	6	5	0	0	0	6	1	18
Maromiza		1	—	1b	—	—	—	—	2	3
		2	1c	1b	—	—	—	1	—	3
		12	1c	1b	—	—	—	1	—	3
		3	1c	1b	—	—	—	1^b	—	3
		4	1c	1c	1(♀c)	1(♂c)	—	—	—	4
		5	1c	1b	—	—	—	1	—	3
	Total	6	5	6	1	1	0	4	2	19

[a]Key: Sex determination: c, certain; b, behavioral (see text). Assumed age categories (see text): A, adult aged 9 yr or more; A*, adult-sized, age 6–9 yr; J2, old juvenile aged 3–6 yr; J1, juvenile aged 1–3 yr; UK, animals of unknown age and sex probably including a high proportion of females, especially those without I.

[b]Infant first seen as a young juvenile and assumed to have been present as an infant in that group in June 1972.

support Petter's (1962) contention that *Indri* societies are "small family groups."

The study of changes in relative spatial displacement around their habitat remains of crucial importance in understanding the group compositions and social behavior of primate species. Rigorous proof of a disciplined state of monogamy would crucially affect our interpretation of the dynamics and behavior of family-living primates, but to date there is knowledge only in the Hylobatidae of specific pair-bond retention over several breeding seasons (Chivers, 1973). The sexual integrity of a single adult pair of animals must always be partially in doubt.

If *Indri* live in monogamous, exclusive families, a stable population may be thought to contain at least approximately equal numbers of group sizes ranging from two to a maximum depending on the maturation period and/or tolerance to subadults by other group members. *Indri* groups, it can be reasonably argued from the results presented above, normally contain a single adult pair with reproductive status and other individuals who are infants or are clearly young, presumably offspring from previous breeding seasons.

Socionomically, *Indri* represents a departure from the extreme behavioral cohesion formerly thought to be characteristic of Indriidae (Petter, 1962); the larger groups of *Propithecus verreauxi* frequently produce two infants in a single birth season (Jolly, 1966; Richard, 1973). The indriids sympatric with *Indri* are *Propithecus diadema* and *Avahi laniger*, the former personally encountered in groups numbering 2, 3, 1, 2, 3, and 5. Albignac (personal communication) reports sighting a group of six animals including two infants, making it possible that no real social difference exists between the two species of *Propithecus* outside the observed tendency of rain forest-living lemurs to have smaller group sizes, larger home ranges, and therefore, lower densities than western forms (Pollock, in preparation). The potentially close genetic

TABLE III. *Indri* Group Composition in June 1973[a]

Location		Group	A♀	A♂	A*	J2	J1	I	UK	Total
Perinet		P	1c	1c	1(♂c)	1(♂c)	1(♀c)	—	—	5
		V	1c	1c	—	—	1(♂c)	—	—	3
		nb								
		ne								
		w								
		U	—	1c	—	—	1	—	>1	>3
		uii								
		s								
	Total	3	2	3	1	1	3	0	>1	>11
Fierenana		i								
		ii								
		iii								
		4	—	—	—	—	—	—	>2	>2
		v								
		6	1c	—	—	—	1	—	1	3
	Total	2	1	0	0	0	1	0	>3	>5
Maromiza		i								
		2	—	—	—	—	1	—	2	3
		12	1c	1b	—	—	1	—	—	3
		3	1c	1b	—	—	1	—	—	3
		4	1c	1c	1(♀c)	1(♂c)	—	1	—	5
		5	1c	—	—	—	1	—	1	3
	Total	5	4	3	1	1	4	1	3	17

[a]Key: As for Table II; lowercase letters and Roman numerals designate groups not recensused.

relationships between *Propithecus verreauxi* group members are apparently diluted by the mating season mobility of males (Richard, 1974). Such plasticity in group complement was not observed in *Indri*. *Avahi laniger* was encountered in groups of 1, 1, 2, 2, 2, 2, 3, and 4 in these same forests.

In the continuum of primate societies, the formation of groups possessing more than one adult female in reproductive condition may require a fundamentally different system for the maintenance of social cohesion than that operating in true families. Unless an ancestral condition has been retained, which seems unlikely (Charles-Dominique and Martin, 1970), the discrete family has evolved in at least three major divisions of the primate order: Ceboidea (Mason, 1966), Hominoidea (Carpenter, 1940; Ellefson, 1967), and Lemuroidea. It is likely that an extensive analysis and comparison of groups of *Callicebus moloch*, *Hylobates* sp., and *Indri indri* will accelerate an understanding of the functional significance of family social systems.

POPULATION DENSITY

Introduction

A loud, continuously modulating call or "song" (see below) is frequently heard in forests inhabited by *Indri indri*. Petter (1974) estimated population density from measures of calling frequency, necessarily assuming two attributes of *Indri* behavior: that all groups called over the study period, and that neighboring groups were vocally different and could be acoustically distinguished.

During the present observations it was discovered that the tendency to call was a function of season, weather, proximity of adjacent groups, and other undetermined conditions. Calling frequency was highly variable; in extreme cases, a silence of several days would be interrupted by four or five vocalizing sessions within a few hours. Figure 5 illustrates the distribution of calling frequency in the two intensively studied groups at Perinet over the period September 1972 to July 1973. The similarly extensive ranges associated with each study session (10 days per circuit for Group P, 6 days for Group V) for each group emphasize the danger in regarding *Indri* singing patterns as consistent. The Circuit IV call frequency for Group V was affected by a 4-day cyclone with continuous hard rain reducing all activities. Over the whole year, the number of silent days was approximately equal to the number of one-song days: for Group P, 27% and 30%, respectively; for Group V, 33% and 29%. Two-song days for the two groups occurred in 23% and 12% of the total days for Groups P and V, respectively. The maximum recorded number of songs in a single day was seven for Group P and six for Group V.

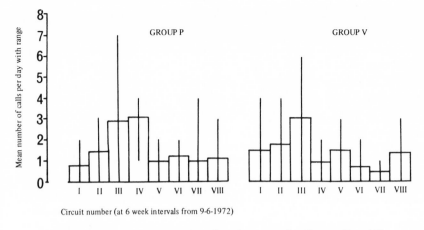

FIG. 5. Seasonal variation in calling frequency for Groups P and V.

It is reasonable to assume that over a lengthy period all groups will call many times, since most inter- and much intragroup communication is effected by this means. Conclusions about population density made solely from vocalization data exclude the essential discrimination between many groups calling and multiple-calling groups, and ignore the presence of consistently quiet individuals.

Results

Population density estimates in the three study areas are shown in Table IV. At Fierenana and Maromiza, the groups found and/or heard calling over the whole study period within an estimated radius of 1000 m were either counted or estimated. This enabled maximum density limits to be assigned to each area, as the audibly enscribed circle included only parts of the territories (see below) of some groups. At Perinet, the forest region was more accurately measured from maps and aerial photographs, and the densities lie within the stated limits. An adjustment in parentheses is included which corrects for the presence of *Eucalyptus* plantations avoided by the animals. The results from Maromiza should not be directly compared with those from Fierenana, as the former site was specifically chosen for optimal acoustic sensitivity. However, in 1500 m, the access path passed through the territories of six groups, a fact suggestive of a relatively high density.

The measured territories of the two intensive study groups, at 17 and 18 hectares, together make an adjacent group density of 22.9 animals/km². These territories were, however, artificially bordered by paths or a road for

TABLE IV. Population Density of *Indri indri* in Three Study Areas[a]

Location	1	2	3	4	5	6	7	8	9
Fierenana	9	314[e]	6	18	24	33	8–10	9	9
Perinet	9	282	6	18	24	33	8–11	10	11
		(264)					(9–12)	(11)	(12)
Maromiza	12–15	314[e]	6	20	32	65	11–21	12–15	13–16

[a]Key: 1, number of groups; 2, area in hectares ([e] indicates estimated area of 100 hectares); 3, number of groups with known complements; 4, number of animals in groups with known complements; 5, minimum number of animals in all groups based on the assumption that all groups with unknown complements contained two animals; 6, maximum number of animals possible in all groups based on the assumption that all groups with unknown complements contained five animals; 7, limits of population density for cases 5 and 6 (Number of animals/km2); 8, population density calculated on local mean group size; 9, population density calculated on overall mean group size. Overall mean population density for all three study areas 9–15 animals/km2. Mean population density based on overall mean group sizes in all three study areas 10–11 animals/km2.

much of their perimeter. The overlap of about 50 m in the case of three group ranges becomes extremely important in reducing (1) a square and (2) a circular territory each of, for example, 17 hectares, by 43.4% and 38.4%, respectively, leaving only 9.6 and 10.5 hectares for a group's exclusive use. If the assumption is made that no overlap area is used by more than two groups, a population density correction factor of one half these percentages must be employed. For a circular territory in this case the figure of 22.9 individuals/km² is reduced to 18.5. It seems unlikely that *Indri* territory size is a function of group size (Pollock, in preparation), and this relatively high figure could be accounted for either by the artificial bordering noted, if it restricted range expansion, or by the difference between actual group size totals (8) compared to that of the mean (6.2). This last factor would further reduce the animal density to 14.3/km². The inaccuracies of the subjective estimation of 100–hectare areas or the presence of unoccupied patches of forest could, additionally, account for the discrepancy. Finally, the overlap area increment magnifies as the territorial geometry departs from a circular condition.

It should be emphasized that these data manipulations are utilized neither to present absolute, accurate measurements of *Indri* population density nor to illustrate regional variation in animal concentration, but rather to characterize the sort of population limits within which *Indri* lies. The figures given are rough guides to densities in highly selected areas. At Betampona, a warm, well-protected forest near Tamatave, five calls were heard in 3 days in early August 1972, all well over 1000 m distant, indicating far lower numbers than those in the three study areas which were chosen partially for their high animal densities.

These figures enable a biomass estimate to be calculated. This lies, for the three study areas combined, between 0.8 and 1.9 kg/hectare. This is comparable only with that recorded for the spider monkey, other primate folivore/frugivores ranging from 4.0 to 15.0 kg/hectare (Hladik and Hladik, 1969; 1972).

Conclusions

Animal density in three populations of *Indri indri* lay between 8 and 16 individuals/km². This corresponds to from 3 to 5 groups, a figure greatly in excess of Petter's (1974) estimate of 1 group/km² in one of the present study areas (Fierenana). Petter suggested that the population density in degraded areas increased as disturbed animals fled to quiet parts of the forest where large concentrations accumulated. In this study, comparison of the population structures in primary and selectively degraded forests exposed no clear differences, though the sample size was small.

The population density of *Propithecus verreauxi verreauxi* has been calculated at 244 animals/km² (Jolly, 1966) in the gallery forest of Berenty near Fort Dauphin, but Richard's (1973) home ranges for the same species indicated far lower numbers in the *Didierea* bush nearby. Population nuclei, which may be widespread throughout the Lemuroidea, have been suggested for *Lepilemur mustelinus* (Petter, 1962) and *Microcebus murinus* (Martin, 1972). Martin, using mark, release, and recapture techniques, concluded that the mouse lemur population in the 1.5 hectare Mandena study areas consisted of between 20 and 40 highly localized individuals (1300–2700 animals/km²). Petter and Petter-Rousseaux (1964) puts *Microcebus murinus murinus* density at Mahambo in the coastal rain forest at 20 animals/8 hectares (250 animals/km²). *Lepilemur mustelinus leucopus* was calculated by Charles-Dominique and Hladik (1971) to be 200, 220, and 350 animals/km² in *Didierea* bush, and 270, 450, and 810 animals/km² at Berenty. The same authors claim that Perinet densities for *Lepilemur* and *Avahi* combined numbered 200/km², concluding that there may be comparable population densities of island-wide forms in dry and wet forests but acknowledging the difficulties of animal location in the latter.

In this study, a recognizable group of *Propithecus diadema diadema*, observed irregularly over the study period ranged consistently in an area with one minimum dimension of 500 m. If this was the diameter of a circular territory or home range, it would approximate 20 hectares in size. The frequency with which this species was found in all study areas suggested very low densities. *Lemur fulvus fulvus*, in groups of six to eight in the same forest, occupied large areas and was infrequently encountered compared to the same species in the thick deciduous forest of Ampijoroa (personal observation) and other regions in the west (Sussman, personal communication, and Chapter 13).

Mapping accurately all parts of a forest utilized by a single lemur group probably requires several years' research, and even the most rigidly defended territory may be impermanent. Although the variation found in animal density between the several areas was high, it appears that those species inhabiting the rain forest may live in lower concentrations than those of the west. It seems likely that such trends towards smaller group sizes and greater intergroup distances, resulting in lower animal densities, may derive from a common cause.

A SYNOPSIS OF THE BEHAVIOR OF *INDRI INDRI*

Activity patterns

Indri indri, locally known as the "babakoto," "amboanala," or "endrina," has a short, strictly diurnal activity period (Fig. 6) lasting from 5 to 11 daylit hours. The exact length of this period is dependent upon season, temperature, and atmospheric conditions. The animals sleep at a height of between 30 and 100 ft in nonspecific trees. Except for the presence of sleeping partners, nocturnal group dispersion is similar to that occurring while feeding, entailing a maximum dyadic spread of over 100 m. *Indri* young sleep with their mother every night for the first year, initially enclosed inside the round folded "bowl" of her body, and thereafter boxed in the "locomotive" described by Jolly (1966) for *Propithecus verreauxi*. During the second year the frequency of this sleeping arrangement diminishes and becomes irregular. The next youngest of the family occasionally sleeps with the adult male. No more than two animals ever sleep together.

The animals in a group synchronize the initiation of activity over a few minutes, often stretching by hanging suspended by the arms or standing bipedally on their horizontal sleeping support. A brief feeding bout on the nearest available food (sometimes the sleeping trees themselves) precedes a short progression and descent; the members of a group then urinate and defecate in concert, often in a previous defecation area and often repeatedly taking identical supports. There are many such areas in a group's territory, the disposition to travel to a particular site appearing to be related to its proximity to the sleeping trees.

Feeding on leaf buds, young leaves, leaf stalks, adult leaves, flowers, and fruit then commences, interspersed with occasional progressions to new feeding trees, desirable species of fruit being consumed for about 2 h daily. During the course of the study, 63 plant species from 39 genera and 19 families were recognized as forming the basic food repertoire. The greatest number of food species were contained in the Lauracae (14 species), Guttiferacae (9), Ano-

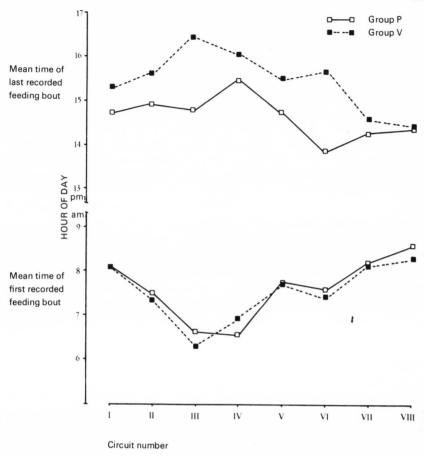

FIG. 6. Seasonal activity variation in two *Indri* groups.

nacae (4), Monimiacae (4), Euphorbiacae (3), Moracae (3), and Sapindacae (3). Between them, the Lauracae and Guttiferacae provide the most commonly eaten species and probably the bulk of food consumed. In addition, they comprise the most common plants of the rain forest (Abraham, personal communication; personal observation).

Fruits were eaten from 25 (39%) of the 63 species, and 13 (21%) were chosen exclusively for fruit during the study period. However, with irregular and polyennial phenological rhythms predominant, it is very likely that these percentages would change with time.

No animal matter, alive or dead, was consumed by *Indri*.

Feeding continues throughout the day in constant activity. This contrasts with both all-season studies on lemurs conducted in dry forest, where a bimodal, diurnal feeding behavior distribution is generally interrupted by mid-day rest periods. In the rain forest, *Lemur fulvus* also follows this pattern typical of tropical primates, whereas the sympatric *Propithecus diadema* probably feeds continuously.

Indri usually feed each day momentarily on the earth exposed by an up-rooted tree trunk.

All levels of the forest are utilized in feeding, small saplings in new leaf being pulled toward a major supporting vertical trunk. Leaves are always ea-ten off the tree, but fruit is picked orally, transferred to the hand, and gnawed out of the clenched fist. As a rule, three limbs are used in support dur-ing a feeding bout, the remaining arm either pulling terminal branches to the mouth or holding fruit.

Although most time is spent feeding on some of the commonest plants of the forest, the animals demonstrate a great intraspecific choice of certain in-dividual trees, especially when these are in new leaf or fruit.

A grooming session announces the end of feeding and the animals rest, often for 18 h, without moving and only occasionally looking up at strange noises, to groom, or to scratch with the toilet claw. The group's daily range of about 350 m does not appear to be organized on a regular territory circum-scription principle.

Territoriality

Indri indri is territorial in that groups use a large central portion of their ranging area exclusively and actively defend this against intruding groups in overlap perimeters. In over 2000 h of observation, mainly on three groups, on only six occasions were other groups seen; these were always in an area just inside the ranging limits of the study group. In this overlap region, a "singing" battle ensued on each occasion (see below). No male fighting was seen but two adult male *Indri* showed trauma: one had most of the left ear missing and the other lacked an eye.

The marking behavior of males (see below) was particularly conspicuous after border encounters; they fed little, remaining in vigilance for the sub-sequent active hours. No obvious effect of these intergroup encounters on ranging behavior was anywhere evident, and it remained unclear exactly how territorial limits were defined. Scent marking appeared to be equally distrib-uted throughout the territory. It was evident, however, that particular groups exclusively occupied a region of forest (in two measured cases, 17 and 18 hec-tares) which was defended by display and perhaps by brief physical in-volvement. Regular patterns in the movement of individuals or groups to pa-trol border areas, as suggested by Chivers (1973) for the siamang, were not observed.

Vocalizations

Marler (1968) suggested that lateral visibility in a primate's habitat determines to some extent the tendency of the animals to include auditory signals as accompaniment to visual displays. Perhaps the most characteristic feature of *Indri's* biology is the loud song choruses ("longue plainte modulée" of Petter, 1962) that resound in the montane forest. Any group's call may induce a nearby group to reply (Fig. 7); this often develops into 30-min singing sessions. Each call normally lasts between 60 and 150 s and consists of a series of cries or howls varying in frequency between 500 and 6000 Hz it can be heard by the human observer under ideal conditions from 1500 to 2000 m away. These calls are emitted throughout the territory[1] of each group, and seem to serve a number of functions. They comprise the majority of intergroup communicatory acts within a population. The auditory acuity of *Indri* is such that they were able to perceive calls at least 3000 m from source.[2] In regions of high population density, this places any group in potentially direct communication with at least 100 others in surrounding areas. The call probably declares the occupation of a territory, functions to reunite temporarily dispersed individuals in a group, and is sometimes produced in response to aerial predators, aeroplanes, and thunder. The call may also have a significant function in the transmission of information concerning the reproductive status of calling individuals, due to the characteristic partial participation of developing adolescents in a group's song.

The frequency of *Indri* song is highly variable, both from day to day and between seasons. In addition, the patterns of calling within a population change according to the time of year, varying from successions of single calls to the predominance of long calling sessions.

A loud barking "roar" ("hurlements" of Petter, 1962) by all group members sometimes introduces a song but is normally the aerial predator alarm call. Ground predators and observers receive an exhaled "hoot" ("coup de klaxon" of Petter, 1962) from the adult male and occasionally from his oldest male offspring.

Other vocalizations are quiet and include a soft "hum" announcing imminent movement, and a variety of discrete disturbance noises.

---

[1] There was no obvious correlation between singing location and specific parts of the study groups' territories. It is possible, however, that detailed analysis of the data will reveal a more subtle relationship.

[2] Auditory sensitivity in *Indri* was evident from the experience of induced calls commencing immediately (allowing for the time spent for sound transmission) after a distant group finished singing (i.e., "grouped?"), where both calls were extremely faint to the human observer situated approximately midway between them.

Study time : 3 successive days

Total time over which all calls were heard : 1408 minutes (23 hours 28 min)

Total number of calls : 85

The percentage of single calls:	28%

The percentage of calls grouped in 2 or more:	72%
3	55%
4	38%
5	33%
6	15%
7	8%
8	0%

Maximum number of grouped calls : 7

FIG. 7. Call clustering at Maromiza in the forest of Vohidrazana in March 1973. A "grouped" call is defined as one either overlapping a previous call, or beginning either immediately after or within 10 seconds of the end of a previous call. The 10 second limit is a natural one. Inter-call intervals beyond this delayed response period consisted of many minutes or hours. The occasional simultaneous occurrence of geographically distinct calling regions causes the above percentages to *overestimate* the numbers of natural groups of calls. The poor auditory sensitivity of the observer certainly contributed to an *underestimation* of the numbers of calls in a calling session as distant calls were missed. The tendency to call at certain times of the day was observed in all study areas, otherwise, conceivably, a whole population might have been in continous calling communication.

Social Behavior

The intragroup social activities of the Indriidae are limited in variety and constitute a minute proportion of the day's activity. *Indri's* overt social behaviors can be concisely listed:

Allogrooming: repeated alternate grooming bouts concentrated on the partner's head and neck. Unilateral allogrooming sessions are rare, and simultaneous mutual allogrooming, common in other lemurs, nonexistent.

Social displacement: the (1) submissive and (2) seemingly ritualized and voluntary quitting of a resting or feeding position in response to (1) agonistic or (2) nonagonistic behavior of another individual. Agonistic behavior in this context is defined by an aggressive approach with attempts to bite, wrestle, and kick the displaced individual, sometimes accompanied by "disturbance" vocalizations from one or both animals. These sounds are quiet "grunts," "kisses," or "wheezes" (in order of increasing intensity) and are made whenever the animals are frightened or anxious.

Play/wrestling: the silent, contorted struggles of animals, usually young. Both affiliative (allogrooming) and agonistic (social displacement) actions are continuously involved in many play bouts.

Sexual activities: the following or orientation of the male toward the female, smelling her genitalia, attempted copulations, and copulatory behavior.

Food stealing: permitted fruit stealing from the adult male or female by young offspring. This is extremely rare except among infants.

There remain the broad and less definable quantities, immeasurable or imperceptible, such as categories of social behavior involving some vocalizations, locomotion and other communicatory acts of "observation," and body movement. Also omitted are actions performed synchronously by the whole group and which may have real social significance, such as "singing," "social defecation sessions," and feeding on earth.

The overall low frequency and the easily visible nature of the behaviors listed above are such that a very high proportion of these social activities could be recorded. All these categories were seasonally variable in frequency, occurring relatively often during the warm summer months. Sexual behavior and play/wrestling were totally restricted to this period between November and March.

Sexual Behavior

Attempted copulation was witnessed, in 1973, on four occasions in Group P: January 10, February 18, 19, and 26. Group P was observed from January 10 to January 20 and from February 18 to March 1. The youngest animal in this group was at that time 21 months old. No sexual behavior of any kind was seen in Group V (observed on the 8 days following Group P in each case) which had produced an infant the previous year.

On all occasions, only the adult pair were involved, and the male's advances—smelling and licking the genitalia of the female, followed immediately by mounting—were easily rejected with a cuff. The female did not give birth to an infant that year. The gestation period of *Indri*, calculated from the above information and the presence of a neonate in another group at the end of May, is probably between 17 and 22 wk. It is likely that the full range and intensity of sexual behavior would only be witnessed in a group every third year.

Marking Behavior

Indri appear to mark parts of their environment by means of glands situated on the side and underneath of the muzzle ("cheek marking"). No throat gland such as that found in males of both species of *Propithecus* is visible on living *Indri*. However, Petter (1962) has described paired gular (anterior part of the throat) glands in both sexes of *Avahi laniger*, a position analogous to the cheek region rubbed by *Indri* males (especially) and females on trunks and branches. No glandular structure or stain could be discerned and the presence of glands could only be inferred from behavior.[3]

Genital marking, on occasions accompanied in the males by urination, is also practiced. Auto-endorsing (Richard, 1974) whereby a cheek-marked area was immediately remarked genitally, was observed on two occasions; in both instances, it was performed by adult males.

Marking behavior, of low general frequency compared to that of the sympatric *Lemur fulvus*, *Hapalemur griseus*, and *Propithecus diadema*, is thus performed predominantly by adult males and occurs relatively often during the summer months.

Infant Development

The young infant, born in May, suckles three or four times during the day in well-defined bouts, at least one of which lasts over 30 min. This very likely causes an extension of the activity period of the group (Fig. 6). Suckling always commenced when the female settled down in the midafternoon to her final sleeping position; the axillary nipple was reached from behind by infants more than three months old.

The local environment holds much interest for the infant a few weeks after birth. Initial grasping of the mother's fur, and periodic movement over her chest and neck, is transferred to nearby small horizontal branches at 4–6 wk. The infant supports itself at first solely by the arms and later by all limbs. From the outset, any activity period is composed of constant movement, at first continuous clambering over the mother and adjacent canopy structures, and later, repeated circuits of four or five jumps which mirror the complete locomotor system of the adults.

Vegetable food is casually gnawed from two months of age; thereafter, feeding gradually develops from a toylike interest in the twig or fruit to in-

---

[3] Many mammals, especially felids, rub the sides of the face or muzzle on their environment or each other. In some cases, specialized sebaceous glands have been found in that region of the skin. However, it has been pointed out to me by Charles-Dominique (personal communication) that some lemurs certainly appear to mark by salivation. Very often the lip is partially reversed by marking *Indri*, and saliva would be deposited at such times.

tensive efforts at food choice and consumption. From 4 or 5 months of age, the infant is carried dorsally, and when separated from the mother, alternates the practice of leaping sequences with concentrated feeding. When the infant is 6 months of age, the mother often moves without the now variegated infant, at first only to new feeding positions, but later to different trees. Following her, the infant is forced to make difficult leaps and may fall 30 ft to the ground several times a day.

At 8 months of age, the infant moves alone, though following the mother closely. Through previous difficulties in landing on the large vertical supports taken by the mother, the infant has learned to compose its own routes. This often results in long pauses and indecisions, confining group movements to short and slow progressions. By this age, however, the young *Indri* has learned to land successfully, and falls are very rare. Suckling is restricted to days of fine weather, and the now liberated mother feeds noticeably less.

The 1-yr-old infant/juvenile suckles no more, but feeds extensively and for longer periods than other members of the group. Refined locomotor techniques and an ability to quickly calculate optimal arboreal routes are now present. Significantly, the choice of food species at this age is not totally similar to that of the mother, suggesting that the individual characteristics of edible species are being recognized and memorized.

Over the second year, the mother–infant relationship gradually reduces in intensity as measured by activity synchrony and spatial proximity. The two do, however, usually feed together and are, compared to any other pair of animals, recognizably "close."

ACKNOWLEDGMENTS

This research was supported by a Royal Society Leverhulme Studentship, the Central Research Fund of the University of London, the Emslie Horniman Anthropological Scholarship, the Boise Fund of Oxford, the Explorers Club of America, and a Grant-in-Aid of Research from the Society of the Sigma Xi.

I am most grateful to Drs. R. D. Martin, A. Jolly, I. Tattersall and R. Sussman for offering their criticism and suggestions for improvement of the manuscript.

The field work in Madagascar was only possible with the help of the Department des Eaux et Forêts in Tananarive, in particular, M. Andriamampianina, M. J-P. Abraham, and Prassede Calabi, and the Betsimisaraka, whose profound religious respect for the "babakoto" has enabled an exemplary distance to be maintained between the two competing species.

REFERENCES

Aldrich-Blake, P., 1970, Problems of social structure in forest monkeys, in: *Social Behaviour in Birds and Mammals.* (J. H. Crook, ed.), pp. 79–101, Academic Press, New York.

Cabanis, Y., Chabouis, L., and Chabouis, F., 1970, *Végétaux et groupements végétaux de Madagascar et des Madagascareignes II*, Bureau pour le Développement de la Production Agricole, Tananarive, Madagascar.

Carpenter, C. R., 1940, A field study in Siam of the behavior and social relations of the gibbon (*Hylobates lar*), *Comp. Psychol. Monogr.* **16(5)**:1–212.

Chalmers, N. R., 1968, Group composition, ecology and the daily activities of free-living mangebeys in Uganda, *Folia Primatol.* **8**:247–262.

Charles-Dominique, P., and Hladik, C. M., 1971, Le *Lepilemur* du Sud de Madagascar: Ecologie, alimentation et vie sociale, *Terre Vie* **25**:3–66.

Charles-Dominique, P. and Martin, R. D. (1970) Evolution of Lorises and Lemurs. *Nature, London* **227**:257–260.

Chivers, D., 1973, Siamang in Malaya, Ph.D. thesis, University of Cambridge.

Grand, T. I., and Lorenz, R., 1968, Functional analysis of the hip joint in *Tarsius bancanus* (Horsfield, 1821) and *Tarsius syrichta* (Linnaeus, 1758), *Folia Primatol.* **9**:161–181.

Hill, W. C. O., 1953, *Primates: Comparative Anatomy and Taxonomy, Vol. 1: Strepsirhini*, Edinburgh University Press, Edinburgh.

Hladik, A., and Hladik, C. M., 1969, Rapports trophiques entre végétations et primates dans la forêt de Barro Colorado (Panama), *Terre Vie* **1**:25–117.

Hladik, C. M., and Hladik A., 1972, Disponibilités alimentaires et domaines vitaux des primates à Ceylan. *Terre Vie* **1**:149–215.

Humbert, H., 1936, Flore de Madagascar et des Comores, *Laboratoire de Phanerogamie du Muséum National d'Histoire Naturelle*, Paris.

Humbert, H., and Cours Darne, G., 1965, Notice de la Carte Madagascar. Institut de la Carte Internationale du Tapis Végétal, Toulouse.

Jolly, A., 1966, *Lemur Behavior*, 187 pp., University of Chicago Press, Chicago.

Koechlin, J., 1972, Flora and Vegetation of Madagascar, in: *Biogeography and Ecology in Madagascar*. (Battistini, R., and Richard-Vindard, G., eds.), pp. 145–190, W. Junk, The Hague.

Marler, P., 1968, Communication in monkeys and apes, in: *Primate Behavior* (I. DeVore, ed.), p. 544–584. Holt Rinehart and Winston, New York.

Martin, R. D., 1969, The evolution of reproductive mechanisms in primates, *J. Reprod. Fert. Suppl.* **6**:49–66.

Martin, R. D., 1972, A preliminary field-study of the lesser mouse lemur *Microcebus murinus* J. F. Miller 1777, in: Behaviour and Ecology of Nocturnal Prosimians, *Fortschr. Verhaltensforsch.* **9**:43–89.

Mason, W. A., 1966, Social organisation of the South American monkey *Callicebus moloch*: A preliminary report, *Tulane Stud. Zool.* **13**:23–28.

Napier, J. R., and Walker, A. C., 1967, Vertical clinging and leaping: A newly recognised category of locomotor behaviour of primates. *Folia Primatol.* **7**:204–219.

Perrier de la Bathie, H., 1921, La végétation malgache, *Ann. Mus. Colon. Marseille*, **IX**:1–268.

Petter, J.-J., 1962, Recherches sur l'écologie et l'éthologie des lémuriens malgaches, *Mém. Mus. Natl. Hist. Nat. n.s.* **27**:1–146.

Petter, J.-J., and Peyrieras, A., 1974., A study of population density and home range of *Indri indri* in Madagascar, in: *Prosimian Biology*, (R. D. Martin, G. A. Doyle, and A. C. Walker, eds.), Duckworth, London.

Petter, J.-J., and Petter-Rousseaux, A., 1964. Première tentative d'estimation de peuplement des lémuriens Malgaches, *Terre Vie* **4**:427–435.

Petter-Rousseaux, A., 1964, Reproductive physiology and behavior of the Lemuroidea. in: *Evolutionary and Genetic Biology of the Primates II.* (J. Buettner-Janusch, ed.), pp. 92–132, Academic Press, New York.

Petter-Rousseaux, A., 1968, Cycles génitaux saisonniers des lémuriens malgaches, in: *Cycles Génitaux Saisonniers de Mammifères Sauvages* (R. Canivenc, ed.), pp. 11–22, Masson, Paris.

Petter-Rousseaux, A., 1969, Day length influence on breeding season in mouse lemurs, Presented at Eleventh Ethological Conference, Rennes.

Pollock, J. I., In preperation. Ph.D. thesis, University College, London.

Richard, A. F., 1973, Social organisation and ecology of *Propithecus verreauxi* Grandidier 1867, Ph.D. thesis, Queen Elizabeth College, London.

Richard, A. F., 1974, Patterns of mating in *Propithecus verreauxi*, In: *Prosimian Biology*, (R. D. Martin, G. A. Doyle, and A. C. Walker, eds.), Duckworth, London.

Walker, A. C., 1967, Locomotor adaptations in recent and fossil Madagascan lemurs, Ph.D. thesis, University of London.

An Analysis of the Social   17
Behavior of Three Groups of
Propithecus verreauxi

ALISON F. RICHARD AND
RAYMOND HEIMBUCH

INTRODUCTION

An 18-month study of the social organization and ecology of *Propithecus verreauxi* was carried out in Madagascar between May 1970 and September 1971. Two groups were studied in the mixed deciduous and evergreen forests of the Ankarafantsika in the northwest of Madagascar and two in the *Didierea* forest which covers large tracts in the south. The two study areas and the ecology of groups of *P. verreauxi* occupying them have been described by Richard (1973, 1974*a*, and in press).

In this chapter, interindividual relationships within three groups of *P. verreauxi* are considered in detail, together with the results of a statistical analysis of patterns of dispersion in one of these groups.

ALISON F. RICHARD Department of Anthropology, Yale University, New Haven, Connecticut RAYMOND HEIMBUCH   Department of Anthropology, Yale University, New Haven, Connecticut

FIG. 1. Adult female in Group IV suckling her four-month-old infant.

METHODS

Two habituated groups (I and II) in the northern study area were studied intensively during the dry season months of July and August 1970 and July 1971, and during the wet season months of October, November, and December 1970. The two habituated groups in the south (III and IV) were studied during the dry season months of April, May, and June 1971, and during the wet season months of January, February, and March 1971. Observations on group III were also made in September 1970. The original composition of and subsequent changes in these four groups are given in Table I.

Data were collected for 72 h per month on each group. Observations were divided into six 12-h periods lasting from 0600 h until 1800 h. One ani-

FIG. 2. Adult female in Group III grooming subadult male.

mal was followed throughout each observation period, and an individual of a different age and sex class was observed each day. Thus observations were evenly distributed between the different classes and between different times of day. Overt interactions between group members were infrequent, and it was possible to keep a record of all interactions that visibly involved the subject. Each interaction was described separately, although they have been grouped into more general categories for purposes of analysis.

Data on other aspects of behavior were recorded at 1-min intervals during observation periods (see also Richard, 1973). The quantitative record pertinent to a consideration of group dispersion was that made of the proximity and identity of the subject's nearest neighbor. Five categories of proximity were used: (1) physical contact, (2) 1–2 m to nearest animal, (3) 2–3 m to nearest animal, (4) 3–6 m to nearest animal, and (5) over 6 m to nearest ani-

TABLE I. Initial Composition of the Four Study Groups and Subsequent Changes

Group	July 1970	Oct. 1970	Dec. 1970	July 1971
I	5 A♀♀ A♂ GOP A♂ N	5 A♀♀ A♂ STR	4 A♀♀ A♂ BE	4 A♀♀ A♂ BE 2 Infs.
II	A♀ A♂ SA ♂ J Inf.	A♀ A♂ SA ♂ J	No change	A♀ A♂ SA ♂ J Inf.
	Sept. 1970	March 1971[a]	June 1971	Sept. 1971
III	2 A♀♀ 2 A♂♂ SA ♂ J 2 Infs.	2 A♀♀ 2 A♂♂ SA ♂ J	No change	2 A♀♀ 2 A♂♂ SA ♂ J 2 Infs.
IV	2 A♀♀ A♂ R SA♂ 2 Infs.	2 A♀♀ A♂ INT SA ♂	No change	2 A♀♀ A♂ INT SA ♂ Inf.

[a] A♂ P's brief detachment from group III and association with group IV is not recorded here. Where group members are not individually named, it can be assumed that their identity remained the same.

mal. All animals could be recognized individually; a record of "identity unknown" was entered only when all group members other than the subject were out of sight. These data were subsequently analyzed using the multivariate analysis of variance. The use of this technique in analyzing such data is described in detail elsewhere (Heimbuch and Richard, 1974), and only the conclusions drawn from this analysis are presented here.

RESULTS

Intragroup Interactions

Description of Agonistic Behavior
Aggressive Behavior. An animal that displaced, threatened, bit, or cuffed another animal in its own group was considered to be the aggressor in an agonistic encounter. Cuffs, given with the hand, and bites, generally administered

on the back of the neck or limb extremities, were both usually accompanied by a "cough" or "hack" vocalization. Animal A was said to have displaced animal B if, as it approached, animal B glanced at it rapidly and leaped off at once, or if animal B adopted a submissive posture (see below) and subsequently moved away. Staring and lunging toward an animal were considered as threats. They resulted in the threatening animal's displacing the recipient of the threat or in the recipient's adopting a submissive posture.

Submissive Behavior. Submissive gestures included baring the teeth with the lips drawn back tightly, rolling up the tail between the hind legs, and hunching the back. They were usually accompanied by the "spat" vocalization described by Jolly (1966) as "a series of high squeaks, given in quick succession ..." (p. 60). At low levels of intensity, these vocalizations were unvoiced and had the quality of a cat's purr. After giving these signals of submission, the aggressee would remain in his original position or move away, depending on the persistence of the aggressor and the nature of the previous relationship between them.

Situations in which Agonistic Behavior Occurred

In all groups, most agonistic encounters occurred in a feeding situation (Table II). Typically, the aggressor was feeding and the aggressee approached too closely, or the aggressor supplanted the aggressee from the latter's feeding station. This contrasts with Jolly's (1966) results; she reports practically no aggressive interactions between animals at Berenty, and only once did she see an aggressive encounter take place with reference to a food source. In the present study, however, spatial displacements without reference to a food source were rarely seen and occurred only at the onset of resting periods, when there was competition for what were presumably preferred resting places.

TABLE II. Frequency of Agonistic Encounters and Situations in which They Were Observed

Group	Total number of agonistic encounters	Number per animal hour	Situations in which aggression occurred		
			Access to food	Access to resting site	Other
I	107	0.29	98	9	—
II	187	0.42	153	10	24
III	109	0.25	84	11	14
IV	191	0.44	128	23	40

[a]Figures are based on equal numbers of hours of observation equally distributed throughout the day for all animals.

In addition to disputes over access to food and resting sites, there were five other situations in which intragroup aggression was occasionally observed:

1. The aggressive reaction of group II's adult female (A♀) to the constant approaches of other group members after the birth of her infant.
2. Her rejection of the group II juvenile male (J) following the disappearance of her infant when, on three occasions, he tried to suckle her.
3. Group IV's adult females' aggressive reactions to adult male R's (A♂ R) attentions in the period preceding the mating season;
4. Agonistic encounters involving more than two animals. Three instances were observed in group II. In the first encounter, A♂ H cuffed J, who gave a "spat" call and the tense-lipped grin expression but did not move off; A♀ immediately leaped at A♂ H and pushed him down a vertical trunk until he leaped off. On the second occasion, a similar pattern was seen, but this time A♂ H had cuffed Y♂. Third, an apparent case of "redirected aggression" was noted: A♀ leaped at A♂ H, who spat-called and leaped at Y♂, who immediately left the tree. No triadic interactions were seen in group I or III, but an apparent case of redirection was once seen in group IV when A♀ FI approached A♂ R; he exhibited submissive gestures and moved away from A♀ FI toward A♀ FNI, whom he snapped at and displaced.
5. Forced grooming. This was observed only between A♂ P and A♂ F in group III. A♂ F frequently approached A♂ P, put his arm around A♂ P's neck, and thrust A♂ P's head into the fur on his (A♂ F's) shoulder, forcing A♂ P to groom him. This gesture is referred to as "collaring." A♂ P always responded by grooming A♂ F, with a full exhibition of submissive gestures.

Frequency of Agonistic Behavior

The frequency of agonistic behavior varied considerably between the four groups, but not consistently between areas (see Table II). This variation appeared to be related to the internal dynamics of each individual group rather than to external, environmental factors.

Patterns of Aggression

Separate analysis of agonistic encounters in the various situations described above give very similar results, so the data are considered together. Tables III, IV, and V show the direction and frequency of aggression between

TABLE III. Frequency of Aggression Between Members of Group II[a]

Aggressor	Recipient ♀	♂	Y♂	J	Total
♀		13	34	26	73
		(13)	(30)	(17)	(60)
♂			11	5	16
			(4)	(4)	(8)
Y♂				10	10
				(16)	(16)
J			4		4
			(0)		(0)
Total		13	49	41	103
		(13)	(34)	(37)	(84)

[a]Figures in parentheses represent number of aggressive encounters recorded during the dry season. Those without parentheses represent those recorded during the wet season.

members of each of groups II, III, and IV, respectively.[1] The significance of these results is considered in the Discussion and Conclusions section.

Seasonal Variation in Agonistic Behavior[2]

Aggressive encounters occurred more frequently in the wet season than in the dry season. Most animals contributed to this increase, and in not one was a decrease in frequency of aggression recorded. However, the frequency of aggression increased more in some group members than in others. These differential increases were not consistent with any particular age/sex class.

The difference between the two seasons is highly significant (Mann-Whitney U test using data from Table VI: $N_1 = N_2 = 4$, $U = 1$, $p \leq 0.01$, for group II; $N_1 = N_2 = 8$, $U = 0$, $p \leq 0.001$, for groups III and IV). Table VI shows the number of aggressive interactions initiated by members of the study groups in the two seasons. These figures are also expressed as a percentage of the total number of aggressive encounters seen in each group during each season. This measures each animal's relative contribution to all recorded aggression. Although most animals contributed to the wet season increase, the index of seasonal change shows that the proportionate increase was higher in some animals than in others. This index is the ratio between the number of aggressive incidents an animal initiates in each season. Where the dry season value is zero, the value of the index is infinity and is consequently not shown.

---

[1]Not all members of group I could be consistently recognized, so it has not been included in this analysis.

[2]Observations made during the wet season preceded those made during the dry season. An "increase" or "decrease" in the wet season is thus a change relative to, although not following upon, the dry season.

TABLE IV. Frequency of Aggression Between Members of Group III[a]

Aggressor	Recipient						
	♀ NFD	♀ FD	♂ F	♂ P	Y ♂	J	Total
♀ NFD		(1)	4	4 (1)	9 (1)	7 (2)	24 (5)
♀ FD	1 (2)		3 (5)	1 (3)	2	6 (3)	13 (13)
♂ F				17 (5)	9 (1)	12 (1)	38 (7)
♂ P					1	1 (2)	2 (2)
Y ♂				1		2	3
J					2		2
Total	1 (2)	(1)	7 (5)	23 (9)	23 (2)	28 (8)	82 (27)

[a]Figures in parentheses represent number of aggressive encounters recorded during the dry season. Those without parentheses represent those recorded during the wet season.

Description of Nonagonistic Social Behavior

Allogrooming. Allogrooming was the most commonly observed non-agonistic interaction. It could be initiated by the groomer without the groomee necessarily presenting to be groomed. When presentation occurred, the prospective groomee held out an arm toward the prospective groomer, who approached, grasped the arm, and began grooming it. In only 10.4% of bouts was grooming concentrated on the limbs. On all other occasions, grooming either started on or was transferred almost at once to the head,

TABLE V. Frequency of Aggression Between Members of Group IV[a]

Recipient	Aggressor						
	♀ FI	A♂ R	♀ FNI	SA♂ Q	♂ INT	Inf.	Total
♀ FI		46	9 (3)	24 (20)	(1)	6	85 (24)
♂ R	5[b]		27	17			49
♀ FNI				10 (6)	(3)	2	12 (9)
SA♂ Q	1[b]	2[b]	3[b]				6
♂ INT				(6)			(6)
Inf.							
Total	6	48	39 (3)	51 (32)	(4)	8	152 (39)

[a]Figures in parentheses represent number of aggressive encounters recorded during the dry season. Those without parentheses represent those recorded during the wet season.
[b]These encounters were recorded in March 1971.

TABLE VI. Contribution of Each Animal to Total Group Aggression in Wet and Dry Seasons, and Index of Seasonal Change

Group	Initiator	Contribution to group aggression				Index of seasonal change
		Wet season		Dry season		
		N	%	N	%	
II	♀	73	71	60	83	1.2
	♂	16	15	8	11	2.0
	Y♂	10	10	4	6	2.5
	J	4	4	—	—	—
III	♀ FD	13	17	13	48	1.0
	♀ NFD	24	29	5	19	4.8
	♂ F	38	46	7	26	5.4
	♂ P	2	2	2	7	1.0
	Y♂	3	4	—	—	—
	J	2	2	—	—	—
IV	♀ FI	85	56	24	62	3.5
	♀ FNI	12	8	9	23	1.3
	♂ R/INT	49	32	6	15	8.2
	SA♂ Q	6	4	—	—	—
	Inf.	—	—	Not present		—

face, and back of the groomee. These areas are inaccessible for self-grooming. "Collaring" of the groomer by the groomee was observed only between adult males F and P in group III and always resulted in an exhibition of submissive gestures by the groomer, A♂ P, as he groomed A♂ F.

All grooming was done by licking with the tongue and scraping the tooth comb over the fur. When grooming the head, the groomer frequently clamped his hand around the groomee's muzzle. 70% of grooming bouts were unidirectional: throughout the bout one animal groomed another. In 4% of bouts, grooming was reciprocal: The two animals involved alternately groomed each other. Reciprocal and simultaneous grooming occurred in 26% of bouts: The animals sat in physical contact, each grooming the other's shoulders or back. This reciprocity developed only after one or the other animal had initiated the bout.

Play Behavior. Play behavior generally involved subadults and juveniles, although adults did occasionally join in, and was characterized by the relaxed, open-mouth play face of the participants. In 73% of bouts, participants played on or within 2 m of the ground. Bouts more than 2 m above the ground usually consisted of two or three animals wrestling together, often hanging by their arms or legs alone, each trying to dislodge the other(s). Wrestling also occurred on or near the ground, together with chases. Play bouts on the ground frequently culminated in one animal lying on his back with the other repeatedly jumping onto him.

Nose-Touching. Nose-touching was observed as a form of greeting behavior, as noted by Jolly (1966), but it constituted only 5.7% of all nonagonistic social behavior. The participants approached each other, touched noses briefly, and moved apart again. No other form of greeting behavior was seen.

Situations in which Grooming and Play Occurred

Most grooming occurred during rest periods. The main exception to this was the A♂ F/A♂ P grooming pair. A♂ F might force A♂ P to groom him in any situation if the latter was close to him.

Play behavior was observed only during the wet season, both in the north and in the south (see Discussion and Conclusions), and took place mainly at the beginning and end of feeding and rest periods. Nose-touching was seen occasionally when animals were reunited after prolonged dispersal of the group during feeding bouts.

Patterns of Grooming and of Play and Nose-Touching

Tables VII–XII show the direction and frequency of grooming and of play and nose-touching between members of each of groups II, III, and IV. These results are considered below.

Seasonal Variation in Nonagonistic Social Behavior

The only seasonal variation in the frequency of grooming was found in group II. However, seasonal variation in the frequency of play behavior was marked in all three groups.

In group II, out of a total of 136 grooming bouts, only 48 occurred in the dry season. The increase in the wet season was not due to a uniform increase

TABLE VII. Frequency of Grooming in Group II[a]

Groomer	Groomee				
	♀	♂	Y♂	J	Total
♀		2	1	18	21
		(1)	(1)	(1)	(3)
♂	3		2	14	19
	(16)		(2)	(2)	(20)
Y♂	3	4		12	19
	(12)			(2)	(14)
J	10	11	8		29
	(7)	(4)			(11)
Total	16	17	11	44	88
	(35)	(5)	(3)	(5)	(48)

[a]Figures in parentheses represent grooming bout frequencies in the dry season. Those without parentheses represent frequencies in the wet season.

TABLE VIII. Frequency of Grooming in Group III[a]

Groomer	Groomee						Total
	♀ FD	♀ NFD	♂ F	♂ P	Y♂	J	
♀ FD		2			2	6	10
		(5)		(1)	(1)	(3)	(10)
♀ NFD	2		2	1		1	6
	(3)		(1)				(4)
♂ F	2				1	3	6
	(1)	(1)		(1)		(5)	(8)
♂ P			4		1	2	7
		(1)	(9)			(1)	(11)
Y♂	2	1	2			4	9
	(1)		(1)			(1)	(3)
J	2	2	2		1		7
			(5)	(1)	(1)		(7)
Total	8	5	10	1	5	16	45
	(5)	(7)	(16)	(3)	(2)	(10)	(43)

[a]Figures in parentheses represent grooming bout frequencies in the dry season. Those without parentheses represent frequencies in the wet season.

in grooming activity by all animals (see Table XIII). In the wet season, 57% of all grooming was initiated by A♀ and J, and they were responsible for 90% of the increase over dry season frequencies. The index of seasonal change shows that A♂ was the only animal who initiated grooming less frequently in the wet season than in the dry. (The index of seasonal change is the ratio between the number of bouts observed in the wet season and the number of bouts observed in the dry season.)

In the dry season, A♀ and J groomed less than the other two members of the group, and A♀ initiated only 6% of dry season grooming. The reverse

TABLE IX. Frequency of Grooming in Group IV[a]

Groomer	Groomee				Total
	♀ FI	♂ R/INT	♀ FNI	SA♂ Q	
♀ FI					
		(3)			(3)
♂ R/INT	2		1	2	5
♀ FNI	10	1			11
	(5)				(5)
SA♂ Q		3			3
	(6)	(1)	(3)		(10)
Total	12	4	1	2	19
	(11)	(1)	(6)		(18)

[a]Figures in parentheses represent grooming frequencies in the dry season. Those without parentheses represent frequencies in the wet season.

TABLE X. Frequency of Play and Nose-Touching in Group II[a]

Partner	Initiator ♀	♂	Y♂	J	Total
♀			1	1	2
			(1)	(1)	(2)
♂			3	3	6
Y♂		3		17	20
J	7	7	25		39
			(1)		(1)
Total	7	10	29	21	67
			(2)	(1)	(3)

[a]Figures in parentheses represent frequency of nose-touching, those without parentheses frequency of play.

TABLE XI. Frequency of Play and Nose-Touching in Group III[a]

Partner	Initiator ♀ FD	♀ NFD	♂ F	♂ P	♂ Y	J	Total
♀ FD			1				1
			(2)			(1)	(3)
♀ NFD	2				1		3
				(1)		(2)	(3)
♂ F					1		1
♂ P					3	1	4
		(1)					(1)
♂ Y	1			1		12	14
						(1)	(1)
J			3	1	11		15
			(2)	(2)			(4)
Total	3	1	3	2	16	13	38
		(1)	(4)	(3)		(4)	(12)

[a]Figures in parentheses represent frequency of nose-touching, those without parentheses frequency of play.

situation held for A♀ with respect to being groomed: 73% of all grooming was directed at her in the dry season, this figure falling to 18% in the wet season.

Play behavior was never seen during the dry season in either study area.

Dispersion Patterns among Members of Group III

A number of behavioral inferences could be drawn from the multivariate analysis of variance of "nearest-neighbor" data collected on group III. These are presented below in the following form: the general dispersion tendencies of age/sex classes are described, and departures from this norm with respect both to seasonal variation and to preferences for particular age/sex classes as nearest neighbor are then noted. It should be emphasized, however, that the analysis of age/sex class dyads does not demonstrate which age/sex class established the observed proximity. This causal factor can be considered only speculatively.

TABLE XII. Frequency of Play and Nose-Touching in Group IV[a]

Partner	♀ FI	♂ R/INT	♀ FNI	SA♂ Q	Inf.	Total
			Initiator			
♀ FI						
		(1)				(1)
♂ R/INT			2	1	1	4
			(1)			(1)
♀ FNI					1	1
	(1)	(4)		(1)		(6)
SA♂ Q						
Inf.				1		1
Total			2	2	2	6
	(1)	(5)	(1)	(1)		(8)

[a]Figures in parentheses represent frequency of nose-touching, those without parentheses frequency of play.

TABLE XIII. Contribution of Each Animal to Total Frequency of Grooming in Group II in Wet and Dry Seasons, and Index of Seasonal Change

	Contribution to grooming				Index of seasonal change
	Wet season		Dry season		
	%	N	%	N	
Groomer					
♀	24	21	6	3	7.0
♂	21.5	19	42	20	0.95
Y♂	21.5	19	29	14	1.35
J	33	29	23	11	2.63
Groomee					
♀	18	16	73	35	0.46
♂	19	17	10.5	5	3.40
Y♂	13	11	6	3	3.67
J	50	44	10.5	5	8.80

Adult Male

Adult males spent the majority of their time in distance categories (see p. 146) 2, 3, and 4 relative to other members of the group. Male–male dyads spent more time in distance category 1 and less time in category 4 relative to their general dispersion tendencies. From this it is inferred that each male "preferred" the other over other group members as nearest neighbor. There was a high level of overt aggression between the two males in the group, however, so this proximity may have been due to a reduction of interanimal distance during agonistic encounters rather than to mutually high tolerance. Some "preference" was also shown for the young male: in adult male–young male dyads, there was an increase in the time spent in distance category 2. The distancing of the adult males relative to the young male differed between seasons; during the wet season, there was an increase in time spent in distance category 5. In adult male–adult female dyads, there was a decrease in time spent in distance category 1 during the dry season.

Adult Female

The adult females generally spent most of their time in distance categories 2, 3, and 4. When the two females were each other's nearest neighbor, there was a considerable increase in the time each spent in categories 1 and 2. This suggests that there was a high level of tolerance between the females. The low frequency of overt aggression between them confirms this inference. In the wet season, the females tended to distance themselves from other group members and from each other. However, in the case of the adult female–young male dyad, there was actually an increase in the time spent in categories 2 and 3 during the wet season, although the females spent almost no time in physical contact with the young male during that period. In contrast, although the general wet season distancing effect was seen in the adult female–adult male dyad, adult females did spend significant amounts of time in physical contact with the males.

Young Male

There were not enough observations of the one young male in group III to establish any clear pattern using him as the subject, or focus, in dyads.

Juvenile

The juvenile's general pattern of dispersion showed much the same tendencies as that of the adult males and females; the majority of his time, irrespective of season or the identity of his nearest neighbor, was spent in distance categories 2, 3, and 4. However, when the young male was his nearest neighbor, there was an increase in the time the juvenile spent in distance category 1 and a decrease in that in category 5. The inference is that the juvenile and the young male generally "preferred" each other's company to that of any other member of the group. Since they frequently played together, this habitual close proximity was to be expected. However, despite the absence of play in the dry season, there was no significant change in the distancing of this dyad between seasons.

Unknown Neighbor

Results for the "unknown neighbor" class show, predictably, that when the neighbor was unknown the individual was at some distance from other group members.

DISCUSSION AND CONCLUSIONS

The Interpretation of Social Structure

In early primate studies, the all-inclusive theory of "social dominance" was put forward as the basis of primate social organization (Zuckerman, 1932). This unitary motivational theory postulated that the individuals constituting a social group were ranked in a linear hierarchy, and that the outcome of all interactions was determined by the relative ranks of the animals participating in the interaction. The criteria by which this hierarchy was established included frequency and direction of aggression, preferential access to food sources and feeding station, and preferential access to receptive females. More recent work has shown that a unitary theory of social structure underestimates the complexity of primate social organizations (e.g., Jolly, 1966; Stoltz and Saayman, 1970). In an extensive discussion of the concept of social dominance, Gartlan (1968) pointed out that "those behavior patterns which are traditionally associated with dominance . . . often show no correlation with one another . . . " (p. 92). He proposed instead to describe the social structure of *Cercopithecus aethiops* in terms of functional roles played by different members of the group. These roles included territorial display, social vigilance, social focus, friendly approach, territorial chasing, punishing, and leading.

Few of the roles envisaged by Gartlan could be differentially applied to members of the groups of *P. verreauxi* studied. However, it would be an oversimplification to revert to the unitary theory of social dominance and to see all the social relationships of *P. verreauxi* solely in the light of a simple linear hierarchy. While the unidirectionality of aggression and displacements in the groups could be used to define a hierarchy in each, there was no consistent correlation between the rank of individuals in a hierarchy established on this criterion and their ranks in hierarchies established according to the frequency of aggression, the direction and frequency of grooming, or preferential access to receptive females.

Feeding Hierarchy

The majority of aggressive interactions occurred with reference to access to a feeding station. In the groups in both study areas, a clear-cut hierarchy existed within the context of feeding and, specifically, priority of access to food; this hierarchy is henceforth called the "feeding hierarchy." In all three groups, the highest-ranking animal in the feeding hierarchy was an adult female. However, dominance in this hierarchy was not necessarily a function of sex: in group IV, A♂ R always displaced A♀ FNI.

There were only two exceptions to the general rule that agonistic encounters in this context were unidirectional:

1. The two adult females in group III, who were both unchallenged by other group members, occasionally engaged in protracted aggressive interactions with each other, of unpredictable outcome. This contrasted with the situation in group IV, where A♀ FI always displaced A♀ FNI.
2. Subadult and juvenile animals were always displaced by adults in agonistic encounters, but no stable relationship appeared to exist between the subadults and juveniles themselves. Aggression between them was two-way.

Outside the context of food, the rigidity of structuring broke down, and other factors probably regulated interindividual relationships. For example, a complete breakdown in structuring occurred in group IV when the two adult females were in estrus in March 1971. (For a detailed discussion see Richard, 1974*b*.)

The Role of the Mother and Her Infant

The presence of an infant and associated changes in the behavior of both its mother and other members of the group probably play an important part in determining the nature of social relationships within that group. For example, it is postulated that the birth of an infant in group II during the 1970 dry season was responsible for changes in interindividual relationships in that group at that time. The newly born infant became a focus of attention within the group, and animals repeatedly approached A♀ and tried to groom or handle her infant. She prevented this by snapping or cuffing at approaching animals. In 41 out of a total of 60 aggressive interactions initiated by A♀ during the dry season, her aggression was directed toward another member of the group who was trying to gain access to the infant. In contrast, 67 out of 73 aggressive interactions initiated by her in the wet season were over access to a feeding station. Thus although the absolute frequency with which A♀ initiated aggression changed little between seasons, the predominant context for these encounters did change radically with the presence of a neonate in the group.

The Role of Idiosyncratic Relationships

Only in a few field studies have most, or all, individuals within a group been consistently identified (e.g., Jolly, 1966; Van Lawick-Goodall, 1968; Mizuhara, 1964). Therefore, it has rarely been possible to consider interactions as a function of the characteristics of two individual animals rather than as relatively stereotyped encounters between members of age/sex classes. It is possible that social structure even among prosimians may be partly determined by individual idiosyncrasies in addition to predictable patterns of age/sex class interactions.

In group IV, the frequency of aggression was positively correlated with the direction of aggression in the feeding hierarchy, but the data from groups II and III show that frequency of aggression was not necessarily a correlate of rank. In group III, A♂ F was the most frequently aggressive animal, although he ranked third in the feeding hierarchy. Forty-nine percent of his encounters were with A♂ P. This frequent and often apparently gratuitous aggression toward A♂ P could be interpreted as an idiosyncratic assertion of dominance in a stable situation or as an incipient attempt to drive the second male out of the group. The latter interpretation is in line with Petter and Peyrieras' (1974) hypothesis that the family group is the basic unit in the social organization of *P. verreauxi*, so that symptoms of stress in the form of heightened aggression might be expected when group composition departs from the "norm." However, from evidence collected on group size and composition in this study, it has been postulated that the group should rather be considered as a foraging party whose age/sex composition varies widely from group to group (Richard, 1973). Thus A♂ F's behavior cannot necessarily be considered as a response to the abnormal presence of a second adult male in the group. A detailed study of a much wider sample of relationships between known individuals belonging to different age/sex classes is necessary before it can be definitively established whether the A♂ F/A♂ P relationship reflects individual idiosyncrasies or a certain pattern of adult male–adult male interaction.

The Dynamics of Social Structure

Social structure should be considered as a dynamic process and not as a static framework. Some indication of the constantly changing nature of social groups has been given by long-term studies of baboon, chimpanzee, and Japanese macaque populations (Rowell, 1969; Van Lawick-Goodall, 1968; Itani, 1963; Mizuhara, 1964). Thus, in a given group, certain relationships may in the short-term appear stable, although they are in fact undergoing a gradual change. This may have been true of the Y♂/J relationship in group II. Y♂ was more frequently aggressive than the adult male, A♂ H, although the latter ranked higher in the feeding hierarchy. However, all of Y♂'s, aggression was directed at J, whereas J initiated aggression toward, and displaced, Y♂ only four times. It is possible that Y♂, the older of these two immature animals, was in the process of changing the existing two-way relationship, involving reciprocal initiation of aggression and frequent play bouts in the wet season, into the dominant–subordinate relationship characteristic of the adult feeding hierarchy. It is postulated that A♂ H was less often aggressive because his position had only to be maintained rather than established.

Grooming Behavior

Adult females generally initiated less grooming and were more commonly groomed than other members of their groups. This suggests a positive correlation between frequency of being groomed and rank in the feeding hierarchy. In a review paper, Sparks (1967) notes that "the majority of the allo-grooming bouts in Old World primates are against the dominance slope of the hierarchy prevailing in these communities . . . " (p. 167). However, despite this tendency for dominant animals to groom least and be groomed most, the existence of frequent grooming due to maternity or harassment of one animal by another complicated relationships and removed the simple linearity of the feeding hierarchy. Seventy-nine percent of the grooming initiated by A♀ in group II was directed at J, and in group III many of the numerous bouts initiated by A♀ FD were directed at J (and at the codominant A♀ NFD). A♂ P and Y♂ in group III were, predictably, rarely groomed by other animals. However, most of A♂ P's grooming was directed not at the two most dominant females, but at A♂ F.

Predictions based on the analysis of data from the other groups are confirmed in group IV. In this group, with no juvenile and only one fully adult male present at any one time, the simple linearity of the feeding hierarchy was reflected in the frequency and direction of grooming bouts.

Play Bouts

Play bouts were most common between subadults and juveniles, and rare, or absent between adults. Although no adult–adult play bouts were observed in group II, there were nonetheless many more bouts between immature animals than between any animals in either of the southern groups. The absence of play between adults was a group-specific occurrence in that it was frequently recorded in the wet season among group I adults.

The apparent discrepancy between play frequencies in groups III and IV was probably due to the lack of peers for the group IV infant. Comparison of only adult–adult and adult–infant figures for group III with group IV figures, corrected for group size, reveals little difference in bout frequencies. More difficult to understand was the discrepancy in the frequency of play between seasons in both study areas: in neither area did animals play during the dry season. The reason for this is not known, but may be related to restrictions in activity due to reduced food availability and hence a reduced energy budget (Richard, 1974a).

Patterns of Group Dispersion

A few general conclusions and many areas for further investigation emerged from the analysis of patterns of dispersion in group III. Individuals generally spent much of their time closest to an individual of the same age/sex class. No trend was identified toward peripheralization of adult males. In fact, if inferences had been possible concerning the unknown neighbor category, the evidence would suggest that males were the least peripheralized members of the group. However, additional data are necessary on this particular aspect of their behavior before even tentative conclusions can be drawn.

The young male occupied a unique position in that he was in a phase of transition from one age/sex class to another. Both the analysis of group dispersion and the observation of overt interactions provide evidence of this changing role. If the basic unit of social organization were the family group, the young male would probably be peripheralized, if not ejected, from the group on maturation. An indication of such a process would be an increase in time spent in categories 4 and 5 during the dry season (which followed the wet season chronologically). Quite the contrary occurred during the dry season: the young male increased his close contact with the females and the males. The increase in close contact with the males suggests that the other males in the group were beginning to respond to the young male as if he were an adult.

Mating occurred during the wet season, and the young male's role at that time should be understood as being particularly ambiguous. There was considerable distancing between the adult males and the young male, while this did not occur in the adult male–juvenile dyad (although the juvenile was also male), or in the adult male–adult male dyad. Thus if competition for females *within* the group existed, it appears to have been limited to competition between the adult male class and the young male. (For a discussion of intergroup competition, see Richard, 1974c.) The young male's relationship with the adult females could also be differentiated seasonally. During the wet season, there was virtually no physical contact between the young male and the females. There was, however, an increase in time spent in categories 2 and 3 relative to the dry season. A possible explanation for this is that the young male was consistently attempting to approach the females but this was not permitted by the females.

Juvenile dyads were problematical. The juvenile spent significant amounts of time close to the young male. It is, perhaps, surprising that no similar preference was shown for the females. However, it is possible that such a preference did exist; if there was a tendency to prefer the company of one of the adult females (presumably his mother) and avoidance of the other female, this would cause the net effect to be close to zero.

We believe that a more extensive use of the multivariate analysis of variance could yield further insight into patterns of dispersion, and research into this aspect of group dynamics is being pursued further. The analysis is being reformulated to differentiate between individual animals. This may allow us to clarify certain questions concerning sexual activity and concerning the status of the young male and juvenile. Additional data from other groups would help establish whether responses between age/sex categories shown in group III are characteristic of all groups of *P. verreauxi* or specific to this group.

ACKNOWLEDGMENTS

Fieldwork was supported by a Royal Society Leverhulme Award, the Explorers' Club of America, the Boise Fund, the Society of the Sigma Xi, a NATO Overseas Studentship, the John Spedan Lewis Trust Fund for the Advancement of Science, and the Central Research Fund of London University.

REFERENCES

Gartlan, J. S., 1968, Structure and function in primate society, *Folia Primatol.* **8:**89–120.
Heimbuch, R., and Richard, A., 1974, An analysis of patterns of dispersion in a group of *Propithecus verreauxi*, in preparation.

Itani, J., 1963, The social construction of national troops of Japanese monkeys in Takasakiyama, *Primates* **4**(3):1–42.

Jolly, A., 1966, *Lemur Behavior*, Chicago University Press, Chicago.

Mizuhara, H., 1964, Social changes of Japanese monkey troops in Takasakiyama, *Primates* **4**:27–52.

Petter, J.-J., and Peyrieras, A., 1974, A study of population density and home range of *Indri indri* in Madagascar, in: *Prosimian Biology* (R. D. Martin, G. A. Doyle, and A. C. Walker, eds.), Duckworth, London.

Richard, A., 1973, The social organization and ecology of *Propithecus verreauxi*, Ph.D. thesis, London University.

Richard, A., in press. Feeding behaviour in *Propithecus verreauxi*, in: *Primate Feeding Behaviour* (T. Clutton-Brock, ed.), Academic Press, London.

Richard, A., 1974a, Intra-specific variation in the social organization and ecology of *Propithecus verreauxi*, *Folia Primatol.*, **22**(2–3):178–207.

Richard, A., 1974b, Patterns of mating in *Propithecus verreauxi*, in: *Prosimian Biology* (R. D. Martin, G. A. Doyle, A. C. Walker, eds.), Duckworth, London.

Rowell, T. E., 1969, Long-term changes in a population of Ugandan baboons, *Folia Primatol.* **11**:241–254.

Sparks, J., 1967, Allogrooming in Primates: A review. in: *Primate Ethology* (D. Morris, ed.), Weidenfeld and Nicolson, London.

Stoltz, L. P., and Saayman, G. S., 1970, Ecology and social organization of chacma baboon troops in the northern Transvaal, *Ann. Transvaal Mus.* **26**:499–599.

Van Lawick-Goodall, J., 1968, The behavior of free-living chimpanzees in the Gombe Stream Reserve, *Anim. Behav. Monogr.* **1**:161–311.

Zuckerman, S., 1932, *The Social Life of Monkeys and Apes*, K. Paul, Trench, Trubner and Co., London.

Future of the Malagasy Lemurs: Conservation or Extinction?

18

ALISON F. RICHARD AND
ROBERT W. SUSSMAN

INTRODUCTION

Madagascar, with its unique flora and fauna, is one of the many areas of the world which has become a focus of concern for conservationists. To date, the emphasis (as evidenced by papers presented at the Conférence Internationale sur la Conservation de la Nature et de ses Ressources à Madagascar 1970) has been on what we consider to be chronic but probably not critical aspects of the problem. These are the ongoing hunting and agricultural practices of the Malagasy living in rural areas. Undoubtedly, indigenous exploitation of the forest and animals does have an impact, but in this chapter we shall also attempt to pinpoint some of the factors which we believe are more immediate and crucial threats to the survival of the Malagasy lemurs. We shall also provide information on the present status of the lemurs and address ourselves to some of the practical problems involved in establishing an adequate program of conservation in Madagascar.

ALISON F. RICHARD Department of Anthropology, Yale University, New Haven, Connecticut and ROBERT W. SUSSMAN Department of Anthropology, Washington University, St. Louis, Missouri

IMMEDIATE THREATS TO THE MALAGASY LEMURS

Traditional Forms of Exploitation

Hunting

Hunting of lemurs for food is common in many areas, although the killing or eating of a number of species is taboo to certain tribes and is legally prohibited throughout the island. The power of local taboos is illustrated by the fact that, for example, *Propithecus verreauxi* is neither killed nor eaten by the Sakalava of the northwest or the Antandroy, a tribe extending over large areas of the south. In some areas of the eastern rain forests, the Tanala have taboos against killing or eating *Avahi laniger*, *Daubentonia madagascariensis*, and *Indri indri*, and the Betsimisaraka also have taboos protecting the latter two species.

Traditionally, the Malagasy hunt lemurs using a number of techniques. The small nocturnal species are usually trapped or cornered in their nests, which often involves tree-felling. Other species are caught in traps or snares or are killed using projectiles such as sharpened sticks, stones and slings, blowguns, and possibly spears.[1] Dogs are employed in low, dense forest to hunt *Lemur catta*, the only extant species of lemur that habitually travels on the ground. Guns have only recently been added to this arsenal.

It may be noted that since the arrival of man in Madagascar a number of species of animals, including at least 14 species of lemur, have become extinct. All of the lemurs were large, slow-moving, diurnal and/or ground-living forms, and there is some archeological evidence to suggest that these were hunted to extinction (Walker, 1967). However, natural climatic change (*cf.* Africa) could have altered the ecological balance and thus brought about natural extinction, which may have only been accelerated by hunting.

Deforestation

At least three factors lead to destruction of the forest at a local level. These are slash-and-burn techniques for agriculture (*tavy*), wood-cutting for firewood, and wood-cutting for house construction.

Slash-and-burn techniques are common all over the island. They are used to provide both land for cultivation and grazing pastures for cattle. It has been estimated that 10,000–20,000 ha of forest are destroyed each year by local populations (Chauvet, 1973). However, the long-term effects of this type of exploitation on the environment are still open to debate; Jean Koechlin (1973) comments:

---

[1] We have no evidence of the use of spears for this purpose; in our experience, they are used for killing larger game (e.g., boar) and are unsuitable for hunting smaller animals.

With regard first of all to the balance of the boundaries between the forest and the savanna, it may be noted that the forest areas are nowadays confined to certain edaphic localities (sand, limestone) which seem to be particularly suitable and these boundaries are virtually stabilised. If, therefore, the savannas of the west are to be regarded as recent secondary formations it must be admitted that the destruction of the forest took place under climatic conditions different from those prevailing today, or else that the forests that have disappeared were of another more fragile type. . . . (p. 187)

The population of Madagascar today is around 8 million; in 1966 it was approximately 6 million; at the turn of the century it was only about 2½ million. Archeological evidence suggests that the island was settled in about the fourth century A.D. (Deschamps, 1965), and thus for over 1500 years had only a small, unevenly distributed population. Even today, many areas of the west are almost completely uninhabited. Guichon (1960) has pointed out that despite this fact, woody formations, whether degraded or not, occupy only about 28% of the surface area of a country where the climax is of a forest type throughout: "It may be wondered, therefore, how these (early) men, with their rudimentary tools, and whose agricultural or pastoral interests are often very limited, could have cleared such extensive areas. . . ." (Chauvet, 1973, p. 188).

It is difficult to estimate the amount of destruction brought about by cutting wood for purposes of cooking and heating, although since wood is the major source of energy for the rural Malagasy we must assume that it is not insignificant. In some areas, however, only dead trees are used for firewood (Pollock, personal communication). Similarly, we can supply little systematic information on tree-felling for the construction of houses, although it must necessarily be common in some areas. Among the Antandroy, for example, all houses are made with wooden planks from one species of tree (*Alluaudia ascendens*). Approximately 40 trees are required to build one house, which has a life expectancy of 10–15 years. The Betsimisaraka traditionally use three species of trees for construction, selecting young individuals of these species, although they have recently begun to use eucalyptus instead (Pollock, personal communication).

Exploitation by Modern Technological Means

Hunting

Although guns are used all over the island, the prerequisite permits are expensive to buy and difficult to obtain for most Malagasy. Similarly, ammunition is expensive and supplies sporadic. Few Malagasy living in rural areas appear to possess guns, and the guns that one does see are badly maintained and used mainly for ceremonial purposes rather than hunting.

In our experience, regular hunting with guns is practiced predominantly by Europeans and town-dwelling store owners, and often by those charged, in one capacity or another, with enforcing restrictions on hunting. Although our evidence is based on personal experience only, we suspect that weekend hunting expeditions are a common feature of life for many individuals in these categories. Thus while we can present no overall picture of the effects of this hunting, the impact is potentially great; on one occasion, for example, A. F. R. encountered a hunter who had shot 12 individuals of the species *Propithecus verreauxi* in a single afternoon.

Deforestation

Forest clearance for various purposes by large-scale commercial enterprises has been virtually overlooked in discussions of conservation in Madagascar to date. We believe this to be actually and potentially the most dangerous and the most pressing problem. At this point, however, lack of information makes it difficult to do more than give a general overview of current trends in this sphere.

Concessions were recently granted to foreign companies to log extensive areas of the remaining east coast forest where there are many hardwood species that are subsequently exported at considerable profit to the companies concerned. A concession to carry out "selective" timbering of the coastal forest north of Fort Dauphin was granted while A. F. R. was in the area in 1971. In 1970, R. W. S. found evidence of local businessmen carrying on "opportunistic," large-scale, and indiscriminate logging—both legally and illegally—in the southwest of the island. Timbering companies also have permits for selective timbering in the eastern rain forest near Perinet. Some of the timbered species are important food sources for populations of *Indri indri* in this area (Pollock, personal communication).

Forests are being cleared for the planting of a spectrum of crops all over the island. In the south, clearance is mainly carried out for large-scale production of sisal and cotton. This is particularly significant in the case of those few remaining gallery forests bordering the rivers of the south. For example, along the Mardrare River large areas of *Tamarindus indica* forest have been sacrificed for the expansion of sisal plantations. Since the home ranges of almost all the lemur species found in the south or southwest consist partly or totally of these tamarind forests, their importance cannot be underestimated. A similar situation obtains in areas of the west around Morondava where mixed deciduous forests have been cleared.[2]

---

[2] M. De Heaulme of S.I.A.M., however, has recognized the need for reserves and set aside 100 ha in the south (at Berenty) and 60,000 ha in the west (the area of Analabe) for the protection of the fauna of these regions.

During the past few years, a number of foreign-owned oil companies have been making seismic surveys the whole length of western Madagascar, both offshore and inland. Between Morondava and Tulear, for example, roads have been cut through at 15-km intervals, running in some cases 100 km east from the coast. Preliminary drilling is now under way both to the northwest and to the south of Tulear, the search no doubt intensified by the recently recognized worldwide fuel shortage. Should oil be found in sufficient amounts and permits for exploitation granted, large-scale clearance of the remaining western forests would result as a consequence of both the extraction and transportation process itself and the establishment of supporting industry.

Other forms of mineral exploitation also pose a potentially immense threat to the total environment of Madagascar. For example, recent surveys of the remaining east coast littoral forests north and west of Fort Dauphin have indicated the presence of rich deposits of titanium. A necessary prerequisite for titanium extraction is complete leveling of the forest, and the process leaves in its wake vast areas of denuded, scarred, and sterile land. If this form of exploitation were undertaken, the consequences would be catastrophic for the long-term future of the southeast of the island.

PRESENT STATUS OF THE MALAGASY LEMURS

Table I contains some of the data relevant to the conservation of Malagasy prosimians. As can be seen, approximately half of the species have never been studied. Furthermore, very little is known of the population statistics of any species over its entire geographical range. This table is meant to be only preliminary. As more information becomes available, it will be revised to provide a better-substantiated basis from which to initiate programs of conservation geared to the particular needs and problems associated with each species. The various sections of Table I are explained in more detail below.

Geographical Range

Accurate maps of the geographical range of each species are not available. The data included in the table under this heading are very general: whether the species has a wide or localized range and if it is specific to a particular type of habitat. The latter category does not include animals that are known to have specialized diets or foraging habits but are found in a number of diverse types of habitat (e.g., *Lepilemur mustelinus*). However, it does include *Hapalemur griseus*, whose distribution is known to be closely tied to

TABLE I. Data Relevant to

Species	Geographical range			Found in reserves	Density per kM²	Average group size	Home range size	Population		
	Wide	Local	Specialized					Abundant	Scarce	No information
Lemur catta	X			B, 5, 10, 11	200–300 (canopy forest)	15–18	6–9 (canopy forest) 23 (degraded forest)	X		
Lemur fulvus	X			A, NM, 1, 3, 4, 5, 7, 8, 9, 11, 12	~1000[a] (*L. f. rufus*, canopy forest)	~9.5[a] (*L. f. rufus*)	~1[a] (*L. f. rufus*)	X		
Lemur macaco		X		6	?	~9.5	?			X
Lemur rubriventer		X	?	4(?)	?	?	?		X	X
Lemur mongoz		X	?	7	~350 (reforested area)	2–4 ("family" groups)	~0.5–1	X		
Lemur variegatus	?		?	NM, 1, 3, 5	?	?	?		?	X
Hapalemur griseus	X	X		1, 3, 4, 12	?	3–5(?)	?		?	X
Hapalemur simus		X	?	None	?	?	?	X		
Lepilemur mustelinus	X			A, B, 1, 4, 6, 7, 8, 10	250–450 (canopy forest) 200–350 (desert-like forest)	Solitary	0.18, females; 0.30, males		?	X
Microcebus murinus	X			A, B, 1, 3, 5, NM, 6, 7	~360[c] (bush) ~250 (canopy forest)	Solitary and nesting groups	~0.07–0.2	X		

[a]Sussman (1972, this volume).
[b]Data on *Lemur catta* are for 1959–1963 (Napier and Napier, 1967).
[c]Charles-Dominique and Hladik (1971).
[d]Pollock (this volume).

Survival of Malagasy Lemurs

Population			Major threats					Captivity (outside Madagascar)			Field data (published)		
			Forest area diminishing			Hunting	Export						
Stable	Increasing	Declining	Industry	Timber	Agriculture			Number in captivity (1973)	Successful in zoos	Number bred in zoos	Long term (over 6 months)	Short or survey	Not studied
		X	X	X	X			Many	Yes	51[b]	Jolly (1966) Sussman (1972,1974) Budnitz and Dainis (this volume)		
		X	X	X	X			198	Yes	116	Sussman (1972, this volume)	Harrington (this volume)	
		X	X	X	X			67	Yes	40		Petter (1962)	
		X	X	X	X			1	No	0			X
		X	X	X	X			175	Yes	76		Tattersall and Sussman (1975)	
		X	X	X	X		X	48	Little	8			X
		X	X	X	X	X		17	No	1		Petter and Peyrieras (1970a)	
		X	X	X	X			0	No	0			X
		X	X	X	X			1	No	0		Charles-Dominique and Hladik (1971)	
?	?		X	X	X			32	Yes	7	Martin (1972, 1973)		

TABLE I. Data Relevant to

| | Geographical range | | | | | | | Population | | |
Species	Wide	Local	Specialized	Found in reserves	Density per kM^2	Average group size	Home range size	Abundant	Scarce	No information
Microcebus coquereli		X	?	A, 9.	210 (canopy forest)	Solitary (assumed)	?		X	
Cheirogaleus major		X	?	NM, 1, 3, 4	?	Solitary (assumed)	?		?	X
Cheirogaleus medius	X		?	B, 7	?	Solitary (assumed)	?		X	X
Allocebus trichotis		X	?	None	?	?	?		X	
Phaner furcifer		X	?	A, 4	500–850 (canopy forest)	Solitary (assumed)	?			X
Indri indri		X	?	1, 3	8–16	3.1 (range 2–5) ("family" groups)	~17		X	X
Avahi laniger	X			1, 3, 7	200^c (rain forest)	2.3^d (range 1–4)	?		?	X
Propithecus diadema	?	?	?	1(?), 3, 11, 12	?	3.14^d (range 1–6)	?		X	X
Propithecus verreauxi	X			A, B, 7, 8, 9, 10, 11	140–240 (canopy forest) 160 (semiarid forest)	~6 (range 3–13)	~2–6 (gallery forest) ~5 (desertlike forest)	X		
Daubentonia madagascariensis		X	X	NM	?	Solitary	?		X	

^{a} Sussman (1972, this volume).
^{b} Data on *Lemur catta* are for 1959–1963 (Napier and Napier, 1967).
^{c} Charles-Dominique and Hladik (1971).
^{d} Pollock (this volume).

Survival of Malagasy Lemurs

| | | | Major threats | | | | | Captivity (outside Madagascar) | | | Field data (published) | | |
| Population | | | Forest area diminishing | | | Hunting | Export | Number in captivity (1972) | Successful in zoos | Number bred in zoos | | | |
Stable	Increasing	Declining	Industry	Timber	Agriculture						Long term (over 6 months)	Short or survey	Not studied
		X	X	X	X			4	No	0		Petter et al. (1971)	
	X		X	X	X			11	No	0			X
	X		X	X	X			9	No	0			X
	X		X	X	X			0	No	0			X
	X		X	X	X			0	No	0		Petter et al. (1971)	
	X		X	X	X			0	No	0	Pollock (this volume)		
	X		X	X	X			0	No	0			X
	X		X	X	X			0	No	0			X
	X		X	X	X			6	No	2	Jolly (1966), Richard (1973, 1974)		
	X		X	X	X		X	0	No	0		Petter and Peyrieras (1970b), Petter and Petter (1967)	

that of bamboo, and *Daubentonia madagascariensis* and *Indri indri*, whose
distributions are restricted to portions of the eastern rain forests of Madagascar.

Reserves

There are 11 national forest reserves, one special reserve—Nosy Mangabe—supported by I.U.C.N., and two private reserves—Berenty and Analabe—owned by M. De Heaulme (Fig. 1). We list the reserves in which each
species is found.[3] A, B, and NM refer to Analabe, Berenty, and Nosy Mangabe, respectively.

Density, Group Size, Home Range Size

Where data on density, group size, and home range size are available,
they are from one or at most a few sites within the total range of the species.
In most cases, population density figures are very rough approximations. The
data given are from the field studies cited in Table I unless otherwise stated.

Population

At this time, there can be only tentative estimates of abundance or scarcity of population, based on density figures, extent of geographical range, and
sightings or lack of such during the last 10 years. Only four species might be
considered abundant (*Lemur catta, Lemur fulvus, Microcebus murinus*, and
Propithecus verreauxi), but only in relation to other Malagasy lemurs; all
species would be considered scarce in relation to nonhuman primates in general. The following species are extremely rare and probably on the brink of
extinction: *Allocebus trichotis, Daubentonia madagascariensis, Hapalemur
simus, Lemur rubriventer, Microcebus coquereli*, and *Propithecus diadema*.
However, because of the lack of adequate population data, this list is probably incomplete.

All of the species are considered to be declining except perhaps *Microcebus murinus*, the smallest and most widespread of the lemurs. This is based
on data on the general and continual destruction of forest vegetation in Madagascar. However, even for *Microcebus murinus*, Martin (1973) gives data in-

---

[3] These data are from: National Reserves, Andriamampianina and Peyrieras (1972); Analabe,
  Petter *et al.* (1971); Berenty, Charles-Dominique and Hladik (1971); Nosy Mangabe, Petter
  (1972).

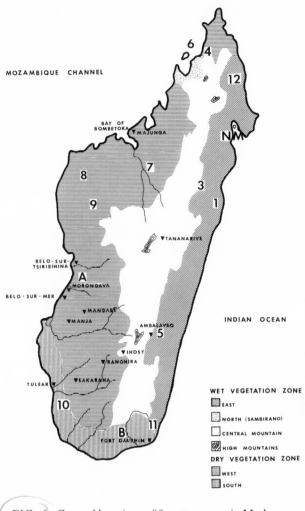

MOZAMBIQUE CHANNEL

BAY OF
BOMBETOKA

MAJUNGA

BELO-SUR-
TSIRIBIHINA

MORONDAVA

BELO-SUR-MER

MANDABE

MANJA

TULEAR

SAKARAHA

FORT DAUPHIN

TANANARIVE

AMBALAVAO

IHOSY

RANOHIRA

INDIAN OCEAN

WET VEGETATION ZONE
EAST
NORTH (SAMBIRANO)
CENTRAL MOUNTAIN
HIGH MOUNTAINS
DRY VEGETATION ZONE
WEST
SOUTH

FIG. 1. General locations of forest reserves in Madagascar

dicating that heavy tree-felling is affecting population densities of this species. An ongoing study of *Microcebus* in the south also suggests that population densities are not as high as previously estimated and that extensive grazing by both cattle and goats is effectively destroying the low bush habitat of the species even in areas where no actual felling is taking place (J. Russell, personal communication).

Major Threats

We believe the major threat to the lemurs of Madagascar to be deforestation caused by timbering and industrial and agricultural exploitation. Although all of the lemurs are hunted, until recently hunting has been documented to be a major threat only to *Hapalemur griseus* (Petter and Peyrieras 1970*a*) and *Daubentonia madagascariensis* (Petter and Peyrieras, 1970*b*). However, in limited areas hunting has traditionally been a common practice and now seems to be on the increase. This, combined with the advent of the "weekend" hunter armed with his gun, means that hunting may soon constitute a major threat to other species.

Commercial exportation of the lemurs could become a threat at any time. For example, at least 27 individuals of the species *Lemur variegatus* were exported to zoos between 1971 and 1973. We can, further, assume that many casualties occurred during the capture of these animals. This venture alone easily could have resulted in the demise of a local population of this species. Furthermore, we have heard several accounts of illegal trade in lemurs from numerous points along Madagascar's coastline.

Captivity

Captivity data for *Lemur catta* are from Napier and Napier (1967). Data for all other species are from Duplaix-Hall (1973).

IS CONSERVATION IN MADAGASCAR FEASIBLE?

It is all too easy to make recommendations for the conservation of the Malagasy flora and fauna. It is all too difficult to reconcile these recommendations with the immediate needs of an underdeveloped country with a population expanding at the rate of at least 3% per annum. As Jolly and Jolly (1970) have pointed out, the general pattern in the past has been to establish reserves without consideration of local needs and often at the expense of the local population. Any effective long-term plans for conservation must, we believe, take into account the needs of the local population as well as those of the environment and its resources, for in the long run these two factors are inseparable. We are ultimately dependent on our environment for survival and, equally, the environment is dependent on the willingness of its human occupants to conserve it. The following comments and tentative suggestions are made with these considerations in mind.

Hunting Legislation

Much of the fauna, including all the lemurs, is protected from hunting by extensive legislation. In reality, however, these laws are insufficiently and inconsistently enforced: in some areas, there is no enforcement at all, in others enforcement is dependent on the identity of the individual breaking the law and/or that of the individual delegated to enforce it. There is no indication that lemur meat is a regular or necessary dietary component for most Malagasy; rather, it is a delicacy eaten infrequently. Thus the enforcement of anti-hunting laws should not entail the enforcement of nutritional deficiencies.

We feel that at this time there is adequate legislation and that the problem is rather one of educating those who must enforce the laws as well as those expected to abide by them. This must include education not only concerning the animals themselves but also concerning the laws protecting them: in many rural areas, the people are not even aware of the existence of these laws. Furthermore, the penalties imposed in the course of law enforcement must be much more closely related to the social realities of the people involved. At present, the penalty for hunting lemurs is a massive fine or extended prison sentence, and officials are understandably reluctant to pursue cases with such potentially dire consequences. We believe that in the social and economic context of Madagascar a much more meaningful and hence effective sanction would be the forfeit of the gun permit and/or goods with social and economic value (e.g., a cow or products from the land). The imposition of large fines is ineffective in an economy in which currency plays a minimal role.

Exportation of Animals

At this time, laws concerning the exportation of live animals are adequate but not adequately enforced. It should, further, be remembered that the lemurs are potentially a valuable financial asset: any significant increase in exportation—prompted by the prospect of considerable financial gain in Madagascar and by the interests of medical research or zoo facilities abroad—will have catastrophic results. In fact, as stated above, at least 27 specimens of *Lemur variegatus* were exported to zoos in 1971 and 1972. Lemurs can also be shot or captured quite legally for "scientific purposes" on purchase of a permit in Madagascar, and can then be imported to other countries without infringing current importation laws. One has only to remember the Archbold Expedition to Madagascar in the 1930s, during which over 400 animals were killed, to realize the possible implications of this loophole in legislation.

Natural Reserves

At present, there are 11 national reserves and at least three other reserves (two privately owned, the third—Nosy Mangabe—funded by I.U.C.N.). Unfortunately, many are reserves in name only; because of a shortage of finance and hence manpower, many of them are insufficiently patrolled. The human populations found (sometimes in considerable numbers) within the boundaries of so-called reserves cut trees for firewood, practice *tavy* (slash-and-burn agriculture), and graze cattle. In these cases, the question of conservation becomes very much a human problem: protection of the environment involves inevitable deprivation of livelihood to people living in that environment, and either a compromise solution must be reached or the people or the reserve moved elsewhere.

Before any such radical steps are taken, however, careful cost–benefit analyses should be undertaken. In most cases, even the most rudimentary items of information are lacking: does the content of each reserve justify an intensive effort at conservation which will probably be costly in both financial and social terms? Although inventories are available of the animal species present in the national reserves (Andriamampianina and Peyrieras, 1972), we have few data on the density of these species or their environmental requirements. Thus we propose that before costly programs of conservation are instigated an intensive survey project be carried out to document the status of areas currently referred to as national reserves as well as other areas which might be more valuable areas for conservation. For example, R. W. S. carried out a study in an essentially unprotected region in the west (Antserananomby) where one area of 10 ha contained over 250 lemurs, including all seven species of lemur found in this region. Unlike many of the African mammals, those of Madagascar that have been studied to date live in small home ranges. It is thus possible that a combination of large and small reserves would be an efficient way of maximizing conservation efforts while minimizing social costs. However, it remains an open question as to what constitutes the minimum area compatible with the maintenance and self-perpetuation of a stable plant and animal community. A reserve that is too small could have dangerous genetic implications and be more susceptible to changing climatic conditions.

In addition to answering some of the more fundamental ecological questions, a reevaluation of the present reserves and other areas should also include economic considerations. Some of the East African Safari Package Tours are beginning to include long weekend stopovers in Madagascar. We believe that this development should be encouraged in order to make a limited number of reserves economically profitable. This in turn would subsidize the maintenance of more remote reserves which could be kept undisturbed by the inevitable disruption caused by even a limited degree of commercialization and would also provide a possible alternative livelihood for local populations deprived of grazing rights, etc., in these areas.

In summary, it is our view that, at present, limited financial resources are thinly spread over poorly known ground, and the first step to be taken should be exploration of the ecological and economic actualities and potentialities of that ground.

Education

The success or failure of conservation efforts depends ultimately on the willingness of people to implement measures undertaken. We believe that willingness to be dependent on education. A real understanding and appreciation of the aims of conservation can be brought about only by teaching people at every level of the educational system about the uniqueness and value of their island's flora and fauna. Biology textbooks currently used all over the island give students a fine grounding in the mammals of northwest Europe. We found none that even referred to the flora and fauna of Madagascar, and we were, indeed, amazed at the frequency with which we encountered well-educated Malagasy who had no idea that much of the island's plant and animal life is unique. An immediate necessity is an inexpensive textbook-cum-encyclopedia, written in Malagasy and distributed throughout the school system, which stresses the uniqueness of the island's flora and fauna and the importance of the lemurs as one of the most primitive living primates.

REFERENCES

Andriamampianina, J., and Peyrieras, A., 1972, Les réserves naturelles intégrales de Madagascar in: *Comptes Rendus de la Conférence Internationale sur la Conservation de la Nature et de ses Ressources à Madagascar*, pp. 102–123, U.I.C.N., Morges, Switzerland.

Charles-Dominique, P., and Hladik, C. M., 1971, Le *Lepilemur* du sud de Madagascar: Ecologie, alimentation et vie sociale, *Terre Vie* 25:3–66.

Chauvet, B., 1973, The forests of Madagascar, in: *Biogeography and Ecology of Madagascar*. (R. Battistini and G. R. Vindard, eds.), pp. 191–199, W. Junk, The Hague.

Deschamps, H., 1965, *Histoire de Madagascar*, 3rd ed., Berger-Levrault, Paris.

Duplaix-Hall, N. (ed.), 1973, *International Zoo Yearbook*, Vol. 13, Zoological Society of London, London.

Guichon, A., 1960, La superficie des formations forestières à Madagascar, *Rev. For. Fr. (Nancy)* 6:408–411.

Humbert, H., and Darne, G. C., 1965, *Notice de la Carte Madagascar*, Institut de la Carte Internationale du Tapis Végétal, Toulouse.

Jolly, A., 1966, *Lemur Behavior*, University of Chicago Press, Chicago.

Jolly, R., and Jolly, A., 1970, The costs and benefits of conservation—Who pays and who benefits, Paper delivered at the Conférence Internationale sur la Conservation de la Nature et de ses Ressources à Madagascar, October 7–11, 1970, Tananarive, Madagascar. (Not included in *Comptes Rendus*.)

Koechlin, J., 1973, Flora and vegetation of Madagascar, in: *Biogeography and Ecology of Madagascar* (R. Battistini and G. R. Vindard, eds.), pp. 145–190, W. Junk, The Hague.

Martin, R. D., 1972, A preliminary field-study of the lesser mouse lemur (*Microcebus murinus*, J. F. Miller 1777), *Z. Tierpsychol* Beiheff **9**:43–89.

Martin, R. D., 1973, A review of the behaviour and ecology of the lesser mouse lemur (*Microcebus murinus* J. F. Miller 1777), in: *Comparative Ecology and Behaviour of Primates* (R. P. Michael and J. H. Crook, eds.), pp. 1–68, Academic Press, London and New York.

Napier, J. R., and Napier, P. H., 1967, *A Handbook of Living Primates*, Academic Press, London.

Petter, J.-J., 1962, Recherches sur l'écologie et l'éthologie des Lémuriens malgaches, *Mém. Mus. Natl. Hist. Nat. Sér. A Zool.* **27**:1–146.

Petter, J.-J., 1972, Intérêt des recherches démographiques sur les Lémuriens malagasy—Importance des réserves de faune—Aménagement de la réserve de Nosy-Mangabe, in: *Comptes Rendus de la Conférence Internationale sur la Conservation de la Nature et de ses Ressources à Madagascar*, pp. 167–169, U.I.C.N., Morges, Switzerland.

Petter, J.-J., and Petter, A., 1967, The aye-aye of Madagascar, in: *Social Communication Among Primates.* (S. A. Altmann, ed.), pp. 195–205, University of Chicago Press, Chicago.

Petter, J.-J., and Peyrieras, A., 1970a, Observations éco-éthologiques sur les Lémuriens malgaches du genre *Hapalemur, Terre Vie* **24**:356–382.

Petter, J.-J., and Peyrieras, A., 1970b, Nouvelle contribution à l'étude d'un Lémurien malgache, le aye-aye (*Daubentonia madagascariensis* E. Geoffroy), *Mammalia* **34**:167–193.

Petter, J.-J., Schilling, A., and Pariente, G., 1971, Observations éco-éthologiques sur deux Lémuriens malgaches nocturnes: *Phaner furcifer* et *Microcebus coquereli, Terre Vie* **25**:287–327.

Richard, A., 1973, Social organization and ecology of *Propithecus verreauxi* Grandidier, Ph.D. thesis, University College, London.

Richard, A., 1974, Intra-specific variation in the social organization and ecology of *Propithecus verreauxi, Folia Primatol.,* **22**(2–3):178–207.

Sussman, R. W., 1972, An ecological study of two Madagascan primates: *Lemur fulvus rufus* Audebert and *Lemur catta* Linnaeus, Ph.D. thesis, Duke University.

Sussman, R. W., 1974, Ecological distinctions in sympatric species of *Lemur*, in: *Prosimian Biology* (R. D. Martin, G. A. Doyle, and A. C. Walker, eds.), Duckworth, London.

Tattersall, I., and Sussman, R. W., 1975, Preliminary observations on the ecology and behavior of the mongoose lemur, *Lemur mongoz mongoz* Linnaeus (Primates, Lemuriformes), at Ampijora, Madagascar, *Anthropol. Papers Am. Mus. Nat. Hist.*

Walker, A. C., 1967, Patterns of extinction among the subfossil Madagascan lemuroids, in: *Pleistocene Extinctions: The Search for a Cause* (P. S. Martin and H. E. Wright, eds.), pp. 425–432, Yale Univ. Press, New Haven.

Index